Bergman Spaces and Related Topics in Complex Analysis

Titles in This Subseries

Israel Mathematical Conference Proceedings (IMCP) is a publication, part of the Contemporary Mathematics Series, devoted to the proceedings of conferences, symposia and seminars. Collections of papers focusing on a certain subject will also be published. Prospective volumes may be submitted to any member of the editorial board. Each volume has an editor (or editors) responsible for its preparation. In order to ensure inexpensive and timely distribution, authors are requested to submit to the Editor of the volume an electronic TEX file of their manuscript in $\mathcal{A}_{\mathcal{M}}\mathcal{S}$-LATEX, using the Contemporary Mathematics style file which can be downloaded at http://www.ams.org/tex/author-info.html. For further information, contact the Managing Editor, IMCP, Department of Mathematics, Bar-Ilan University, Ramat-Gan 52900, Israel; e-mail: rowen@macs.biu.ac.il.

404 **Alexander Borichev, Håkan Hedenmalm, and Kehe Zhu, Editors,** Bergman spaces and related topics in complex analysis, 2006

402 **Zvi Arad, Mariagrazia Bianchi, Wolfgang Herfort, Patrizia Longobardi, Mercede Maj, and Carlo Scoppola, Editors,** Ischia group theory 2004, 2006

387 **Michael Entov, Yehuda Pinchover, and Michah Sageev, Editors,** Geometry, spectral theory, groups, and dynamics, 2005

382 **Mark Agranovsky, Lavi Karp, and David Shoikhet, Editors,** Complex analysis and dynamical systems II, 2005

364 **Mark Agranovsky, Lavi Karp, David Shoikhet, and Lawrence Zalcman, Editors,** Complex analysis and dynamical systems, 2004

Published Earlier as IMCP

15 **Vitali Milman, Iossif Ostrovskii, Mikhail Sodin, Vadim Tkachenko, and Lawrence Zalcman, Editors,** Entire functions in modern analysis: Boris Levin memorial conference, 2001

14 **Robert Brooks and Mikhail Sodin, Editors,** Lectures in memory of Lars Ahlfors, 2000

13 **Yuri Brudnyi, Michael Cwikel, and Yoram Sagher, Editors,** Function spaces, interpolation spaces, and related topics, 1999

12 **Mina Teicher, Editor,** The heritage of Emmy Noether, 1999

11 **Lawrence Zalcman, Editor,** Proceedings of the Ashkelon workshop on complex function theory (May 1996), 1997

10 **Jean-Pierre Fouque, Kenneth J. Hochberg, and Ely Merzbach, Editors,** Stochastic analysis: random fields and measure-valued processes, 1995

9 **Mina Teicher, Editor,** Proceedings of the Hirzebruch 65 conference on algebraic geometry, 1995

8 **Ilya Piatetski-Shapiro and Stephen Gelbart, Editors,** The Schur lectures (1992), 1995

7 **Anthony Joseph and Steven Shnider, Editors,** Quantum deformations of algebras and their representations, 1993

6 **Haim Judah, Editor,** Set theory of the reals, 1992

5 **Michael Cwikel, Mario Milman, and Richard Rochberg, Editors,** Interpolation spaces and related topics, 1992

4 **Simson Baron and Dany Leviatan, Editors,** Approximation interpolation and summability, in honor of Amnon Jakimovski, 1991

3 **Stephen Gelbart, Roger Howe, and Peter Sarnak, Editors,** Festschrift in honor of I. I. Piatetski-Shapiro, part II: Papers in analysis, number theory and automorphic L-Functions, 1990

2 **Stephen Gelbart, Roger Howe, and Peter Sarnak, Editors,** Festschrift in honor of I. I. Piatetski-Shapiro, part I: Papers in representation theory, 1990

1 **Louis Rowen, Editor,** Ring theory, in honor of S. A. Amitsur, 1989

CONTEMPORARY MATHEMATICS

404

Israel Mathematical Conference Proceedings

Bergman Spaces and Related Topics in Complex Analysis

Proceedings of a Conference in Honor of Boris Korenblum's 80th Birthday

November 20–22, 2003
Barcelona, Spain

Alexander Borichev
Håkan Hedenmalm
Kehe Zhu
Editors

American Mathematical Society
Providence, Rhode Island

Bar-Ilan University
Ramat Gan, Israel

2000 *Mathematics Subject Classification.* Primary 30-06, 46E15, 32A36; Secondary 30D50, 30D55.

Library of Congress Cataloging-in-Publication Data

Bergman spaces and related topics in complex analysis : proceedings of a conference in honor of Boris Korenblum's 80th birthday, November 20–22, 2003, Barcelona, Spain / Alexander Borichev, Håkan Hedenmalm, Kehe Zhu, editors.
    p. cm. — (Contemporary mathematics, ISSN 0271-4132 ; 404) (Israel mathematical conference proceedings)
    Includes bibliographical references.
    ISBN 0-8218-3712-5 (alk. paper)
    1. Bergman spaces.   2. Functions of several complex variables.   I. Korenblum, Boris. II. Borichev, Alexander A., 1963–   III. Hedenmalm, Håkan.   IV. Zhu, Kehe, 1961–   V. Series. VI. Contemporary mathematics (American Mathematical Society) ; v. 404.

QA331.7.B465   2006
515′.9—dc22   2006040697

Contents

Preface   vii

A Word from the Organizers   ix

List of Participants   xi

Conference Program   xv

Opening Words by Kristian Seip   xvii

Publications of Boris Korenblum   xix

Invariant Subspaces for the Backward Shift on Hilbert Spaces
of Analytic Functions with Regular Norm
    ALEXANDRU ALEMAN, STEFAN RICHTER, and CARL SUNDBERG   1

Surjectivity and Invariant Subspaces of Differential Operators
on Weighted Bergman Spaces of Entire Functions
    AHARON ATZMON and BRUNO BRIVE   27

Exceptional Values and the MacLane Class $\mathcal{A}$
    KARL F. BARTH and PHILIP J. RIPPON   41

Operators on Weighted Bergman Spaces
    OSCAR BLASCO   53

A Wiener Tauberian Theorem for Weighted Convolution Algebras
of Zonal Functions on the Automorphism Group of the Unit Disc
    ANDERS DAHLNER   67

Domination on Sets and in Norm
    WALTER K. HAYMAN   103

Extensions of the Asymptotic Maximum Principle
    CHARLES HOROWITZ and BERNARD PINCHUK   111

Singularity Resolution of Weighted Bergman Kernels
    STEFAN JAKOBSSON   121

Blaschke Sets for Bergman Spaces
    BORIS KORENBLUM   145

Phragmén-Lindelöf-type Problems for $A^{-\alpha}$
    XAVIER MASSANEDA and PASCAL J. THOMAS   153

A Representation Formula for Reproducing Subharmonic Functions
in the Unit Disc
ANDERS OLOFSSON   165

On Dominating Sets for Bergman Spaces
FERNANDO PÉREZ-GONZÁLEZ and JULIO C. RAMOS   175

Trigonometric Obstacle Problem and Weak Factorization
SERGUEI SHIMORIN   187

A Sharp Norm Estimate of the Bergman Projection on L^p Spaces
KEHE ZHU   199

Preface

This volume contains fourteen research papers on Bergman spaces and related topics. It grew out of a conference in honor of Boris Korenblum's 80th birthday, held on November 20–22, 2003, at the University of Barcelona, organized by J. Bruna, H. Hedenmalm, B. Pinchuk, K. Seip, and K. Zhu. On a personal note, we would like to say that we appreciate Boris Korenblum's deep research contributions to Complex Analysis, as well as the generous support he has extended to colleagues, PhD students, and post-docs.

The volume includes also the list of participants, the list of lectures, and the opening words delivered by Kristian Seip on November 20, 2003, welcoming Boris Korenblum.

Finally, we add the list of publications of Boris Korenblum.

We thank the authors for their contributions and the referees for their careful and patient work.

Alexander Borichev
Håkan Hedenmalm
Kehe Zhu

A Word from the Organizers

The conference honoring Boris Korenblum was made possible by generous support from the following sources:

- Departament de Matemàtiques, Universitat Autònoma de Barcelona
- Facultat de Matemàtiques, Universitat de Barcelona
- European Research Training Network in Classical Analysis, Operator Theory, Geometry of Banach spaces, their interplay and their applications (contract No. HPRN-CT-2000-00116)
- Gelbart Research Institute for the Mathematical Sciences at Bar-Ilan University.

We thank these organizations for making the conference possible. We also thank the city of Barcelona for offering a pleasant background.

Joaquim Bruna
Håkan Hedenmalm
Bernard Pinchuk
Kristian Seip
Kehe Zhu

List of Participants

Abakumov, Evgeny
University Marne-la-Vallee

Abuzyarova, Natalia
KTH, Stockholm

Aleman, Alexandru
Lund University

Anderson, J. Milne
University College London

Armitage, David
Queens University, Belfast

Ascensi, Gerard
Universitat Autònoma de Barcelona

Atzmon, Aharon
Tel Aviv University

Bieche, Camille
Université de Provence

Blasco, Oscar
Universidad de Valencia

Blasi Babot, Daniel
Universitat Autònoma de Barcelona

Borichev, Alexander
CNRS – Université de Bordeaux I

Bruna, Joaquim
Universitat Autònoma de Barcelona

Buckley, Stephen
NUI Maynooth

Burgués, Josep M.
Universitat Autònoma de Barcelona

Cantón Pire, Alícia
Universitat Autònoma de Barcelona

Carlsson, Linus
CRM

Carmona, Joan Josep
Universitat Autònoma de Barcelona

Carro, María J.
Universitat de Barcelona

Cascante, Carme
Universitat de Barcelona

Cerdà, Joan
Universitat de Barcelona

Clop, Albert
Universitat Autònoma de Barcelona

Cufí, Julià
Universitat Autònoma de Barcelona

Demange, Bruno
Universitat Autònoma de Barcelona

Domanski, Pawel A.
Adam Mickiewicz University, Poznan

Donaire Benito, Juan Jesús
Universitat Autònoma de Barcelona

Dussau, Xavier
Université de Bordeaux I

Dyakonov, Konstantin
Universitat de Barcelona

Fedorovskiy, Konstantin
Moscow State University (MGU)

Fernández Arias, Arturo
Universidad Nacional de Educación a
Distancia

Gallardo, Eva A.
Universidad de Zaragoza

Gardiner, Stephen
University College Dublin

González Fuente, María José
Universidad de Cádiz

González Llorente, José
Universitat Autònoma de Barcelona

Granados Sanandres, Ana
University of British Columbia

Gulisashvili, Archil
Ohio University & CRM

Hayman, Walter K.
Imperial College

Hedenmalm, Håkan
KTH Stockholm

Kellay, Karim
Université de Provence

Korenblum, Boris
SUNY at Albany

Lemmers, Oscar
Universitat Autònoma de Barcelona

Louhichi, Issam
Université de Bordeaux I

Lovera, Stéphanie
Université de Provence

Lyubarskii, Yurii
NTNU Trondheim

Marshall, Donald
University of Washington

Martin Pedret, Joaquim
Universitat Autònoma de Barcelona

Marzo Sanchez, Jordi
Universitat de Barcelona

Massaneda, Xavier
Universitat de Barcelona

Mateu, Joan
Universitat Autònoma de Barcelona

Melnikov, Mark
Universitat Autònoma de Barcelona

Monreal Galan, José Ignacio
Universitat Autònoma de Barcelona

Montes Rodríguez, Alfonso
Universidad de Sevilla

Nicolau, Artur
Universitat Autònoma de Barcelona

Nikolski, Nikolai
Université de Bordeaux I

O'Farrell, Anthony G.
NUI Maynooth

Olofsson, Anders
KTH Stockholm

Orobitg, Joan
Universitat Autònoma de Barcelona

Ortega, Joaquim M.
Universitat de Barcelona

Pascuas, Daniel
Universitat de Barcelona

Pau, Jordi
Universitat Autònoma de Barcelona

Perdomo Gallípoli, Yolanda
Lund University

Pereyra, Cristina
University of New Mexico

Perez-Gonzalez, Fernando
Universidad de La Laguna

Pinchuk, Bernard
Bar-Ilan University and Netanya
Academic College

Ponce Escudero, Manuel
Universidad de Sevilla

Prat Baiget, Laura
Universitat de Barcelona

Richter, Stefan
University of Tennessee

Rippon, Philip J.
Open University

Saksman, Eero
University of Jyväskylä

Saludes, Jordi
Universitat Politècnica de Catalunya

Segura Manzano, María Dolores
Universidad de Sevilla

Seip, Kristian
NTNU Trondheim

Shapiro, Harold
KTH Stockholm

Shimorin, Serguei
KTH Stockholm

Sodin, Mikhail
Tel Aviv University

Soria, Javier
Universitat de Barcelona

Stray, Arne
Bergen University

Strouse, Elizabeth
Université de Bordeaux I

Sundberg, Carl
University of Tennessee

Thomas, Pascal J.
Université Paul Sabatier, Toulouse

Tolsa, Xavier
Universitat Autònoma de Barcelona

Verdera, Joan
Universitat Autònoma de Barcelona

Walsh, David
NUI Maynooth

Youssfi, Hassan
Université de Provence

Zalcman, Lawrence
Bar-Ilan University

Zhu, Kehe
SUNY at Albany

Conference Program

THURSDAY, NOVEMBER 20, 2003

16.45:	KRISTIAN SEIP	Opening speech
17.00–17.50:	W. HAYMAN	Domination in sets and in norm
18.15-19.05:	L. ZALCMAN	Normal and quasinormal families of meromorphic functions
19:30-20:20	N. NIKOLSKI	Condition numbers of multipliers and Toeplitz operators

FRIDAY, NOVEMBER 21, 2003

17.00–17.50:	D. MARSHALL	Convergence of the Zipper algorithm for conformal mapping
10.30-11.20	Y. LYUBARSKII	Interpolation in generalized Paley-Wiener spaces
12.00-12.50	A. NICOLAU	Regularity of measures, entropy and the law of the iterated logarithm
15.40-16.30:	K. ZHU	The Bergman projection and related integral operators
17.10-18.00:	S. RICHTER	The index of invariant subspaces in spaces of analytic functions
18.40-19.30:	C. SUNDBERG	Nontangential limits in $L^p(\mu)$ spaces
21.30:	*Official Banquet*	

SATURDAY, NOVEMBER 22, 2003

9.00-9.50:	A. ALEMAN	A Korenblum type estimate for Moebius invariant spaces of analytic functions
10.30-11.20	H. HEDENMALM	Bergman spaces and differential geometry
12.00-12.50:	S. SHIMORIN	Weighted Bergman spaces and the estimates of derivatives of univalent functions
15.00-16.00:	*Open problems*	
16.30-17.20:	H. SHAPIRO	Algebraic aspects of the Dirichlet problem
18.00-18.50:	M. SODIN	How often do analytic functions visit an angle?
19.30-20.20:	A. ATZMON	Banach spaces of entire functions of zero exponential type

Opening words by Kristian Seip at the "Korenblum Fiesta", Barcelona, November 20–22, 2003

Dear Boris and dear friends of Boris!

On behalf of the organizers, I am happy to welcome you all here, for this celebration of Boris's 80th birthday and his contribution to our science. I think Barcelona is close to perfect for what we want this meeting to be, a rather informal gathering of close scientific and personal friends of Boris in a relaxed atmosphere, with excellent food, wine, and pleasant surroundings. I know that you, Boris, as many others in this room, have enjoyable memories from time spent here, and this certainly makes Barcelona no less appropriate as the place for this meeting. So I am very happy that Joaquim took on the job of organizing the meeting; and since I did very little myself, I dare say he has taken care of it in an excellent way.

I will not give complete review of Boris's scientific life—we will learn more about that during the meeting. But I will mention two highlights. I believe the first may come as a surprise to you. It is from CT—computed tomography. I quote from the introductory chapter entitled "In the Beginning" by Steve Webb from the 1988 volume *The Physics of Medical Imaging*: "It is perhaps less known that a CT (Computed Tomography) scanner was built in Russia in 1958. Korenblyum et al. [Tetelbaum, Tyutin] (1958) published the mathematics of reconstruction from projections together with experimental data and wrote: 'At the present time in Kiev Polytechnic Institute, we are constructing the first experimental apparatus for getting X-ray images of thin sections by the scheme described in this article'." May I remind you that G. N. Hounsfield received the 1979 Nobel Prize for Physiology and Medicine for his construction of a machine used to X-ray computed tomography in a clinical environment. I suspect that a neutral observer may find Boris's achievements in this field more significant than our precious theory of Bergman spaces. Let me add that Boris has kept up his interest in physics. As late as last year, he published a paper entitled "Classical Properties of Low-Dimensional Conductors: Giant Capacitance and Non-Ohmic Potential Drop," with Emmanuel Rashba, in *Physical Review Letters*!

Let me make a big jump to something we all know well: the two famous *Acta* papers. There are probably not many people who have really penetrated all aspects of those papers. The reviewer, Walter Hayman, ended his review of the first of them in *Mathematical Reviews* with the following words: "The above sketch must suffice to give some idea of this extremely complicated but profound paper." I have personally been very much inspired by those papers, which contain amazing and deep ideas. One of the most striking aspects is the way linear programming enters the study of zero sets for analytic functions. I am not able to guess how you got the necessary insight, but it is clear that it is based on a broad knowledge and understanding. As far as I know, there is no other way of getting such precise estimates for zero sequences for functions in Bergman spaces.

Boris played a decisive role in the development of the theory of Bergman spaces since around 1990, both through his papers and as a mentor. I asked Håkan about Boris's role as a mentor, and got the following words from him: "I think Boris is one of the truly passionate mathematicians. He really believes that Mathematics is important to the real world, and is willing to discuss it at length at any time, not just during working hours. He gave a lot of support at a time when I felt my mathematical ideas met with little or no response, and I was not sure that I wanted to continue doing mathematics. He also gave me a whole new (to me) field to study: the Bergman spaces. The maximum principle he was then working on suggested that really new phenomena could appear here. I got the idea to introduce extremal problems, in a very simple setting with a single zero at first; the buzz-word we used in these early discussions with Boris was the "envelope." Boris was always very generous, and refused co-authorship when he felt that his contribution was not quite up to the mark he set. The work on the "envelope" gave rise to the factorization theory you now know. The use of Green's formula and the like was stimulated by a paper of Boris, "Transformation of zero sets by contractive operators in the Bergman space," which should be appreciated better than at present."

The last statement is certainly interesting; I believe Håkan is right. There are probably ideas in that paper that should be pursued and that could give more insight into the zero sets of Bergman spaces. I know Boris himself has thought and still thinks about that.

Mathematics has obviously meant a lot to you, Boris, but we know there have been hardships in your life of a different caliber than most of us have experienced, such as your service as a soldier in the Red Army during World War II and your painful procedure for emigration as well as immigration to Israel and the US. Most of us know little about these sides of your life. What we know is your passion for mathematics, your generosity with ideas and willingness to help fellow mathematicians. I asked another friend of mine about Boris and got the following response: "Boris always works independently of others. He usually finds new questions that are of interest to him regardless of whether they are fashionable or not. Yet pretty often those questions happen to be the key ones in new areas." The theory of Bergman spaces certainly is a good example in this respect.

Publications of Boris Korenblum

1. *On the representation of functions of L^p-classes by singular integrals at Lebesgue-points,* Dokl. Akad. Nauk SSSR **58** (1947), 973–976.

2. *On the convergence of singular integrals for some general classes of summable functions,* Sbornik Trudov Mat. Inst. Akad. Nauk Ukr. SSR (1948) no. 11, 60–82.

3. (with S. G. Krein and B. Ya. Levin) *On certain non-linear problems in the theory of singular integrals,* Dokl. Akad. Nauk SSSR **62** (1948), 17–20.

4. *On certain special commutative normed rings,* Dokl. Akad. Nauk SSSR **64** (1949), 281–284.

5. *On theorems of Tauberian type,* Dokl. Akad. Nauk SSSR **64** (1949), 449–452.

6. *On the convergence theory of Fourier series,* Dopovidi Akad. Ukrain. RSR **1951** (1951), 320–323.

7. *On two theorems from the theory of absolutely monotone functions,* Uspekhi Matem. Nauk **6** (1951) no. 4 (44), 172–175.

8. *On lacunary Laplace-Stieltjes integrals,* Dokl. Akad. Nauk SSSR **76** (1951), 779–782.

9. *Theorems of Tauberian type for a class of Dirichlet series,* Dokl. Akad. Nauk SSSR **81** (1951), 725–727.

10. *On a problem of interpolation,* Dokl. Akad. Nauk SSSR **81** (1951), 991–994.

11. (with M. G. Krein and S. I. Tetelbaum) *On the mathematical theory of the optimal amplitude-phase modulation method,* Sbornik Trudov Inst. Elektrotekhn. Akad. Nauk Ukr. SSR **7** (1951), 96–104.

12. *A general Tauberian theorem for the ratio of functions,* Dokl. Akad. Nauk SSSR **88** (1953), 745–748.

13. *A general Tauberian theorem for the ratio of functions,* Uspekhi Mat. Nauk **8** (1953) no. 3, (55), 151–153.

14. *On the asymptotic behavior of Laplace integrals near the boundary of the region of convergence,* Dokl. Akad. Nauk SSSR **104** (1955), 173–176.

15. *Generalization of Wiener's Tauberian theorem and the spectrum of fast growing functions,* Dokl. Akad. Nauk SSSR **111** (1956), 280–282.

16. *Harmonic analysis of fast growing functions,* Uspekhi Mat. Nauk **12** (1957) no. 1 (73), 201–203.

17. *On a normed ring of functions with convolution,* Dokl. Akad. Nauk SSSR **115** (1957), 226–229.

18. *A generalization of Wiener's Tauberian theorem and harmonic analysis of rapidly growing functions,* Trudy Moskov. Mat. Obshch. **7** (1958), 121–148.

19. (with S. I. Tetelbaum and A. A. Tyurin) *On a tomography pattern,* Izv. Vuzov, Radiophysics **1** (1958) no. 3, 151–157.

20. *A generalization of Wiener's Tauberian theorem and spectrum of rapidly growing functions.* Proceedings of the 3rd All-Union Mathematical Congress, 4 (Moscow, 1959), 56–58.

21. *Methods of the theory of functions of a complex variable in generalized harmonic analysis on the line,* Issledovaniya po sovremennym problemam teorii funkciĭkompleksnogo peremennogo, Gosudarstv. Izdat. Fiz.-Mat. Lit., Moscow, 1960, pp. 526–531.

22. *Phragmen-Lindelöf-type theorems for quasi-analytic classes of functions.* Moscow, 1961; in the same collection as no. 21 (Markushevich, ed.), pp. 510–514.

23. *The Weierstrass theorem in spaces of infinitely differentiable functions,* Dokl. Akad. Nauk SSSR **150** (1963), 1214–1217.

24. *A theorem of Weierstrass type in spaces of infinitely differentiable functions,* Studies Contemporary Problems Constructive Theory of Functions (Proc. Second All-Union Conf., Baku, 1962), Izdat. Akad. Nauk Azerbaĭdzhan. SSR, Baku, 1965, pp. 555–561.

25. *Quasi-analytic classes of functions in a circle,* Dokl. Akad. Nauk SSSR **164** (1965), 36–39.

26. *Non-triviality conditions for certain classes of functions analytic in an angle and problem of quasianalyticity,* Dokl. Akad. Nauk SSSR **166** (1966), 1046–1049.

27. *A generalized Watson problem and some of its applications,* Abstracts of Communications, section 4, Moscow, ICM, 1966, p. 59.

28. (with V. I. Rybalskiĭ and B. I. Hatzet) *On an extremum problem linked to the network graph of a project,* Abstract of a report, Conference on Mathematical Optimal Programming, Novosibirsk, 1965, pp. 41–43.

29. *An estimate of the type of an entire function and a time-optimal control problem,* Dokl. Akad. Nauk SSSR **170** (1966), 286–289.

30. (with V. I. Rybalskiĭ) *Optimal distributing networks,* Ekonom. i Mat. Metody **3** (1967), 88–94.

31. (with V. A. Dubovik) *Hadamard-Kolmogorov type inequalities involving constraints,* Mat. Zametki **5** (1969), 13–20.

32. (with Yu. S. Davidovich and B. I. Hatzet) *A certain property of logarithmically concave functions,* Dokl. Akad. Nauk SSSR **185** (1969), 1215–1218.

33. (with E. S. Dekhtiariuk) *Asymptotic uniqueness theorems for certain classes of infinitely differentiable functions,* Ukr. Mat. Zh. **21** (1969), 684–693.

34. (with V. S. Korolevich) *Analytic functions that are regular in a disk and smooth on its boundary,* Mat. Zametki **7** (1970), 165–172.

35. (with V. A. Dubovik) *On accessibility domain for a linear system with bounded phase coordinates,* Proc. 2nd Winter School Math. Programming, no. 2, Moscow, 1969, pp. 496–502.

36. (with Yu. S. Davidovich and B. I. Hatzet) *On a class of unimodal functions,* Ibidem, pp. 470–494.

37. *A certain extremal property of outer functions,* Mat. Zametki **10** (1971), 53–56.

38. *The functions that are holomorphic in the disk and smooth up to its boundary,* Dokl. Akad. Nauk SSSR **200** (1971), 24–27.

39. *Invariant subspaces of the shift operator in certain weighted Hilbert spaces of sequences,* Dokl. Akad. Nauk SSSR **202** (1972), 1258–1260.

40. *Invariant subspaces of the shift operator in a weighted Hilbert space,* Mat. Sb. **89** (1972), 110–137.

41. *Closed ideals of ring A^n,* Funktsional. Anal. i Prilozhen. **6** (1972) no. 3, 38–52.

42. (with V. M. Faĭvyshevskiĭ) *A certain class of compression operators that are connected with the divisibility of analytic functions,* Ukrain. Mat. Zh. **24** (1972), 692–695.

43. *An extension of the Nevanlinna theory,* Acta Math. **135** (1975) no. 3–4, 187–219.

44. (with W. K. Hayman) *An extension of the Riesz-Herglotz formula,* Ann. Acad. Sci. Fenn. Ser. A I Math. **2** (1976), 175–201.

45. *A Beurling-type theorem,* Acta Math. **138** (1977) no. 3–4, 265–293.

46. *Weakly invertible elements in Bergman spaces,* 99 Unsolved Problems in Linear and Complex Analysis, Leningrad, LOMI, 1978, 149–150.

47. (with W. K. Hayman) *A critical growth rate for functions regular in a disk,* Michigan Math. J. **27** (1980) 21–30.

48. *Analytic functions of unbounded characteristic and Beurling algebras,* Proceedings of the International Congress of Mathematicians (Helsinki, 1978), Acad. Sci. Fennica, Helsinki, 1980, pp. 653–658.

49. *Cyclic elements in some spaces of analytic functions,* Bull. Amer. Math. Soc. **5** (1981), 317–318.

50. *Some problems in potential theory and the notion of harmonic entropy,* Bull. Amer. Math. Soc. **8** (1983), 459–462.

51. (with E. Thomas) *An inequality with applications in potential theory,* Trans. Amer. Math. Soc. **279** (1983), 525–536.

52. *A generalization of two classical convergence tests for Fourier series, and some new Banach spaces of functions,* Bull. Amer. Math. Soc. **9** (1983), 215–218.

53. *On a class of Banach spaces of functions associated with the notion of entropy,* Trans. Amer. Math. Soc. **290** (1985), 527–553.

54. *BMO estimates and radial growth of Bloch functions,* Bull. Amer. Math. Soc. **12** (1985), 99–102.

55. (with L. Brown) *Cyclic vectors in $A^{-\infty}$,* Proc. Amer. Math. Soc. **102** (1988), 137–138.

56. (with J. Bruna) *A note on Calderon-Zygmund singular integral convolution operators,* Bull. Amer. Math. Soc. **16** (1987), 271–273.

57. (with J. Bruna) *On Kolmogorov's theorem, the Hardy-Littlewood maximal function and the radial maximal function,* J. Anal. Math. **50** (1988), 225–239.

58. *Une forme plus précise d'un théorème de Kolmogorov,* C.R. Acad. Sci. Paris Sér. I Math. **306** (1988) no. 5, 235–238.

59. *Unimodular Möbius-invariant contractive divisors for the Bergman space,* The Gohberg Anniversary Collection, Vol. II (Calgary, AB, 1988), Oper. Theory Adv. Appl., **41**, Birkhäuser, Basel, 1989, pp. 353–358.

60. *Transformation of zero sets by contractive operators in the Bergman space,* Bull. Sci. Math. **114** (1990), 385–394.

61. *A maximum principle for the Bergman space,* Publ. Mat. **35** (1991), 479–486.

62. (with W. K. Hayman) *Representation and uniqueness theorems for polyharmonic functions,* J. Anal. Math. **60** (1993), 113–133.

63. (with M. Stessin) *On Toeplitz-invariant subspaces of the Bergman space,* J. Funct. Anal. **111** (1993), 76–96.

64. *Outer functions and cyclic elements in Bergman spaces,* J. Funct. Anal. **115** (1993), 104–118.

65. (with K. Richards) *Majorization and domination in the Bergman space,* Proc. Amer. Math. Soc. **117** (1993), 153–158.

66. (with R. O'Neill, K. Richards, and K. Zhu) *Totally monotone functions with applications to the Bergman space,* Trans. Amer. Math. Soc. **337** (1993), 795–806.

67. (with J. E. McCarthy) *Non-attainable boundary values of H^∞ functions,* Extracta Math. **8** (1993), 138–141.

68. (with K. Zhu) *An application of Tauberian theorems to Toeplitz operators,* J. Operator Theory **33** (1995), 353–361.

69. (with C. Horowitz and B. Pinchuk) *Extremal functions and contractive divisors in A^{-n},* Ann. Scuola Norm. Super. Pisa Cl. Sci. **23** (1996), 179–191.

70. (with P. J. Rippon and K. Samotij) *On integrals of harmonic functions over annuli,* Ann. Acad. Sci. Fenn. Ser A I Math. **20** (1995), 3–26.

71. (with H. Hedenmalm and K. Zhu) *Beurling type invariant subspaces of the Bergman space,* J. London Math. Soc. **53** (1996), 601–614.

72. (with J. E. McCarthy) *The range of Toeplitz operators on the ball,* Rev. Mat. Iberoamericana **12** (1996), 47–61.

73. (with C. Horowitz and B. Pinchuk) *Sampling sequences for $A^{-\infty}$,* Michigan Math. J. **44** (1997), 389–398.

74. (with A. Mascuilli and J. Panariello) *A generalization of Carleman's uniqueness theorem and a discrete Phragmén-Lindelöf theorem,* Proc. Amer. Math. Soc. **126** (1998), 2025–2032.

75. (with H. Hedenmalm and K. Zhu) *Theory of Bergman Spaces.* Graduate Texts in Mathematics, **199**, Springer-Verlag, New York, 2000.

76. (with J. Racquet) *Concurrence of uniqueness and boundedness conditions for regular sequences,* Complex Variables Theory Appl. **41** (2000), 231–239.

77. *$A^{-\alpha}$ zero sets: new methods and techniques.* Complex analysis, operators, and related topics, Oper. Theory Adv. Appl. **113**, Birkhäuser, Basel, 2000, pp. 63–178.

78. (with T. L. Lance and M. I. Stessin) *Projective generators in Hardy and Bergman spaces,* Bull. Sci. Math. **124** (2000), 435–445.

79. (with C. Bénéteau) *Jensen type inequalities and radial null sets,* Analysis (Munich) **21** (2001), 99–105.

80. *Asymptotic maximum principle,* Ann. Acad. Sci. Fenn. Math. **27** (2002), 249–255.

81. (with E. I. Rashba) *Classical properties of low-dimensional conductors: giant capacitance and non-Ohmic potential drop,* Phys. Rev. Lett. **89** 096803 (2002).

82. (with C. Bénéteau) *Some coefficient estimates for H^p functions,* Complex Analysis and Dynamical Systems, Contemp. Math. **364** (2004), 5–14.

83. *Blaschke sets for Bergman spaces,* (this volume).

Contemporary Mathematics
Volume **404**, 2006

Invariant Subspaces for the Backward Shift on Hilbert Spaces of Analytic Functions with Regular Norm

Alexandru Aleman, Stefan Richter, and Carl Sundberg

Dedicated to Boris Korenblum on the occasion of his 80th birthday

ABSTRACT. We investigate the structure of invariant subspaces of the backward shift operator $Lf = (f - f(0))/\zeta$ on a large class of abstract Hilbert spaces of analytic functions on the unit disc where the forward shift operator $M_\zeta f = \zeta f$ acts as a contraction. Our main results show that under certain regularity conditions on the norm of such a space, the functions in a nontrivial invariant subspace of L have meromorphic *pseudocontinuations* in the Nevanlinna class of the exterior of the unit disc. We also provide a regularity condition which implies that the subspace itself is contained in the Nevanlinna class of the disc. These results imply that the spectrum of the restriction of L to these subspaces intersects the unit disc in a discrete set and this fact is then applied to prove a general index-one theorem for the forward shift invariant subspaces of the Cauchy dual of the original space. Finally, we give a detailed discussion of the weighted shift operators for which our main results apply.

1. Introduction

Let $\mathbb{D}$ denote the open unit disc in the complex plane and let ζ denote the identity function on $\mathbb{D}$, $\zeta(z) = z$, $z \in \mathbb{D}$. In this paper, we shall consider Hilbert spaces of analytic functions on the open unit disc $\mathbb{D}$ that contain the constants and such that the operator M_ζ of multiplication with the function ζ acts as a contractive operator on $\mathcal{H}$, and such that the backward shift defined by the rule

$$ Lf(z) = \frac{f(z) - f(0)}{z} $$

acts as a bounded linear operator. We will investigate the structure of the non-trivial invariant subspaces of the backward shift. Our goal is to show that there are quite general regularity conditions on the norm of such a Hilbert space which

2000 *Mathematics Subject Classification.* Primary 47B32.

Key words and phrases. Backward shift, pseudocontinuation.

Part of this work was done while the second author visited Lund University. He would like to thank the Mathematics Department for its hospitality. Furthermore, work of the first author was supported by the Royal Swedish Academy of Sciences; work of the second author was supported by the National Science Foundation and the Göran Gustafsson Foundation; and work of the third author was supported by the National Science Foundation.

automatically imply very special properties of the nontrivial invariant subspaces of the backward shift. In order to explain our results, let us begin with the basic properties of the spaces to be considered. Throughout the paper, we shall assume that $\mathcal{H}$ is a Hilbert space consisting of analytic functions on $\mathbb{D}$ which satisfies the following two axioms:

(1.1)   For each $f \in \mathcal{H}$, we have $\zeta f \in \mathcal{H}$ and $||\zeta f|| \leq ||f||$.

(1.2)   The analytic polynomials are dense in $\mathcal{H}$; and for each $\lambda \in \mathbb{D}$,

we have that $(\zeta - \lambda)\mathcal{H}$ is closed in $\mathcal{H}$ with

$$\dim \mathcal{H} \ominus (\zeta - \lambda)\mathcal{H} = \dim \mathcal{H} \cap ((\zeta - \lambda)\mathcal{H})^\perp = 1.$$

These two properties immediately imply that

(1.3)   For each $\lambda \in \mathbb{D}$, the evaluation functional $f \to f(\lambda)$ is continuous on $\mathcal{H}$.

(1.4)   For every $\lambda \in \mathbb{D}$, there is a $c_\lambda > 0$ such that $\left\| \dfrac{\zeta - \lambda}{1 - \bar{\lambda}\zeta} f \right\| \geq c_\lambda ||f||$

for all $f \in \mathcal{H}$.

Moreover, from the closed graph theorem we also see that the backward shift L defined above is a bounded linear operator on $\mathcal{H}$. Many common examples of Hilbert spaces of analytic functions satisfy these axioms. Such are the Hardy space, the weighted Bergman spaces, or, more generally, the closure of analytic polynomials in $L^2(\mu)$ for certain measures μ on $\overline{\mathbb{D}}$. Another relevant class of examples are those spaces where M_ζ acts as a weighted shift, i.e., an operator for which there is an orthonormal basis $\{e_n\}$ such that each e_n is mapped to a positive multiple of e_{n+1}. In the case of the operator M_ζ, this immediately implies that each e_n is a multiple of ζ^n and also, that the norm is a weighted l^2-norm of the sequence of Taylor coefficients of the functions in $\mathcal{H}$.

We shall be concerned with the structure of $\mathrm{Lat}(L, \mathcal{H})$, the lattice of invariant subspaces of the backward shift operator. For specific Banach spaces of analytic functions these subspaces have been intensively studied [**DSS**], [**RiSu**], [**AR**], [**ARR**]. For many examples satisfying (1.1) and (1.2), it turns out that whenever a function belongs to a nontrivial invariant subspace for L, it has a meromorphic pseudocontinuation to the exterior of the unit disc $\mathbb{D}_e = \{z \in \mathbb{C}, |z| > 1\}$.

DEFINITION 1.1. A meromorphic function g in $\mathbb{D}_e$ is a pseudocontinuation of the analytic function f in $\mathbb{D}$ if both functions f and g have the same nontangential limits a.e. on the unit circle $\partial\mathbb{D}$.

Recall also that it follows from the Lusin-Privalov uniqueness theorem that a pseudocontinuation is unique if it exists (see e.g., [**RS1**, Remark 6.2.2]). Even more information is available in many of the cases mentioned above. The values of the pseudocontinuation $\tilde{f}$ of a function f in a nontrivial L-invariant subspace $\mathcal{M}$ satisfies the equality

(1.5) $$\tilde{f}(\lambda) = \frac{\langle \frac{f}{\zeta - \lambda}, h \rangle}{\langle \frac{1}{\zeta - \lambda}, h \rangle}, \quad |\lambda| > 1$$

whenever h is a nonzero function in $\mathcal{M}^\perp$. Following [**RS2**], the right hand side of (1.5) is called the h-prolongation of f. Thus, for all $h \in \mathcal{M}^\perp \backslash \{0\}$ the h-prolongation of f equals the pseudocontinuation of f. By the results proved in [**ARR**] for general

Banach spaces of analytic functions, the remarkable fact that the values of the h-prolongations of f coincide for all $h \in \mathcal{M}^{\perp} \setminus \{0\}$, is actually equivalent to the fact that $(I - \lambda L|\mathcal{M})^{-1}$ exists. Thus, in all of these cases, a point $\alpha \in \mathbb{D} \setminus \{0\}$ belongs to the spectrum of the restriction of L to a nontrivial invariant subspace if and only if the right hand side of (1.5) is not defined at α^{-1}, i.e., $(\zeta - \alpha^{-1})^{-1}$ belongs to that subspace. Clearly, from (1.2) the set of these points is discrete in $\mathbb{D}$, and it easily follows from (1.1) that it actually forms a Blaschke sequence.

The argument outlined here breaks down in the case of an invariant subspace $\mathcal{M} \in \mathrm{Lat}(L, \mathcal{H})$ which, at its turn, contains a nontrivial M_ζ-invariant subspace $\mathcal{N}$. Indeed, in this case the h-prolongation of any $f \in \mathcal{N}$ vanishes identically and thus, it cannot equal a pseudocontinuation of f unless $f = 0$. The existence of such invariant subspaces has been proved recently by J. Esterle [**E**]. His examples are L-invariant subspaces of Hilbert spaces $\mathcal{H}$ as above where $M_\zeta|\mathcal{H}$ is a contractive weighted shift with highly irregular weights which however, decay to zero arbitrarily slow. As it has been pointed out by Borichev [**B**], these invariant subspaces $\mathcal{M}$ not only contain functions that fail to satisfy (1.5), but the spectrum of the restriction of L to $\mathcal{M}$ is the closed unit disc. Further details and open questions related to this phenomenon are discussed in [**RS2**].

The purpose of this paper is to present two abstract conditions on the space $\mathcal{H}$ under which, the invariant subspaces of the backward shift L on $\mathcal{H}$ consist of functions that have of pseudocontinuations to the exterior of the unit disc. Both of our conditions can be seen as regularity assumptions on the norm on $\mathcal{H}$. The first condition is very easy to state and is nothing else than a uniform version of condition (1.4):

(1.6)
$$\text{There is a } c > 0 \text{ such that } \left\| \frac{\zeta - \lambda}{1 - \overline{\lambda}\zeta} f \right\| \geq c \|f\|$$

for all $f \in \mathcal{H}$ and all $\lambda \in \mathbb{D}$.

In Theorem 2.2 below, we prove that given any proper invariant subspace for L on $\mathcal{H}$, every function in the subspace has a meromorphic pseudocontinuation in the Nevanlinna class of $\mathbb{D}_e$ such that (1.5) holds for all nonzero $h \in \mathcal{M}^{\perp}$. In particular, the spectrum of the restriction of L to that subspace intersects $\mathbb{D}$ in a discrete set.

It is well-known that condition (1.6) is satisfied for Bergman spaces with so-called standard weights (also see Lemma 4.2). We mention that it follows from the proof of Proposition 4.10 of [**MR**] that all Hilbert spaces with "Bergman-type kernels" satisfy condition 1.6 (see [**MR**] for definitions). Nevertheless we shall see in Section 4 of this paper that condition (1.6) implies that there are sequences (λ_n) in $\mathbb{D}$ with $|\lambda_n| \to 1$ such that the norms of the evaluation functionals at these points grow slower than a negative power of the distance to the boundary. Hence, the condition does not cover weighted shifts with rapidly decreasing weights. A particular class of examples of this type are weighted Bergman spaces with radial weights that decay exponentially to zero near the boundary. Recall that $\mathcal{H}$ is a weighted Bergman space with a radial weight if the norm on $\mathcal{H}$ can be expressed as

$$\|f\|^2 = \int_{\mathbb{D}} |f|^2 d\mu$$

where μ has the form $d\mu = \omega dr \times dt$, for some positive integrable function ω on $[0, 1)$. However, an even stronger result than the conclusion of the above theorem holds true for all such weighted Bergman spaces. Namely, in addition to the existence of pseudocontinuations, nontrivial invariant subspaces for L are always contained in the Nevanlinna class of the unit disc ([**ARR**, Theorem 1.11]). In order to motivate our second abstract assumption, it is useful to recall briefly the argument in [**ARR**] which leads to the proof of this result. Let $\mathcal{M}$ be invariant for L, $f \in \mathcal{M}$ and h be a nonzero function in $\mathcal{M}^\perp$. Then h annihilates all functions of the form $(f - f(\lambda))/(\zeta - \lambda) = (1 - \lambda L)^{-1} Lf$, $f \in \mathcal{M}$, $\lambda \in \mathbb{D}$, or equivalently,

$$(1.7) \qquad f(\lambda) \int_{\mathbb{D}} \frac{\overline{h}}{\zeta - \lambda} d\mu = \int_{\mathbb{D}} \frac{f\overline{h}}{\zeta - \lambda} d\mu,$$

for all $\lambda \in \mathbb{D}$. Now due to the special form of the measure μ and to the fact that h is analytic, it follows immediately that

$$(1.8) \qquad \int_{|\zeta| < |\lambda|} \frac{\overline{h}}{\zeta - \lambda} d\mu = 0.$$

Given $r \in (0, 1)$, we can split the integrals involved in (1.7) according to the regions $\{|\zeta| < r\}$ and $\{|\zeta| > r\}$ and use (1.8) to obtain an equality of the form

$$f(r\lambda)\overline{F_r(\lambda)} = \overline{G_r(\lambda)} + H_r(\lambda),$$

a.e. on $\partial\mathbb{D}$, where F_r, G_r, H_r have uniformly bounded norms in the usual Hardy spaces H^p, $0 < p < 1$ for almost every $r \in (0, 1)$ and $H_r \to 0$ when $r \to 1^-$ in H^p, $0 < p < 1$. Thus, by letting $r \to 1^-$ we see that there are functions $F, G \in H^p$, $0 < p < 1$ such that $fF \in H^p$ and $f\overline{F} = \overline{G}$ a.e. on $\partial\mathbb{D}$, which leads to the desired result. This special representation of the functions in nontrivial L-invariant subspaces is essentially based on the equality (1.8). It turns out that it is possible to formulate an abstract regularity condition for the norm on a Hilbert space of analytic functions which leads to a similar situation.

Let $\mathcal{H}$ be a Hilbert space of analytic functions satisfies the conditions (1.1) and (1.2). The following condition will be the hypothesis of Theorem 2.4.

(1.9)   There exists a set $A(\mathcal{H}) \subseteq (0, 1)$ with $\sup A(\mathcal{H}) = 1$ such that for every $r \in A(\mathcal{H})$, the scalar product on $\mathcal{H}$ can be written as a sum of two scalar products $\langle \cdot, \cdot \rangle = \langle \cdot, \cdot \rangle_r + \langle \cdot, \cdot \rangle_{\frac{1}{r}}$ such that $\frac{1}{r} M_\zeta$ and rL are contractive with respect to the norms induced by $\langle \cdot, \cdot \rangle_r$ and $\langle \cdot, \cdot \rangle_{\frac{1}{r}}$, respectively.

It is not difficult to verify that weighted Bergman spaces with radial weights satisfy this condition if $\| \cdot \|_r, \| \cdot \|_{\frac{1}{r}}$ are defined as integrals over $\{|z| < r\}$ and $\{|z| \geq r\}$ respectively. As we shall see in Section 4 of this paper, (1.9) applies to many other examples, especially to weighted shifts. For spaces $\mathcal{H}$ which satisfy (1.9), we prove in Theorem 2.4 that any nontrivial invariant subspace for L on $\mathcal{H}$ is contained in the Nevanlinna class of the unit disc. Moreover, every function in the subspace has a meromorphic pseudocontinuation in the Nevanlinna class of $\mathbb{D}_e$ and the spectrum of the restriction of L to that subspace intersects $\mathbb{D}$ in a discrete set.

One application of Theorems 2.2 and 2.4 concerns the index of ζ-invariant subspaces in the Cauchy dual $\mathcal{H}'$ of $\mathcal{H}$. This is presented in Section 3 of the paper. $\mathcal{H}'$ consists of analytic functions of the form

$$f(\lambda) = \langle g, (1 - \overline{\lambda}\zeta)^{-1} \rangle_{\mathcal{H}}, \quad \lambda \in \mathbb{D},$$

where $g \in \mathcal{H}$ and $\|f\|_{\mathcal{H}'} = \|g\|_{\mathcal{H}}$. The space $\mathcal{H}'$ satisfies (1.2)-(1.4), and M_ζ is a bounded expansive operator on this space. Given a ζ-invariant subspace $\mathcal{N}$ of $\mathcal{H}'$, the index of $\mathcal{N}$ is defined by

$$\operatorname{ind} \mathcal{N} = \dim \mathcal{N}/\zeta\mathcal{N} = \dim \mathcal{N} \cap (\zeta\mathcal{N})^\perp.$$

From the results in [**Ri**], we will derive the fact that whenever the space $\mathcal{H}$ satisfies (1.1), (1.2) and either one of the assumptions (1.6) or (1.9), then every M_ζ-invariant subspace $\mathcal{N}$ of $\mathcal{H}'$ has index one (Corollary 3.1). This is no longer true for the Cauchy duals of the spaces considered by Esterle in [**E**] which contain ζ-invariant subspaces of arbitrary index. This was pointed out by Borichev, [**B**]. Since no explicit proof of this fact is available in published form, we have included an outline of his argument, see Proposition 3.2.

In the last section of this paper, we present several examples of weighted shift operators that satisfy our conditions. We will consider spaces $\mathcal{H}_w$ where the norm has the form

$$\left\| \sum_{n=0}^\infty a_n \zeta^n \right\|^2 = \sum_{n=0}^\infty |a_n|^2 w_n,$$

with $w = (w_n)_{n \geq 0}$ a fixed nonincreasing sequence of positive numbers with

$$\lim_{n \to \infty} \frac{w_{n+1}}{w_n} = 1.$$

In [**AB**, p. 24], it is pointed out that if $\frac{w_{n+1}}{w_n}$ is nondecreasing with n then the norm on $\mathcal{H}_w$ is equivalent to a Bergman space norm with a radial weight and that norm will satisfy (1.9). Hence the conclusion of Theorem 2.4 holds for the space $\mathcal{H}_w$. The conditions we consider in Section 4 go beyond this regularity condition. For example, we show that if we set

$$\alpha_- = \liminf_{n \to \infty} (n+1)\left(1 - \frac{w_{n+1}}{w_n}\right), \qquad \alpha_+ = \limsup_{n \to \infty} (n+1)\left(1 - \frac{w_{n+1}}{w_n}\right),$$

then the space $\mathcal{H}$ satisfies (1.6), whenever $\alpha_+ - \alpha_- < 1$ (see Corollary 4.3), while if

$$\sup\left\{r \in (0,1) : \sum_{n=0}^\infty \frac{w_n}{r^{2n+2}} \max\left\{r^2 - \frac{w_{n+1}}{w_n}, 0\right\} < 1\right\} = 1$$

then $\mathcal{H}$ satisfies (1.9) (Corollary 4.6). This last condition applies, for example, to all strictly decreasing weight sequences w with the property that

$$\alpha_+ \left(\ln \frac{\alpha_+}{\alpha_-} - \left(1 - \frac{\alpha_-}{\alpha_+}\right)\right) e^{\alpha_+ - \alpha_-} < 1,$$

and weight sequences satisfying $1 - \frac{w_{n+1}}{w_n} = \frac{\alpha}{n^\gamma} + o(1/n)$ as $n \to \infty$, for some $\alpha > 0$ and $0 < \gamma < 1$ (Examples 4.8 and 4.9). It also is satisfied for strictly decreasing sequences $\{w_n\}$ such that there is a k_0 such that $\frac{w_{m+1}}{w_m} > \frac{w_{n+1}}{w_n}$ whenever $m \geq n + k_0$ (Lemma 4.7).

2. Main results

This section contains the proofs of the main results announced in the introduction. We begin with some basic estimates of contractive operators on separable Hilbert spaces which are needed in our arguments. The result below can be found in [**ARS**] together with a more detailed analysis and further estimates for contractions. For the sake of completeness, we have included an outline of the proof.

For a complex-valued function u on $\mathbb{D}$, we shall denote the nontangential limit of u at at a point $\zeta \in \partial\mathbb{D}$ by nt-$\lim_{\lambda \to z} u(\lambda)$. Analogous notations will be used for the nontangential limit superior and limit inferior of real-valued functions at points of $\partial\mathbb{D}$.

LEMMA 2.1. *Let $\mathcal{K}$ be a separable Hilbert space and T be a contraction on $\mathcal{K}$. Then for every $x \in \mathcal{K}$ we have*

$$\text{nt-}\limsup_{\lambda \to z}(1 - |\lambda|^2)\|(1 - \lambda T)^{-1}x\|^2 < \infty$$

a.e. on $\partial\mathbb{D}$. If $\lim_{n \to \infty} \|T^n x\| = 0$ then

$$\text{nt-}\lim_{\lambda \to z}(1 - |\lambda|^2)\|(1 - \lambda T)^{-1}x\|^2 = 0,$$

a.e. on $\partial\mathbb{D}$.

PROOF. The first part follows immediately from the inequality

$$(1 - |\lambda|^2)\|(1 - \lambda T)^{-1}x\|^2 \leq \|(1 - \lambda T)^{-1}x\|^2 - \|\lambda T(1 - \lambda T)^{-1}x\|^2 = u_x(\lambda).$$

Indeed, a direct calculation shows that the right hand side of this inequality is a harmonic function of $\lambda \in \mathbb{D}$ which is also nonnegative. Such functions u_x have finite nontangential limits $u_x(z)$ for a.e. $z \in \partial\mathbb{D}$. This proves the first part of the Lemma.

Since $(I - \lambda T)^{-1}x = x + \lambda(I - \lambda T)^{-1}Tx$ one verifies inductively that for all $n \in \mathbb{N}$ we have nt-$\limsup_{\lambda \to z}(1-|\lambda|^2)\|(1-\lambda T)^{-1}x\|^2 =$ nt-$\limsup_{\lambda \to z}(1-|\lambda|^2)\|(1-\lambda T)^{-1}T^n x\|^2$. Thus, for all $n \in \mathbb{N}$ and a.e. $z \in \partial\mathbb{D}$ we obtain nt-$\limsup_{\lambda \to z}(1-|\lambda|^2)\|(1-\lambda T)^{-1}x\|^2 \leq u_{T^n x}(z)$, where $u_{T^n x}$ is the harmonic function as in the first paragraph of the proof, but associated with $T^n x$. Note that $\|u_{T^n x}\|_{L^1(\partial\mathbb{D})} \leq u_{T^n x}(0) = \|T^n x\|^2 \to 0$ as $n \to \infty$, hence a subsequence $u_{T^{n_j} x}$ converges to 0 a.e., and this completes the proof. $\qquad\square$

With these preparations we can now turn to our first result about invariant subspaces for the backward shift.

THEOREM 2.2. *Let $\mathcal{H}$ be a Hilbert space of analytic functions satisfying conditions (1.1), (1.2) and (1.6) and let $\mathcal{M} \in \text{Lat}(L, \mathcal{H})$ be a nontrivial invariant subspace. Then every $f \in \mathcal{M}$ has a meromorphic pseudocontinuation that belongs to the Nevanlinna class of the exterior of the closed unit disc and satisfies (1.5) for all $h \in \mathcal{M}^{\perp} \setminus \{0\}$. In particular, $\sigma(L|\mathcal{M}) \cap \mathbb{D}$ is a discrete subset of $\mathbb{D}$.*

PROOF. We begin with the simple observation that for $f, h \in \mathcal{H}$ the function

$$F_{f,h}(\lambda) = \bar{\lambda}\langle f(1 - \bar{\lambda}\zeta)^{-1}, h\rangle, \quad \lambda \in \mathbb{D}$$

is the complex conjugate of the Cauchy transform of a finite measure on $\partial\mathbb{D}$. Indeed, this follows by an application of the spectral theorem to the minimal unitary dilation of M_ζ. Consequently, $F_{f,h}$ has nontangential limits a.e. on $\partial\mathbb{D}$ which are the same as the nontangential limits of the function $\lambda \mapsto F_{f,h}(\frac{1}{\lambda})$, $|\lambda| > 1$. Clearly, this last function is again a Cauchy transform of a finite measure on $\partial\mathbb{D}$. In particular, it belongs to the Nevanlinna class of the exterior of the closed unit disc. Of course, the reasoning applies to the function $F_{1,h}(\lambda) = \bar{\lambda}\langle (1 - \bar{\lambda}\zeta)^{-1}, h\rangle$, $\lambda \in \mathbb{D}$ and in addition, if $h \neq 0$ the function $F_{1,h}$ does not vanish identically because polynomials are dense in $\mathcal{H}$. Our goal is to prove that each $f \in \mathcal{M}$ has nontangential limits a.e. on $\partial \mathbb{D}$ which satisfy

$$(2.1) \qquad \text{nt-} \lim_{\lambda \to z} f(\lambda) = \text{nt-} \lim_{\lambda \to z} \frac{F_{f,h}(\lambda)}{F_{1,h}(\lambda)}$$

for every $h \in \mathcal{M}^{\perp} \setminus \{0\}$. By the above considerations we then obtain that f has the pseudocontinuation $\lambda \mapsto F_{f,h}(1/\lambda)/F_{1,h}(1/\lambda)$, $|\lambda| > 1$ which is just (1.5). As we have already shown, this function belongs to the Nevanlinna class of the exterior of the closed unit disc.

We start with the fact that for $|\lambda| < 1$

$$\langle (f - f(\lambda))(\zeta - \lambda)^{-1}, h \rangle = \langle (1 - \lambda L)^{-1} L f, h \rangle = 0$$

which implies

$$\bar{\lambda} \langle (f - f(\lambda))(1 - \bar{\lambda} \zeta)^{-1}, h \rangle = F_{f,h}(\lambda) - f(\lambda) F_{1,h}(\lambda)$$

$$= (1 - |\lambda|^2) \left\langle \frac{f - f(\lambda)}{(\zeta - \lambda)(1 - \bar{\lambda} \zeta)}, h \right\rangle.$$

Let us now estimate the last expression above. By (1.6) we have

$$\left\| \frac{f - f(\lambda)}{\zeta - \lambda} \right\| \leq \frac{1}{c} \left\| \frac{f - f(\lambda)}{1 - \bar{\lambda} \zeta} \right\|$$

which leads to

$$\left| \left\langle \frac{f - f(\lambda)}{(\zeta - \lambda)(1 - \bar{\lambda} \zeta)}, h \right\rangle \right| \leq \frac{1}{c} \left\| \frac{f - f(\lambda)}{1 - \bar{\lambda} \zeta} \right\| \left\| (1 - \lambda M_{\zeta}^*)^{-1} h \right\|.$$

Thus,

$$(2.2) \quad |F_{f,h}(\lambda) - f(\lambda) F_{1,h}(\lambda)| \leq \frac{1}{c} (1 - |\lambda|^2) \left\| \frac{f - f(\lambda)}{1 - \bar{\lambda} \zeta} \right\| \left\| (1 - \lambda M_{\zeta}^*)^{-1} h \right\|$$

$$\leq \frac{1}{c} (1 - |\lambda|^2) \left(\left\| \frac{f}{1 - \bar{\lambda} \zeta} \right\| + |f(\lambda)| \left\| \frac{1}{1 - \bar{\lambda} \zeta} \right\| \right) \left\| (1 - \lambda M_{\zeta}^*)^{-1} h \right\|$$

From the first part of Lemma 2.1, it follows that

$$\text{nt-} \limsup_{\lambda \to z} \sqrt{1 - |\lambda|^2} \left\| \frac{g}{1 - \bar{\lambda} \zeta} \right\| < \infty$$

a.e. on $\partial \mathbb{D}$ and for all $g \in \mathcal{H}$. Also, we claim that

$$(1 - |\lambda|^2) \|(I - \lambda M_{\zeta}^*)^{-1} h\|^2 \to 0$$

as λ approaches a.e. $z \in \partial \mathbb{D}$ nontangentially. This follows immediately from the second part of Lemma 2.1 once we show that $\|(M_{\zeta}^*)^n h\| \to 0$ when $n \to \infty$. But this last condition is verified whenever h is a finite linear combination of reproducing kernels for $\mathcal{H}$ and since this set is dense in $\mathcal{H}$ and M_{ζ}^* is a contraction, the condition holds for all $h \in \mathcal{H}$.

From these observations, we conclude that there is a subset E of $\partial \mathbb{D}$ of full measure such that for each $z \in E$ we have:

1) $F_{f,h}$ has a nontangential limit at z,
2) $F_{1,h}$ has a nonzero nontangential limit at z,
3) $\text{nt-} \lim_{\lambda \to z} (1 - |\lambda|^2) \|(I - \lambda M_{\zeta}^*)^{-1} h\|^2 = 0$.
4) $\text{nt-} \limsup_{\lambda \to z} \sqrt{1 - |\lambda|^2} \|(1 - \bar{\lambda} \zeta)^{-1} f\| < \infty$

5) nt-$\limsup\limits_{\lambda\to z} \sqrt{1-|\lambda|^2}\|\frac{1}{1-\bar{\lambda}\zeta}\| < \infty$.

We will now use (2.2) to show that

$$\text{nt-}\limsup_{\lambda\to z} |f(\lambda)| < \infty$$

for each point $z \in E$. Indeed, if λ_n converges nontangentially to $z \in E$ such that $|f(\lambda_n)| \to \infty$, then the left hand side of (2.2) is $O(|f(\lambda_n)|)$, while the right hand side is $o(|f(\lambda_n)|)$. But then from (2.2) we see that (2.1) holds at every point of E.

To prove the assertion about the spectrum of $L|\mathcal{M}$ we use Corollary 2.3 and Proposition 2.8 of [**ARR**]. According to these results we have for $|\lambda| > 1$ that $((1-\lambda L)|\mathcal{M})^{-1}$ exists if and only if the value

$$c_\lambda(f,g) = \langle \zeta f(\zeta-\lambda)^{-1}, g\rangle/\langle\lambda(\zeta-\lambda)^{-1}, g\rangle$$

does not depend on the choice of $g \in \mathcal{M}^\perp$, $g \neq 0$. Since $\frac{\zeta}{\zeta-\lambda} = 1 + \frac{\lambda}{\zeta-\lambda}$ we see that $c_\lambda(f,g) = \frac{F_{f,g}(1/\bar{\lambda})}{F_{1,g}(1/\bar{\lambda})}$, whenever $F_{1,g}(1/\bar{\lambda}) \neq 0$. But the above reasoning shows that whenever λ is not a pole of the meromorphic pseudocontinuation of f to the exterior disc, $c_\lambda(f,g)$ equals the value of this pseudocontinuation at λ. Since the pseudocontinuation is unique the result follows. □

The assumption that polynomials are dense in $\mathcal{H}$ is not really necessary for the above argument. In fact, the same reasoning applies whenever $\mathcal{M}$ is a nontrivial L-invariant subspace that does not contain all polynomials.

Let us now turn to our second result. As pointed out in the Introduction, Theorem 2.2 can be improved for Hilbert spaces of analytic functions that satisfy (1.9). In fact, in addition to the existence of pseudocontinuations we shall show that the subspaces under consideration are always contained in the Nevanlinna class of the unit disc.

Let us begin with some simple observations derived from the condition (1.9).

PROPOSITION 2.3. *Let $\mathcal{H}$ be a Hilbert space of analytic functions in $\mathbb{D}$ satisfying* (1.1), (1.2) *and* (1.9). *Then*

(i) *For all $r \in A(\mathcal{H})$ the operator $\frac{1}{r}M_\zeta$ is expansive with respect to the norm induced by $\langle\cdot,\cdot\rangle_{\frac{1}{r}}$.*

(ii) *There exists an equivalent Hilbert-space norm $\|\;\|_1$ on $\mathcal{H}$ such that $(\mathcal{H}, \|\;\|_1)$ satisfies* (1.9) *for all r in some set $A_1(\mathcal{H}) \subset (0,1)$, $\sup A_1(\mathcal{H}) = 1$ such that*

$$\liminf_{\substack{r\to 1^-\\r\in A_1(\mathcal{H})}} \langle f,f\rangle_{1,r} > 0$$

for all $f \in \mathcal{H}$, $f \neq 0$.

PROOF. (i) Since rL is contractive with respect to the norm induced by $\langle\cdot,\cdot\rangle_{\frac{1}{r}}$, we have

$$\frac{1}{r^2}\langle \zeta f, \zeta f\rangle_{\frac{1}{r}} \geq \langle L\zeta f, L\zeta f\rangle_{\frac{1}{r}} = \langle f,f\rangle_{\frac{1}{r}}.$$

(ii) Let $r_0 \in (0,1)$ be fixed and set

$$\|f\|_1^2 = \|f\|^2 + \int_{\partial\mathbb{D}} |f(r_0 z)|^2 \frac{|dz|}{2\pi}.$$

By (1.3) it follows immediately that $\| \cdot \|_1$ is equivalent to the original norm and clearly, (1.9) is satisfied with $A_1(\mathcal{H}) = A(\mathcal{H}) \cap (r_0, 1)$ and

$$\langle f, g \rangle_{1,r} = \langle f, g \rangle_r + \int_{\partial \mathbb{D}} f(r_0 z) \overline{g(r_0 z)} \frac{|dz|}{2\pi}, \quad \langle f, g \rangle_{1,\frac{1}{r}} = \langle f, g \rangle_{\frac{1}{r}}.$$

The fact that this new norm satisfies the condition in the statement is straightforward. $\qquad \Box$

THEOREM 2.4. *Let $\mathcal{H}$ be a Hilbert space of analytic functions satisfying conditions (1.1), (1.2) and (1.9). Let $\mathcal{M} \in \mathrm{Lat}(L, \mathcal{H})$ be a nontrivial invariant subspace for L. Then $\mathcal{M}$ is contained in the Nevanlinna class and every $f \in \mathcal{M}$ has a meromorphic pseudocontinuation that belongs to the Nevanlinna class of the exterior of the closed unit disc. Furthermore, (1.5) holds for all $h \in \mathcal{M}^\perp \setminus \{0\}$ and $\sigma(L|\mathcal{M}) \cap \mathbb{D}$ is a discrete subset of $\mathbb{D}$.*

PROOF. We shall first show that $\mathcal{M}$ is contained in the Nevanlinna class of the unit disc. Let $f \in \mathcal{M}$ and $h \in \mathcal{M}^\perp, h \neq 0$ and denote by $\| \cdot \|_r, \| \cdot \|_{\frac{1}{r}}$ the norms induced by $\langle \cdot, \cdot \rangle_r$ and $\langle \cdot, \cdot \rangle_{\frac{1}{r}}$ respectively. By Proposition 2.3 (ii) we can assume without loss of generality that $\inf\{\|h\|_r, \ r \in A(\mathcal{H})\} > 0$. For $|\lambda| < 1$ we have

$$\langle (f - f(\lambda))(\zeta - \lambda)^{-1}, h \rangle = \langle (1 - \lambda L)^{-1} L f, h \rangle = 0.$$

If $r = |\lambda| \in A(\mathcal{H})$, use (1.9) to write

$$(2.3) \qquad 0 = \langle (f - f(\lambda))(\zeta - \lambda)^{-1}, h \rangle_r + \langle (f - f(\lambda))(\zeta - \lambda)^{-1}, h \rangle_{\frac{1}{r}}.$$

We claim that

$$\langle (f - f(\lambda))(\zeta - \lambda)^{-1}, h \rangle_r = \lim_{\xi \to \lambda} \langle (f - f(\xi))(\zeta - \xi)^{-1}, h \rangle_r.$$

Indeed, if we denote by $\mathcal{H}_r$ the completion of $\mathcal{H}$ with respect to $\langle \cdot, \cdot \rangle_r$ then (1.9) together with von Neumann's inequality imply that every function h, analytic in a disc $\{|z| < r'\}$ with $r' > r$, multiplies $\mathcal{H}$ into $\mathcal{H}_r$. Moreover, the norm of the induced operator does not exceed the supremum norm of h on the disc of radius r. Thus, since f is analytic in $\mathbb{D}$ and $(f - f(\xi))(\zeta - \xi)^{-1}$ converges uniformly to $(f - f(\lambda))(\zeta - \lambda)^{-1}$ on $\{|z| \leq r\}$ when $\xi \to \lambda$, we obtain that $(f - f(\xi))(\zeta - \xi)^{-1} \cdot 1$ converges in $\mathcal{H}_r$ to $(f - f(\lambda))(\zeta - \lambda)^{-1} \cdot 1$ when $\xi \to \lambda$.

Using the claim and (2.3), we see that

$$\langle (f - f(\lambda))(\zeta - \lambda)^{-1}, h \rangle_{\frac{1}{r}} = \lim_{\xi \to \lambda} \langle (f - f(\xi))(\zeta - \xi)^{-1}, h \rangle_{\frac{1}{r}};$$

and thus, we can write

$$(2.4) \qquad \lim_{\rho \to 1^+} \langle (f - f(\rho\lambda))(\zeta - \rho\lambda)^{-1}, h \rangle_r + \lim_{t \to 1^-} \langle (f - f(t\lambda))(\zeta - t\lambda)^{-1}, h \rangle_{\frac{1}{r}}$$

$$= \lim_{\rho \to 1^+} \langle f(\zeta - \rho\lambda)^{-1}, h \rangle_r - f(\lambda) \lim_{\rho \to 1^+} \langle (\zeta - \rho\lambda)^{-1}, h \rangle_r$$

$$+ \lim_{t \to 1^-} \langle (1 - t\lambda L)^{-1} L f, h \rangle_{\frac{1}{r}} = 0.$$

Let F_r, G_r, H_r be defined in the unit disc by

$$\overline{F_r(z)} = \langle f(\zeta - \frac{r}{\bar{z}})^{-1}, h \rangle_r, \quad \overline{G_r(z)} = \left\langle \left(\zeta - \frac{r}{\bar{z}}\right)^{-1}, h \right\rangle_r,$$

$$H_r(z) = \langle (1 - rzL)^{-1} L f, h \rangle_{\frac{1}{r}}.$$

From the hypothesis (1.9) and the general theory of unitary dilations applied to $M_{\zeta/r}$ and rL in the respective inner product spaces, it follows that F_r, G_r, H_r are the Cauchy transforms of finite measures μ_r, ν_r, ω_r on $\partial\mathbb{D}$ whose total variations satisfy

$$|\mu_r| \leq c\|f\|_r\|h\|_r, \quad |\nu_r| \leq c\|1\|_r\|h\|_r \quad |\omega_r| \leq c\|f\|_{\frac{1}{r}}\|h\|_{\frac{1}{r}}$$

for some absolute constant $c > 0$. This implies (see [**D**, Theorem 3.5]) that $F_r, G_r, H_r \in H^p$ for every $0 < p < 1$ and that there exists an absolute constant $c_p > 0$ depending only on p with

$$(2.5) \qquad \max\{\|F_r\|_p^p, \|H_r\|_p^p\} \leq c_p\|f\|^p\|h\|^p \quad \text{and} \quad \|G_r\|_p^p \leq c_p\|h\|^p.$$

With these considerations, we can rewrite (2.4) as

$$\overline{F_r(z)} - f(rz)\overline{G_r(z)} + H_r(z) = 0$$

a.e. on $\partial\mathbb{D}$, and from (2.5) we deduce that

$$\int_{\partial\mathbb{D}} |f(rz)G_r(z)|^p \frac{|dz|}{2\pi} \leq \|F_r\|_p^p + \|H_r\|_p^p \leq 2c_p\|f\|^p\|h\|^p$$

whenever $0 < p < 1$ and $r \in A(\mathcal{H})$. Also from (2.5) we see that $\{G_r, \ r \in A(\mathcal{H})\}$ is a normal family. Note that from our assumptions it follows that there is a sequence $\{r_n\}$ in $A(\mathcal{H})$ converging to 1 such that G_{r_n} converges uniformly on compacts to a nonzero function $G \in H^p$. Otherwise, we would have $G_r(z) \to 0$ uniformly on compacts when $r \to 1^-, r \in A(\mathcal{H})$ which implies that $\langle q, h\rangle_r \to 0$, $r \to 1^-$, $r \in A(\mathcal{H})$, for every polynomial q. Since polynomials are dense in $\mathcal{H}$, this contradicts the fact that $\inf\{\|h\|_r, \ r \in A(\mathcal{H})\} > 0$. Thus, if G is as above, then for $0 < \rho < 1$ fixed but arbitrary and for all $0 < p < 1$ we deduce from (2.5) that

$$\int_{\partial\mathbb{D}} |f(\rho z)G(\rho z)|^p \frac{|dz|}{2\pi} \leq \lim_{n\to\infty} \int_{\partial\mathbb{D}} |f(\rho r_n z)G_{r_n}(\rho z)|^p \frac{|dz|}{2\pi} \leq 2c_p\|f\|^p\|h\|^p$$

which shows that $fG \in H^p$, and hence f belongs to the Nevanlinna class.

To prove the remaining part of the statement, we proceed precisely as in the proof of the preceding theorem. We are going to show first that for $f \in \mathcal{M}$, $h \in \mathcal{M}^\perp$, $h \neq 0$, we have

$$(2.6) \qquad \text{nt-}\lim_{\lambda \to z} \langle (f - f(\lambda))(1 - \bar{\lambda}\zeta)^{-1}, h\rangle = 0$$

a.e. on $\partial\mathbb{D}$. Recall from the considerations made at the beginning of the proof of Theorem 2.2 that the functions

$$F_{f,h}(\lambda) = \bar{\lambda}\langle (1 - \bar{\lambda}\zeta)^{-1}f, h\rangle, \quad F_{1,h}(\lambda) = \bar{\lambda}\langle (1 - \bar{\lambda}\zeta)^{-1}, h\rangle, \quad \lambda \in \mathbb{D}$$

are complex conjugates of Cauchy transforms of finite measures on $\partial\mathbb{D}$ and hence, they have nontangential limits a.e. on $\partial\mathbb{D}$. Then (2.6) will imply that the nontangential boundary values of f (which exist a.e. by the previous argument) satisfy

$$f(z) = \text{nt-}\lim_{\lambda \to z} F_{f,h}(z)/F_{1,h}(z)$$

a.e. on $\partial\mathbb{D}$. The same arguments as in the proof of Theorem 2.2 show that the right hand side equals the nontangential limit from the outside of the unit disc of a quotient of two Cauchy transforms of finite measures on $\partial\mathbb{D}$ such that the denominator does not vanish identically, because h cannot annihilate all polynomials. Consequently, f has a pseudocontinuation that belongs to the Nevanlinna class of the exterior of $\mathbb{D}$ and satisfies (1.5). Finally, the fact that $\sigma(L|\mathcal{M}) \cap \mathbb{D}$ is a discrete subset of $\mathbb{D}$ follows with the same argument as in the proof of Theorem 2.2. Thus, it remains to prove that (2.6) holds.

It follows immediately from what we have already proved that the nontangential limit in (2.6) exists for almost every $z \in \partial\mathbb{D}$. In order to prove (2.6), it will be sufficient to show that there exists a sequence $\{r_n\}$ tending to 1 from below such that

$$(2.7) \qquad \lim_{n \to \infty} \langle (f - f(r_n z))(1 - r_n \bar{z}\zeta)^{-1}, h \rangle = 0$$

for a.e. $z \in \partial\mathbb{D}$. As in the proof of Theorem 2.2 we use the fact that $h \in \mathcal{M}^{\perp}$ to conclude that

$$\langle (f - f(\lambda))(1 - \bar{\lambda}\zeta)^{-1}, h \rangle = (1 - |\lambda|^2)\langle (f - f(\lambda))(\zeta - \lambda)^{-1}, (1 - \lambda M_\zeta^*)^{-1} h \rangle.$$

For $r \in A(\mathcal{H}), r > 1/3$ and $|\lambda| = \rho = \frac{3r-1}{2}$ we use (1.9) to write

$$(2.8) \quad |(1 - |\lambda|^2)\langle (f - f(\lambda))(\zeta - \lambda)^{-1}, (1 - \lambda M_\zeta^*)^{-1} h \rangle|$$
$$\leq |(1 - |\lambda|^2)\langle (f - f(\lambda))(\zeta - \lambda)^{-1}, (1 - \lambda M_\zeta^*)^{-1} h \rangle_r|$$
$$+ |(1 - |\lambda|^2)\langle (f - f(\lambda))(\zeta - \lambda)^{-1}, (1 - \lambda M_\zeta^*)^{-1} h \rangle_{\frac{1}{r}}|$$
$$= \tau_1(\lambda) + \tau_2(\lambda).$$

We shall conclude the proof by showing that there is a nonzero H^∞-function u such that

$$(2.9) \qquad \liminf_{\rho \to 1^-} \int_{|\lambda|=\rho} [\tau_1(\lambda) + |u(\lambda)|\tau_2(\lambda)]|d\lambda| = 0.$$

For $r \in A(\mathcal{H}), r > 1/3$, we can use the Cauchy-Schwarz inequality to obtain for $\rho = \frac{3r-1}{2}$

$$(1 - \rho^2) \int_{|\lambda|=\rho} |\langle (f - f(\lambda))(\zeta - \lambda)^{-1}, (1 - \lambda M_\zeta^*)^{-1} h \rangle_{\frac{1}{r}}| \frac{|d\lambda|}{2\pi\rho}$$

$$\leq (1 - \rho^2) \left(\int_{|\lambda|=\rho} \left\| \left(1 - \frac{\lambda}{r}rL\right)^{-1} Lf \right\|_{\frac{1}{r}}^2 \frac{|d\lambda|}{2\pi\rho} \int_{|\lambda|=\rho} \left\| (1 - \lambda M_\zeta^*)^{-1} h \right\|^2 \frac{|d\lambda|}{2\pi\rho} \right)^{1/2}.$$

Since rL is contractive with respect to $\|\cdot\|_{\frac{1}{r}}$ as in the proof of Lemma 2.1, we have by Parseval's formula

$$\int_{|\lambda|=\rho} \left\| \left(1 - \frac{\lambda}{r}rL\right)^{-1} Lf \right\|_{\frac{1}{r}}^2 \frac{|d\lambda|}{2\pi\rho} \leq (r^2 - \rho^2)^{-1}\|f\|_{\frac{1}{r}}^2 \leq 9(1 - \rho)^{-1}\|f\|^2;$$

and with the same reasoning, we obtain

$$\lim_{\rho \to 1^-} (1 - \rho^2) \int_{|\lambda|=\rho} \|(1 - \lambda M_\zeta^*)^{-1} h\|^2 \frac{|d\lambda|}{2\pi\rho} = 0.$$

Thus, we conclude from above that

$$\int_{|\lambda|=\rho} (\tau_1(\lambda)d|\lambda| = (1 - \rho^2) \int_{|\lambda|=\rho} |\langle (f - f(\lambda))(\zeta - \lambda)^{-1}, (1 - \lambda M_\zeta^*)^{-1} h \rangle_{\frac{1}{r}}| \frac{|d\lambda|}{2\pi\rho} \to 0$$

when $r \in A(\mathcal{H})$ and $r \to 1^-$.

The estimation of the other term in (2.9) is a little more subtle, but very similar. Let $u \in H^\infty$, $u \neq 0$ with $uf \in H^\infty$ and $\|u\|_\infty, \|uf\|_\infty \leq 1$. For $r \in A(\mathcal{H}), r > 1/3$, let $\rho' = \frac{1+r}{2}$ and use the fact that the function $\lambda \mapsto |u(\lambda)|^2\|(f - f(\lambda))(\zeta - \lambda)^{-1}\|_r^2$ is subharmonic to obtain

$$\int_{|\lambda|=\rho} |u(\lambda)|^2\|(f - f(\lambda))(\zeta-\lambda)^{-1}\|_r^2 \frac{|d\lambda|}{2\pi\rho}$$

$$\leq \int_{|z|=\rho'} |u(z)|^2\|(f - f(z))(\zeta - z)^{-1}\|_r^2 \frac{|dz|}{2\pi\rho'}$$

$$\leq 2\int_{|z|=\rho'} \|f(\zeta - z)^{-1}\|_r^2 + \|(\zeta - z)^{-1}\|_r^2 \frac{|dz|}{2\pi\rho'}$$

$$\leq 2\frac{\|f\|^2 + \|1\|^2}{\rho'^2 - r^2},$$

where the last step follows again by Parseval's formula and from the fact that $M_{\zeta/r}$ is contractive with respect to $\|\cdot\|_r$. Another use of the Cauchy Schwarz inequality yields the estimate

$$(1 - \rho^2)\int_{|\lambda|=\rho} |u(\lambda)\langle(f - f(\lambda))(\zeta - \lambda)^{-1}, (1 - \lambda M_\zeta^*)^{-1}h\rangle_r|\frac{|d\lambda|}{2\pi\rho}$$

$$\leq \sqrt{2}(\|f\| + \|1\|)\frac{1 - \rho^2}{(\rho'^2 - r^2)^{1/2}}\left(\int_{|\lambda|=\rho} \|(1 - \lambda M_\zeta^*)^{-1}h\|^2\frac{|d\lambda|}{2\pi\rho}\right)^{1/2} \to 0$$

as $r \to 1^-$ with the same reasoning as above. As pointed out before, this implies (2.6) and Theorem 2.4 is completely proved. $\qquad\Box$

3. The index of ζ-invariant subspaces in the Cauchy dual

Both of our main theorems can be applied to study invariant subspaces for the operator of multiplication by ζ in the Cauchy dual space of $\mathcal{H}$. The Cauchy dual of $\mathcal{H}$ was defined in the Introduction. An alternate way to define it is as the completion of the set of polynomials with respect to the norm

$$\|p\|' = \sup\left\{\left|\int_{\partial\mathbb{D}} p(z)\overline{q(z)}\frac{|dz|}{2\pi}\right|, \ q \text{ is a polynomial and } \|q\|_{\mathcal{H}} \leq 1\right\},$$

that is, $\mathcal{H}'$ is the dual of $\mathcal{H}$ with respect to the H^2-pairing

$$(3.1) \qquad\qquad \langle f, g\rangle_2 = \lim_{r\to1^-}\int_{\partial\mathbb{D}} f(rz)\overline{g(rz)}\frac{|dz|}{2\pi}.$$

A third equivalent way to define the space $\mathcal{H}'$ is to use the Cowen-Douglas model for the operator L (see [**Ri**]). More precisely, one checks that each $\lambda \in \mathbb{D}$ is an eigenvalue of multiplicity one for L. Since the corresponding eigenvectors $(1 - \bar{\lambda}\zeta)^{-1}$ span $\mathcal{H}$, the general theory of such operators implies that L^* is unitarily equivalent to multiplication by ζ on a Hilbert space of analytic functions and with the correct normalization it turns out that this space coincides with $\mathcal{H}'$. In many cases, the space $\mathcal{H}'$ can be found explicitly. For example, if $M_\zeta|\mathcal{H}$ is a weighted shift, i.e., when $\mathcal{H}$ can be identified with a weighted l^2-space, its Cauchy dual is a weighted l^2-space as well, whose weights are just the reciprocals of the original ones. Moreover, under our hypothesis, the space $\mathcal{H}'$ is a much smaller space that is usually contained in the Hardy space H^2. This happens, for example, if the constant functions belong to

$\zeta \mathcal{H}^{\perp}$ (see also [**ARR**, Section 5] for a more detailed discussion of the Cauchy dual). The most relevant fact for our purposes is that the operators $L|\mathcal{H}$ and $M_{\zeta}^{*}|\mathcal{H}'$ are unitarily equivalent.

Given $\mathcal{N} \in \mathrm{Lat}(M_{\zeta}, \mathcal{H}')$, the lattice of M_{ζ} invariant subspaces of $\mathcal{H}'$, such that $\mathcal{N} \neq \{0\}$, we define the index of $\mathcal{N}$ as

$$\mathrm{ind}\,\mathcal{N} = \dim \mathcal{N}/\zeta \mathcal{N} = \dim \mathcal{N} \cap (\zeta \mathcal{N})^{\perp}.$$

For many examples it is shown in [**ARR**] that the index of such an invariant subspace is always one. The same holds true for the Cauchy duals of the abstract spaces $\mathcal{H}$ considered in the previous section, and this can be deduced from the results in [**Ri**] and the unitary equivalence of the operators $L|\mathcal{H}$ and $M_{\zeta}^{*}|\mathcal{H}'$.

COROLLARY 3.1. *Suppose that $\mathcal{H}$ satisfies the hypotheses of either Theorem 2.2 or Theorem 2.4. Then every nonzero invariant subspace for M_{ζ} on the Cauchy dual $\mathcal{H}'$ has index one.*

PROOF. If $\mathcal{N}$ is such a subspace of $\mathcal{H}'$, the restriction of $(M_{\zeta}|\mathcal{H}')^{*}$ to $\mathcal{N}^{\perp}$ is unitarily equivalent to the restriction of L to a nontrivial invariant subspace denoted by $\mathcal{M}$. Now if $\mathcal{H}$ satisfies the hypothesis of Theorem 2.2 or the one of Theorem 2.4 we can conclude that the set $\sigma(L|\mathcal{M}) \cap \mathbb{D}$ is discrete in $\mathbb{D}$ and hence, the same holds for the spectrum of the restriction of M_{ζ}^{*} to $\mathcal{N}^{\perp}$. By Theorem 4.5 in [**Ri**], this implies that $\mathcal{N}$ has index one. $\qquad\square$

A completely different situation occurs in the spaces constructed recently by Esterle in [**E**]. In order to discuss in more detail some consequences of his work, let us introduce the following notation. For $f \in \mathcal{H}, g \in \mathcal{H}'$ let $f * g$ be the sequence obtained by the convolution of the sequences of Taylor coefficients of f and g. In other words, the n-th term $(f * g)_{n}$, $n \in \mathbb{Z}$, of $f * g$ is given by

$$(f * g)_{n} = \langle M_{\zeta}^{n} f, g \rangle_{2}, \text{ for } n \geq 0 \quad \text{and} \quad (f * g)_{n} = \langle L^{|n|} f, g \rangle_{2}, \text{ for } n < 0.$$

Now Theorem 4.10 in [**E**] together with its proof implies the existence of Hilbert spaces of analytic functions $\mathcal{H}$ which satisfy (1.1) and (1.2) and share the following remarkable property:

(3.2)   For any array w_{ij} of sequences with finite support there exist

$f_{i} \in \mathcal{H}, g_{j} \in \mathcal{H}'$ such that $f_{i} * g_{j} = w_{ij}$.

In fact, Esterle's theorem shows that $\mathcal{H}$ can be chosen such that $M_{\zeta}|\mathcal{H}$ is a weighted shift and that the weights $\|\zeta^{n}\|^{1/2}$ involved in the definition of the norm, decay arbitrarily slow to zero. Moreover, (3.2) actually holds even for arrays of infinite sequences w_{ij} that belong to a certain weighted l^{1}-space (see Proposition 4.6, and Theorems 4.5 and 4.2 in [**E**]).

The proof of the following result was kindly communicated to us by Alexander Borichev.

PROPOSITION 3.2. *Let $\mathcal{H}$ be a Hilbert space of analytic functions that satisfies (1.1), (1.2) and (3.2). Then $\mathrm{Lat}(M_{\zeta}, \mathcal{H}')$ contains invariant subspaces of arbitrary index.*

PROOF. According to the results in [**E**], there exist functions $f_i \in \mathcal{H}$, $g_j \in \mathcal{H}'$ which solve the system of equations

$$f_i * g_j = \delta_{ij} e_0,$$

where $\delta_{ij} = 0$ if $i \neq j$ and $\delta_{ii} = 1$, and $e_0 = (\delta_{0n})$. Then all functions g_j belong to the invariant subspace

$$\mathcal{N}_0 = \{g \in \mathcal{H}', \ (f_i * g)_n = 0, \ n \leq -1, i = 1, 2, ..\}.$$

The point is that no nontrivial linear combination of the functions g_j can belong to $M_\zeta \mathcal{N}_0$. Indeed, $\sum \alpha_j g_j \in M_\zeta \mathcal{N}_0$ implies that

$$0 = \left(f_i * \sum \alpha_j g_j \right)_0 = \sum \alpha_j \delta_{ij}$$

for all i, that is, all coefficients α_j vanish. This shows that $[g_j]$ are linearly independent in $\mathcal{N}_0 / M_\zeta \mathcal{N}_0$, i.e., ind $\mathcal{N}_0 = \infty$. Also, given any positive integer N, let $\mathcal{N}$ be the invariant subspace generated by the functions g_j, $j = 1, 2, \ldots, N$, that is, the closed span of all polynomial multiples of these functions. Then it is easy to see that ind $\mathcal{N} \leq N$, and by the above argument we have ind $\mathcal{N} = N$ which completes the proof. □

4. Weighted shift operators and conditions (1.6) and (1.9)

Recall that an operator T on a separable Hilbert space is called a weighted shift if there exists an orthonormal basis $\{e_n : n \geq 0\}$ of the space such that $T e_n = \alpha_n e_{n+1}$, $n \geq 0$, where $\alpha_n > 0$. It is clear that for spaces $\mathcal{H}$ of analytic functions in the unit disc the fact that $M_\zeta | \mathcal{H}$ is a weighted shift means that the space $\mathcal{H}$ consists of power series in the unit disc whose coefficients belong to a given weighted l^2-space, i.e., the norm of the function $f = \sum_{n \geq 0} a_n \zeta^n$ is given by

$$\|f\|_w^2 = \sum_{n=0}^{\infty} |a_n|^2 w_n.$$

To stress the dependence on the weight sequence, we will denote this space of analytic functions by $\mathcal{H}_w$. In this paper, we shall assume that the weight sequence $w = (w_n)_{n \geq 0}$ is strictly positive, non-increasing, that it satisfies $\lim_{n \to \infty} \frac{w_{n+1}}{w_n} = 1$, and that $w_0 = 1$. It is then a simple exercise to show that $\mathcal{H}_w$ satisfies the conditions (1.1) and (1.2). If the weights w_n are bounded below then $\mathcal{H}_w$ is just $H^2(\mathbb{D})$ with an equivalent norm, while if $w_n \to 0$, $n \to \infty$ it follows immediately by the dominated convergence theorem that $\zeta^n f \to 0$ for each $f \in \mathcal{H}_w$. We want to discuss some conditions on the weight sequence w under which the assumptions (1.6) or (1.9) hold for the space $\mathcal{H}_w$.

PROPOSITION 4.1. *Let $v = (v_n)_{n \geq 0}$ be a weight sequence as above and suppose that $M_\zeta | \mathcal{H}_v$ satisfies (1.6) with the constant $c \in (0, 1)$, i.e.,*

$$\left\| \frac{\zeta - \lambda}{1 - \overline{\lambda} \zeta} f \right\|_v \geq c \|f\|_v, \qquad f \in \mathcal{H}_v, \ \lambda \in \mathbb{D}.$$

Furthermore, let $w = (w_n)_{n \geq 0}$ be a weight sequence for which there exists a non-negative integer n_0 and a constant $b > 0$ such that

$$1 - \frac{v_{n+1}}{v_n} \leq 1 - \frac{w_{n+1}}{w_n} \leq b \left(1 - \frac{v_{n+1}}{v_n} \right)$$

for all $n \geq n_0$. If $b(1 - c^2) < 1$, then

$$\left\| \frac{\zeta - \lambda}{1 - \overline{\lambda}\zeta} \zeta^{n_0} f \right\|_w^2 \geq (1 - b(1 - c^2)) \|\zeta^{n_0} f\|_w^2 \quad \text{for all } f \in \mathcal{H}_w.$$

In particular, $\mathcal{H}_w$ satisfies (1.6).

PROOF. Write $b_\lambda = \frac{\zeta - \lambda}{1 - \overline{\lambda}\zeta}$, then for $h \in \mathcal{H}_w$

$$\|h\|_w^2 - \|b_\lambda h\|_w^2 = (1 - |\lambda|^2) \|D_w (1 - \overline{\lambda} M_\zeta)^{-1} h\|_w^2,$$

where D_w denotes the positive square root of the operator $1 - M_\zeta^* M_\zeta$ acting on $\mathcal{H}_w$. Of course, a similar identity holds in $\mathcal{H}_v$. Hence we can restate the first hypothesis as

$$(1 - |\lambda|^2) \|D_v (1 - \overline{\lambda} M_\zeta)^{-1} f\|_v^2 \leq (1 - c^2) \|f\|_v^2$$

for all $f \in \mathcal{H}_v$, $\lambda \in \mathbb{D}$, and we must show that

$$(1 - |\lambda|^2) \|D_w (1 - \overline{\lambda} M_\zeta)^{-1} \zeta^{n_0} f\|_w^2 \leq b(1 - c^2) \|\zeta^{n_0} f\|_w^2$$

for all $f \in \mathcal{H}_w$, $\lambda \in \mathbb{D}$.

We write $f = \sum_{n \geq 0} a_n \zeta^n \in \mathcal{H}_w$, then

$$(4.1) \qquad (1 - |\lambda|^2) \|D_w (1 - \overline{\lambda} M_\zeta)^{-1} M_\zeta^{n_0} f\|_w^2$$

$$= (1 - |\lambda|^2) \sum_{n=n_0}^{\infty} (w_n - w_{n+1}) \left| \sum_{k=n_0}^{n} a_{k-n_0} \overline{\lambda}^{n-k} \right|^2.$$

From the second assumption we have for all $n \geq n_0$

$$w_n - w_{n+1} = \frac{w_n}{v_n} v_n \left(1 - \frac{w_{n+1}}{w_n} \right) \leq b \frac{w_n}{v_n} v_n \left(1 - \frac{v_{n+1}}{v_n} \right) = b \frac{w_n}{v_n} (v_n - v_{n+1}).$$

Note that the first inequality of the second assumption says that (w_n/v_n) is nonincreasing for $n \geq n_0$. Hence we can use (4.1) to obtain

$$(1 - |\lambda|^2) \|D_w (1 - \overline{\lambda} M_\zeta)^{-1} M_\zeta^{n_0} f\|_w^2$$

$$\leq b(1 - |\lambda|^2) \sum_{n=n_0}^{\infty} (v_n - v_{n+1}) \left(\sum_{k=n_0}^{n} |a_{k-n_0}| \sqrt{\frac{w_k}{v_k}} |\lambda|^{n-k} \right)^2.$$

If we denote by $g = \sum_{k \geq n_0} |a_{k-n_0}| \sqrt{\frac{w_k}{v_k}} \zeta^k$, then $g \in \mathcal{H}_v$, $\|g\|_v = \|\zeta^{n_0} f\|_w$ and the right hand side of the last inequality can be written as

$$b(1 - |\lambda|^2) \|D_v (1 - |\lambda| M_\zeta)^{-1} g\|_v^2 \leq b(1 - c^2) \|g\|_v^2 = b(1 - c^2) \|\zeta^{n_0} f\|_w^2.$$

This concludes the proof of the first assertion of the proposition. The second assertion follows from the fact that M_ζ is bounded below and contractive on $\mathcal{H}_w$, hence the operator $M_\zeta | \mathcal{H}_w$ is similar to $M_\zeta | \zeta^{n_0} \mathcal{H}_w$. $\square$

In order to obtain simple conditions for a weight sequence $(w_n)_{n \geq 0}$ such that (1.6) holds in $\mathcal{H}_w$, we are going to apply the above result with a special choice of the sequence $(v_n)_{n \geq 0}$. For $\beta > -1$ we let v^β be defined by

$$v_n^\beta = (\beta + 1) \int_0^1 r^n (1 - r)^\beta dr, \quad n \geq 0.$$

Using integration by parts, we find that

$$\frac{n+1}{\beta+1}(1 - \frac{v_{n+1}^\beta}{v_n^\beta}) = \frac{\int_0^1 r^{n+1}(1-r)^\beta dr}{\int_0^1 r^n(1-r)^\beta dr} = \frac{v_{n+1}^\beta}{v_n^\beta} < 1$$

and hence

(4.2) $$1 - \frac{v_{n+1}^\beta}{v_n^\beta} = (1 + \frac{n+1}{\beta+1})^{-1} = \frac{\beta+1}{\beta+n+2}.$$

Finally, a direct computation shows that the norm in $\mathcal{H}_{v^\beta}$ is given by

$$\|f\|_{v^\beta}^2 = (\beta+1)\int_\mathbb{D} |f(z)|^2(1-|z|^2)^\beta dA(z).$$

LEMMA 4.2. *If $f \in \mathcal{H}_{v^\beta}$ and $\lambda \in \mathbb{D}$, then*

$$\left\|\frac{\zeta-\lambda}{1-\overline{\lambda}\zeta}f\right\|_{v^\beta}^2 \geq \frac{1}{\beta+2}\|f\|_{v^\beta}^2.$$

PROOF. Note first that v_{n+1}^β/v_n^β increases with n; hence,

$$\|zf\|_{v^\beta}^2 \geq v_1^\beta\|f\|_{v^\beta}^2 = \frac{1}{\beta+2}\|f\|_{v^\beta}^2.$$

Further, as before, use the notation $b_\lambda(z) = \frac{\lambda-z}{1-\overline{\lambda}z}$ and a straightforward change of variable to obtain

$$\|b_\lambda f\|_{v^\beta}^2 = (\beta+1)\int_\mathbb{D} |(b_\lambda f)(z)|^2(1-|z|^2)^\beta dA(z)$$

$$= (\beta+1)\int_\mathbb{D} |zf(b_\lambda(z))|^2|b_\lambda'(z)|^2(1-|b_\lambda(z)|^2)^\beta dA(z)$$

$$\geq \frac{\beta+1}{\beta+2}\int_\mathbb{D} |f(b_\lambda(z))|^2|b_\lambda'(z)|^2(1-|b_\lambda(z)|^2)^\beta dA(z) = \frac{1}{\beta+2}\|f\|_{v^\beta}^2,$$

which gives the desired result. $\qquad\square$

COROLLARY 4.3. *For a weight sequence $w = (w_n)_{n\geq 0}$, let*

$$\alpha_-(w) = \liminf_{n\to\infty}(n+1)\left(1 - \frac{w_{n+1}}{w_n}\right), \qquad \alpha_+(w) = \limsup_{n\to\infty}(n+1)\left(1 - \frac{w_{n+1}}{w_n}\right).$$

If $\alpha_+(w) < \infty$ and $\alpha_+(w) - \alpha_-(w) < 1$, then $\mathcal{H}_w$ satisfies (1.6).

PROOF. In the case $\alpha_-(w) > 0$, the result follows by a direct application of Proposition 4.1 with $v = v^\beta$, $\beta = \alpha_-(w) - \varepsilon - 1$, where $\varepsilon > 0$ is sufficiently small. Indeed, if $0 < \varepsilon < \alpha_-(w)$ and β as above, then there exists n_0 dependent on ε and such that for all $n \geq n_0$ we have $\frac{\alpha_-(w)-\varepsilon}{n+1} < \varepsilon$ and by (4.2)

$$1 - \frac{v_{n+1}^\beta}{v_n^\beta} = \frac{\alpha_-(w)-\varepsilon}{n+1+\alpha_-(w)-\varepsilon} < \frac{\alpha_-(w)-\varepsilon}{n+1} \leq 1 - \frac{w_{n+1}}{w_n}$$

$$\leq \frac{\alpha_+(w)+\varepsilon}{n+1} = \frac{\alpha_+(w)+\varepsilon}{\alpha_-(w)-\varepsilon}\left(1 + \frac{\alpha_-(w)-\varepsilon}{n+1}\right)\left(1 - \frac{v_{n+1}^\beta}{v_n^\beta}\right)$$

$$\leq \frac{\alpha_+(w)+\varepsilon}{\alpha_-(w)-\varepsilon}(1+\varepsilon)\left(1 - \frac{v_{n+1}^\beta}{v_n^\beta}\right).$$

We set $b = \frac{\alpha_+(w)+\varepsilon}{\alpha_-(w)-\varepsilon}(1+\varepsilon)$ and note that Lemma 4.2 implies that we need to verify that $b(1-c^2) = \frac{\alpha_+(w)+\varepsilon}{\alpha_-(w)-\varepsilon+1}(1+\varepsilon) < 1$. The hypothesis implies that this is true for sufficiently small ε.

Next suppose $\alpha_-(w) = 0$. Then the hypothesis implies $\alpha_+(w) < 1$, and we can choose $\varepsilon > 0$ such that $\delta = 1 - (\alpha_+(w) + 2\varepsilon) > 0$. For $0 < t < \varepsilon$, define w^t by

$$w_n^t = w_n \prod_{k=0}^{n-1}\left(1 - \frac{t}{k+1}\right), \quad n \geq 1, \ w_0^t = 1$$

and note that

$$1 - \frac{w_{n+1}^t}{w_n^t} = \frac{t}{n+1} + \left(1 - \frac{t}{n+1}\right)\left(1 - \frac{w_{n+1}}{w_n}\right).$$

Set $\beta = t - 1$, then by (4.2) we have for all n

$$1 - \frac{v_{n+1}^\beta}{v_n^\beta} < \frac{t}{n+1} < 1 - \frac{w_{n+1}^t}{w_n^t}.$$

Furthermore, there exists an n_0 which depends on ε, but is independent of t such that for all $n \geq n_0$

$$1 - \frac{w_{n+1}^t}{w_n^t} \leq \frac{t + \alpha_+(w) + \varepsilon}{n+1} = \frac{t + \alpha_+(w) + \varepsilon}{t}\left(1 + \frac{t}{n+1}\right)\left(1 - \frac{v_{n+1}^\beta}{v_n^\beta}\right)$$

$$\leq \frac{\alpha_+(w) + 2\varepsilon}{t}(1+t)\left(1 - \frac{v_{n+1}^\beta}{v_n^\beta}\right).$$

Thus, in this case we can use Proposition 4.1 with $b = \frac{\alpha_+(w)+2\varepsilon}{t}(1+t)$ and $c^2 = 1/(t+1)$. We obtain $1 - b(1-c^2) = 1 - (\alpha_+(w)+2\varepsilon) = \delta$. Hence we have for every polynomial f

$$\left\|\frac{\zeta - \lambda}{1 - \overline{\lambda}\zeta}\zeta^{n_0} f\right\|_{w^t}^2 \geq \delta\|\zeta^{n_0} f\|_{w^t}^2.$$

Since n_0 and δ do not depend on t, we can let $t \to 0$ to obtain

$$\left\|\frac{\zeta - \lambda}{1 - \overline{\lambda}\zeta}\zeta^{n_0} f\right\|_w^2 \geq \delta\|\zeta^{n_0} f\|_w^2$$

for all polynomials f and the result follows. $\square$

Before we turn to our second assumption, let us discuss some cases where (1.6) does not hold. We shall use the following simple observation which yields a necessary condition for a space $\mathcal{H}$ to satisfy (1.6). If k_λ is the reproducing kernel in $\mathcal{H}$ and $\mathcal{H}$ satisfies (1.1), (1.2) and (1.6) then there is a positive constant C such that

(4.3) $$\Delta \log k_\lambda(\lambda) \leq C(1 - |\lambda|^2)^{-2},$$

where Δ denotes the Laplace operator. Indeed, the condition (1.6) implies that for every $f \in \mathcal{H}$ with $f(\lambda) = 0$ the function $g = f(1 - \overline{\lambda}\zeta)/(\zeta - \lambda)$ belongs to $\mathcal{H}$ and satisfies $\|g\| \leq C\|f\|$. This shows that

$$\sup\{(1 - |\lambda|^2)|f'(\lambda)|, \ f \in \mathcal{H}, f(\lambda) = 0, \|f\| \leq 1\} \leq Ck_\lambda(\lambda).$$

Now a direct computation shows that the function $h_\lambda = \frac{\partial}{\partial\lambda}k_\lambda - \frac{\partial}{\partial\lambda}k_\lambda(\lambda)\frac{k_\lambda}{k_\lambda(\lambda)}$ satisfies $h_\lambda(\lambda) = 0$ and

$$\langle f, h_\lambda \rangle = f'(\lambda)$$

whenever $f \in \mathcal{H}$ with $f(\lambda) = 0$. Thus, the supremum above equals

$$(1 - |\lambda|^2)\|h_\lambda\| = (1 - |\lambda|^2)\left(\Delta k_\lambda(\lambda) - \frac{|\frac{\partial}{\partial\lambda}k_\lambda(\lambda)|^2}{k_\lambda(\lambda)}\right)^{1/2},$$

and our inequality can be rewritten as

$$\frac{\Delta k_\lambda(\lambda)}{k_\lambda(\lambda)} - \frac{|\frac{\partial}{\partial\lambda}k_\lambda(\lambda)|^2}{k_\lambda(\lambda)^2} \leq C(1 - |\lambda|^2)^{-2}$$

which is just (4.3). It is interesting to note that the quantity $\Delta \log k_\lambda(\lambda)$ involved in (4.3) is a unitary invariant for spaces of analytic functions which is also called the Cowen-Douglas curvature. Note also that

$$(1 - |\lambda|^2)^{-2} = -\frac{1}{4}\Delta \log(1 - |\lambda|^2)$$

so that by (4.3), we can conclude that $-C' \log(1 - |\lambda|^2) - \log k_\lambda(\lambda)$ is subharmonic in $\mathbb{D}$. By the maximum principle, this implies that there are sequences (λ_n) with $|\lambda_n| \to 1$ such that $\limsup_{n\to\infty}(C' \log(1 - |\lambda_n|^2) + \log k_{\lambda_n}(\lambda_n)) < \infty$. Thus, if $\mathcal{H}$ satisfies (1.6) there are sequences (λ_n) as above such that $k_{\lambda_n}(\lambda_n)$ grows slower than a negative power of the distance to the boundary. This cannot happen, for example, if $\mathcal{H} = \mathcal{H}_w$ and $w_n = o(n^{-\alpha})$, $n \to \infty$, for all $\alpha > 0$. The verification of this last statement is immediate and will be omitted.

Our next goal is to examine the condition (1.9). Given a weighted shift, that is, a space $\mathcal{H}_w$ as above, there is a simple attempt to produce a decomposition of the norm as prescribed in (1.9). We set for $f = \sum_{n\geq 0}a_n\zeta^n \in \mathcal{H}_w$

$$\|f\|_r^2 = \sum_{n\geq 0}|a_n|^2r^{2n+2}w_n, \quad \|f\|_{\frac{1}{r}}^2 = \|f\|_w^2 - \|f\|_r^2 = \sum_{n\geq 0}|a_n|^2(1 - r^{2n+2})w_n.$$

Of course, $\|\cdot\|_r$ satisfies the requirement in (1.9) since

$$\|\zeta f\|_r^2 = \sum_{n\geq 0}|a_n|^2r^{2n+4}\frac{w_{n+1}}{w_n}w_n \leq r^2\|f\|_r^2,$$

while $\|\cdot\|_{\frac{1}{r}}$ might not satisfy that (1.9). A sufficient condition for this is given below.

PROPOSITION 4.4. *Suppose that*

$$\alpha_+(w) = \limsup_{n\to\infty}(n + 1)\left(1 - \frac{w_{n+1}}{w_n}\right) < 1.$$

Then there exists an equivalent Hilbert space norm on $\mathcal{H}_w$ that satisfies (1.9).

PROOF. If $\alpha_+(w) < 1$, there is a positive integer n_0 such that

$$\frac{w_{n+1}}{w_n} \geq \frac{n}{n + 1}, \quad n \geq n_0 \geq 1.$$

If we set $v_n = w_{n+n_0}$, then the norms on $\mathcal{H}_w$ and $\mathcal{H}_v$ are equivalent, and $\frac{v_{n+1}}{v_n} \geq \frac{n+n_0}{n+n_0+1} \geq \frac{n+1}{n+2}$. Moreover, if we apply the above decomposition on $\mathcal{H}_v$ it suffices to show that $r\|Lf\|_{\frac{1}{r}} \leq \|f\|_{\frac{1}{r}}$. For $f = \sum_{n\geq 0} a_n \zeta^n \in \mathcal{H}_v$, we have

$$r^2\|Lf\|_{\frac{1}{r}}^2 = r^2 \sum_{n=1} (1 - r^{2n}) v_{n-1} |a_n|^2 \leq \sup_{n\geq 1} \frac{(r^2 - r^{2n+2}) v_{n-1}}{(1 - r^{2n+2}) v_n} \|f\|_{\frac{1}{r}}^2.$$

Recall that $\frac{v_{n-1}}{v_n} \leq \frac{n+1}{n}$ and use also the inequality

$$\frac{r^2 - r^{2n+2}}{1 - r^{2n+2}} = 1 - \left(\sum_{k=0}^{n} r^{2k}\right)^{-1} \leq 1 - \frac{1}{n+1}$$

to obtain that

$$\sup_{n\geq 1} \frac{(r^2 - r^{2n+2}) v_{n-1}}{(1 - r^{2n+2}) v_n} \leq 1$$

and the result follows.   □

Let us now describe an alternative way to produce a decomposition of the form (1.9). If $\mathcal{H}$ is any Hilbert space of analytic functions satisfying (1.1), (1.2) and in addition, if $\mathcal{H} \ominus \zeta\mathcal{H}$ consists of constant functions then the following identity holds for all $f \in \mathcal{H}$ and all $0 \leq r < 1$

$$\|f\|^2 = \int_{|\lambda|=r} |f(\lambda)|^2 \frac{|d\lambda|}{2\pi r} + \int_{|\lambda|=r} \langle (M_\zeta^* M_\zeta - r^2) L(1 - \lambda L)^{-1} f, L(1 - \lambda L)^{-1} f \rangle \frac{|d\lambda|}{2\pi r}.$$

A proof of this equality can be found in [**ARi**, Lemma 2.2]. Now write

$$M_\zeta^* M_\zeta - r^2 = D_r^+ - D_r^-$$

with positive operators D_r^+, D_r^- and set

$$(4.4) \quad \|f\|_{\frac{1}{r}}^2 = c(r) \int_{|\lambda|=r} |f(\lambda)|^2 \frac{|d\lambda|}{2\pi r} + \int_{|\lambda|=r} \langle D_r^+ L(1-\lambda L)^{-1} f, L(1-\lambda L)^{-1} f \rangle \frac{|d\lambda|}{2\pi r},$$

where $0 < c(r) < 1$ and

$$(4.5) \quad \|f\|_r^2 = \|f\|^2 - \|f\|_{\frac{1}{r}}^2.$$

While (4.4) clearly defines a norm on $\mathcal{H}$, this is not necessarily the case with the expression in (4.5).

PROPOSITION 4.5. *Let $A(\mathcal{H})$ be the set of all $r \in (0,1)$ for which there exists $0 < c(r) < 1$ such that the sesquilinear form defined by (4.5) is positive definite. If $\sup A(\mathcal{H}) = 1$, then $\mathcal{H}$ satisfies (1.9).*

PROOF. We use first the equalities

$$L(1 - \lambda L)^{-1} f = Lf + \lambda L^2 (1 - \lambda L)^{-1} f$$

together with the Parseval formula to conclude that

$$\int_{|\lambda|=r} \langle D_r^+ L(1 - \lambda L)^{-1} f, L(1 - \lambda L)^{-1} f \rangle \frac{|d\lambda|}{2\pi r} = \langle D_r^+ Lf, Lf \rangle$$

$$+ r^2 \int_{|\lambda|=r} \langle D_r^+ L(1 - \lambda L)^{-1} Lf, L(1 - \lambda L)^{-1} Lf \rangle \frac{|d\lambda|}{2\pi r}.$$

This immediately implies that

$$r^2 \|Lf\|_{\frac{1}{r}}^2 \le \|f\|_{\frac{1}{r}}^2.$$

Similarly, we have

$$L(1 - \lambda L)^{-1} M_\zeta f = f + \lambda L(1 - \lambda L)^{-1} f;$$

and by Parseval's formula,

$$\int_{|\lambda|=r} \langle D_r^- L(1 - \lambda L)^{-1} M_\zeta f, L(1 - \lambda L)^{-1} M_\zeta f \rangle \frac{|d\lambda|}{2\pi r} = \langle D_r^- f, f \rangle$$

$$+ r^2 \int_{|\lambda|=r} \langle D_r^- L(1 - \lambda L)^{-1} f, L(1 - \lambda L)^{-1} f \rangle \frac{|d\lambda|}{2\pi r}.$$

This gives the inequality

$$\|M_\zeta f\|_r^2 \le r^2 \|f\|_r^2$$

and the result follows.   □

The expressions considered in (4.4) and (4.5) are easily computed in the case of a weighted shift. For example, a direct computation shows that on any space $\mathcal{H}_w$ the operators D_r^+ and D_r^- are diagonal with respect to the orthonormal basis $\{e_n = w_n^{-1/2} \zeta^n : n \ge 0\}$ and

$$D_r^+ e_n = \max\left\{\frac{w_{n+1}}{w_n} - r^2, 0\right\} e_n \quad D_r^- e_n = \max\left\{r^2 - \frac{w_{n+1}}{w_n}, 0\right\} e_n.$$

With this setup, it is now relatively easy to show that if $\frac{w_{n+1}}{w_n}$ is nondecreasing, then w satisfies the hypothesis of Proposition 4.5 and so $\mathcal{H}_w$ satisfies (1.9) (see Corollary 4.10). We shall prove a stronger result.

COROLLARY 4.6. *For a weight sequence w, let*

$$\rho(w) = \sup\left\{r \in (0,1) : \sum_{n=0}^{\infty} \frac{w_n}{r^{2n+2}} \max\left\{r^2 - \frac{w_{n+1}}{w_n}, 0\right\} < 1\right\}.$$

If $\rho(w) = 1$, then $\mathcal{H}_w$ satisfies (1.9).

PROOF. We want to apply Proposition 4.5. In order to estimate $\|f\|_r$ as defined by (4.4) and (4.5), note first that for $f = \sum_{n\ge0} a_n \zeta^n \in \mathcal{H}_w$ we have

$$L(1 - \lambda L)^{-1} f(z) = \frac{f(z) - f(\lambda)}{z - \lambda} = \sum_{k\ge0} z^k \sum_{n\ge k+1} a_n \lambda^{n-k-1}.$$

Then, as pointed out above,

$$\langle D_r^- L(1-\lambda L)^{-1} f, L(1-\lambda L)^{-1} f \rangle = \sum_{n=0}^{\infty} w_k \max\left\{r^2 - \frac{w_{k+1}}{w_k}, 0\right\} \left|\sum_{n\ge k+1} a_n \lambda^{n-k-1}\right|^2,$$

and by the Parseval formula,

$$\int_{|\lambda|=r} \langle D_r^- L(1-\lambda L)^{-1}f, L(1-\lambda L)^{-1}f \rangle \frac{|d\lambda|}{2\pi r}$$

$$= \sum_{k=0}^{\infty} w_k \max\left\{r^2 - \frac{w_{k+1}}{w_k}, 0\right\} \sum_{n \geq k+1} |a_n|^2 r^{2n-2k-2}$$

$$= \sum_{n=1}^{\infty} |a_n|^2 r^{2n} \sum_{k=0}^{n-1} \frac{w_k}{r^{2k+2}} \max\left\{r^2 - \frac{w_{k+1}}{w_k}, 0\right\}$$

$$\leq \left(\sum_{k=0}^{\infty} \frac{w_k}{r^{2k+2}} \max\left\{r^2 - \frac{w_{k+1}}{w_k}, 0\right\}\right) \sum_{n=1}^{\infty} |a_n|^2 r^{2n}.$$

If we now set

$$A(\mathcal{H}) = \left\{r \in (0,1): \sum_{n=0}^{\infty} \frac{w_n}{r^{2n+2}} \max\left\{r^2 - \frac{w_{n+1}}{w_n}, 0\right\} < 1\right\}$$

and

$$c(r) = \frac{1 + \sum_{n=0}^{\infty} \frac{w_n}{r^{2n+2}} \max\{r^2 - \frac{w_{n+1}}{w_n}, 0\}}{2} < 1$$

for $r \in A(\mathcal{H})$, then the conditions in Proposition 4.5 are satisfied. Indeed, by assumption we know that $\sup A(\mathcal{H}) = 1$ and from above we have for $r \in A(\mathcal{H})$

$$\|f\|_r^2 \geq c(r) \sum_{n=0}^{\infty} |a_n|^2 r^{2n} - (2c(r)-1)) \sum_{n=0}^{\infty} |a_n|^2 r^{2n} = (1-c(r)) \int_{|\lambda|=r} |f(\lambda)|^2 \frac{|d\lambda|}{2\pi r}.$$

The result now follows by an application of Proposition 4.5. $\qquad \square$

Some situations where this condition applies are described below. We will fix a strictly decreasing weight sequence w and for $n \geq 0$ set

$$\gamma_n = \inf\left\{1 - \frac{w_{k+1}}{w_k} : k \leq n\right\}$$

$$\delta_n = \sup\left\{1 - \frac{w_{k+1}}{w_k} : k \geq n\right\}.$$

Then $\{\delta_n\}$ and $\{\gamma_n\}$ are nonincreasing sequences which satisfy

$$0 < \gamma_n \leq 1 - \frac{w_{n+1}}{w_n} \leq \delta_n \to 0$$

as $n \to \infty$. Furthermore set $m_n = \sup\{k : \gamma_n \leq \delta_k\}$. Note that the definitions imply that for all n we have $n \leq m_n < \infty$ and $\delta_n - \gamma_n \geq \delta_k - \gamma_n \geq 0$ for all $k = n, ..., m_n$.

LEMMA 4.7. *Let w be a strictly decreasing weight sequence such that*

$$\limsup_{n \to \infty} (m_n - n)\gamma_n < \infty \quad \text{and}$$

$$\limsup_{n \to \infty} e^{(m_n - n)\gamma_n} \sum_{k=n+1}^{m_n} \delta_k - \gamma_n < 1.$$

Then $\mathcal{H}_w$ satisfies (1.9).

In particular, if w is a strictly decreasing weight sequence such that

$$\limsup_{n \to \infty} (m_n - n)\delta_n < 1,$$

then $\mathcal{H}_w$ *satisfies* (1.9).

Clearly this implies that $\mathcal{H}_w$ satisfies (1.9) whenever w is strictly decreasing and there is a $k \geq 0$ such that $m_n \leq n + k$ for all n.

PROOF. We are going to apply Corollary 4.6. Let $0 < R < 1$. We have to show that there is an r, $R < r < 1$ such that

$$(4.6) \qquad \sum_{k=0}^{\infty} \frac{w_k}{r^{2k+2}} \max\left\{r^2 - \frac{w_{k+1}}{w_k}, 0\right\} < 1.$$

The hypothesis implies that there is $\delta > 0$ and a positive integer N_0 such that

$$(4.7) \qquad e^{(1+\delta)(m_n-n)\gamma_n} \sum_{k=n+1}^{m_n} (\delta_k - \gamma_n) < 1 \text{ for all } n \geq N_0.$$

Let $0 < \varepsilon < 1$ such that $-\ln(1-x) \leq (1+\delta)x$ for all $0 \leq x \leq \varepsilon$, and $N_1 \geq N_0$ such that $\gamma_n < \varepsilon$ for all $n \geq N_1$. Then

$$(4.8) \qquad (1-\gamma_n)^{-1} = e^{-\ln(1-\gamma_n)} \leq e^{(1+\delta)\gamma_n} \text{ for all } n \geq N_1.$$

Since $\frac{w_{n+1}}{w_n} < 1$ for all n and $\{\frac{w_{n+1}}{w_n}\} \to 1$, we can find an index $n \geq N_1$ such that $\frac{w_{n+1}}{w_n} > R^2$ and $\frac{w_{k+1}}{w_k} \leq \frac{w_{n+1}}{w_n}$ for all $k \leq n$. We will verify (4.6) for r defined by $r^2 = \frac{w_{n+1}}{w_n} = 1 - \gamma_n$.

We have

$$\sum_{k=0}^{\infty} \frac{w_k}{r^{2k+2}} \max\left\{r^2 - \frac{w_{k+1}}{w_k}, 0\right\}$$

$$= \sum_{k=0}^{n} \frac{w_k}{r^{2k+2}} \max\left\{r^2 - \frac{w_{k+1}}{w_k}, 0\right\} + \sum_{k=n+1}^{m_n} \frac{w_k}{r^{2k+2}} \max\left\{r^2 - \frac{w_{k+1}}{w_k}, 0\right\}$$

$$\leq 1 - \frac{w_{n+1}}{r^{2n+2}} + \frac{w_{n+1}}{r^{2m_n+2}} \sum_{k=n+1}^{m_n} (\delta_k - \gamma_n).$$

Since $r^{-2} = (1-\gamma_n)^{-1}$, we have by (4.8) and (4.7)

$$r^{-2(m_n-n)} \sum_{k=n+1}^{m_n} (\delta_k - \gamma_n) \leq e^{(1+\delta)(m_n-n)\gamma_n} \sum_{k=n+1}^{m_n} (\delta_k - \gamma_n) < 1.$$

This verifies the first part of the lemma. To see the final part, we will verify the hypothesis of the first part. It is clear that the first limsup is finite. Furthermore, we note that the hypothesis implies that for all sufficiently large n we have $(m_n-n)\delta_n \leq a < 1$. Hence

$$e^{(m_n-n)\gamma_n} \sum_{k=n+1}^{m_n} (\delta_k - \gamma_n) \leq e^{a\frac{\gamma_n}{\delta_n}}(m_n - n)(\delta_n - \gamma_n) \leq e^{ax}a(1-x),$$

where $0 \leq x = \frac{\gamma_n}{\delta_n} \leq 1$. It is easy to see that this last expression must be less than 1. $\qquad\square$

EXAMPLE 4.8. Let $\alpha_+(w)$ and $\alpha_-(w)$ be as defined in Corollary 4.3. If $0 < \alpha_- \leq \alpha_+ < \infty$, and if

$$\alpha_+\left(\ln\frac{\alpha_+}{\alpha_-} - \left(1 - \frac{\alpha_-}{\alpha_+}\right)\right)e^{\alpha_+ - \alpha_-} < 1,$$

then $\mathcal{H}_w$ satisfies the conclusion of Theorem 2.4. In fact, there is a weight sequence $\tilde{w}$ such that $\mathcal{H}_{\tilde{w}}$ satisfies (1.9) and the norms on $\mathcal{H}_w$ and $\mathcal{H}_{\tilde{w}}$ are equivalent.

There is some overlap with Corollary 4.3. In particular, we note that if α_- is sufficiently large, then this method applies to more weight sequences than Corollary 4.3 and gives a stronger conclusion.

PROOF. Let $\varepsilon > 0$ such that $\alpha_- - \varepsilon > 0$. There is some N such that for all $k \geq N$, we have

$$0 < \frac{\alpha_- - \varepsilon}{k+1} \leq 1 - \frac{w_{k+1}}{w_k} \leq \frac{\alpha_+ + \varepsilon}{k+1}.$$

Thus, by changing at most finitely many of the w_n we may assume that $\{w_n\}$ is strictly decreasing. The changed weight sequence will produce an equivalent norm on $\mathcal{H}_w$.

Now, if $n \geq N$ and $n \leq k \leq m_n$, then

$$\frac{\alpha_- - \varepsilon}{n+1} \leq \gamma_n \leq \delta_k \leq \frac{\alpha_+ + \varepsilon}{k+1}.$$

Thus, $1 \leq \frac{m_n+1}{n+1} \leq \frac{\alpha_+ + \varepsilon}{\alpha_- - \varepsilon}$ and

$$(m_n - n)\gamma_n \leq (m_n - n)\frac{\alpha_+ + \varepsilon}{m_n + 1}$$
$$= \left(1 - \frac{n+1}{m_n+1}\right)(\alpha_+ + \varepsilon)$$
$$\leq \alpha_+ - \alpha_- + 2\varepsilon.$$

Furthermore,

$$\sum_{k=n+1}^{m_n} (\delta_k - \gamma_n) \leq \int_{n+1}^{m_n+1} \frac{\alpha_+ + \varepsilon}{x} dx - (m_n - n)\frac{\alpha_- - \varepsilon}{n+1}$$
$$= (\alpha_+ + \varepsilon)\ln\frac{m_n+1}{n+1} - (\alpha_- - \varepsilon)\left(\frac{m_n+1}{n+1} - 1\right)$$
$$\leq (\alpha_+ + \varepsilon)\ln\frac{\alpha_+ + \varepsilon}{\alpha_- - \varepsilon} - (\alpha_- - \varepsilon)\left(\frac{\alpha_+ + \varepsilon}{\alpha_- - \varepsilon} - 1\right)$$
$$= \ln\frac{\alpha_+ + \varepsilon}{\alpha_- - \varepsilon} - \left(1 - \frac{\alpha_- - \varepsilon}{\alpha_+ + \varepsilon}\right)$$

since the function $f(x) = (\alpha_+ + \varepsilon)\ln x - (\alpha_- - \varepsilon)(x - 1)$ is increasing for $1 \leq x \leq \frac{\alpha_+ + \varepsilon}{\alpha_- - \varepsilon}$. Thus by choosing ε sufficiently small the hypothesis implies

$$\limsup_{n \to \infty} (m_n - n)\gamma_n < \infty$$

and

$$\limsup_{n \to \infty} e^{(m_n-n)\gamma_n} \sum_{k=n+1}^{m_n} (\delta_k - \gamma_n) < 1.$$

Hence the result follows from Lemma 4.7. □

An example of exponentially decreasing weights is given by the following.

EXAMPLE 4.9. Let $\alpha > 0$ and $0 < \gamma < 1$ such that

$$1 - \frac{w_{n+1}}{w_n} = \frac{\alpha}{n^\gamma} + o(1/n) \ \text{ as } n \to \infty,$$

then $\mathcal{H}_w$ satisfies the conclusion of Theorem 2.4.

We omit the verification as it is similar to Example 4.8. One can use the second part of Lemma 4.7. We mention that one can show that such weights are of the general form $w_n = b_n e^{-\frac{\alpha}{1-\gamma} n^{1-\gamma}}$, $b_n = e^{\sum_{k=0}^n \frac{\varepsilon_k}{k+1}}$, where $|\varepsilon_k| \to 0$.

COROLLARY 4.10. *If $M_\zeta | \mathcal{H}_w$ is a hyponormal weighted shift operator, then $\mathcal{H}_w$ satisfies (1.9).*

PROOF. It is well-known that under our assumptions on the weights w the operator $M_\zeta | \mathcal{H}_w$ is hyponormal, if and only if w_{n+1}/w_n is nondecreasing, [**C**, p. 55]. In the notation of Lemma 4.7 that implies that $m_n(w) = n$ for each n. Thus, if w_n is also strictly decreasing, then the result follows from Lemma 4.7. Having observed this, it becomes clear that the general case follows directly from Corollary 4.6. □

As pointed out in the introduction, an alternative proof of this result can be found in [**AB**].

COROLLARY 4.11. *Let $u = (u_n)_{n\geq 0}$ be a sequence of positive numbers such that the sequence $w = (1/u_n)_{n\geq 0}$ of reciprocals satisfies the hypothesis of Lemma 4.7, then every nonzero invariant subspace of $M_\zeta | \mathcal{H}_u$ has index 1.*

PROOF. This follows immediately from Corollary 3.1 and Lemma 4.7. □

We point out that in the particular case mentioned in Corollary 4.10, it follows that the reproducing kernel for the Cauchy dual of $\mathcal{H}_w$ is a so-called Pick kernel (or complete NP kernel). In this case the conclusion of Corollary 4.11 was known, [**MT**].

References

[AB]   E. Abakumov and A. Borichev, *Shift invariant subspaces with arbitrary indices in l^p spaces*, J. Funct. Anal. **188** (2002), no. 1, 1–26.

[ARi]   A. Aleman and S. Richter, *Simply invariant subspaces of H^2 of some multiply connected regions*, Integral Equations Operator Theory **24** (1996), no. 2, 127–155.

[AR]   A. Aleman and W. T. Ross, *The backward shift on weighted Bergman spaces*, Michigan Math. J. **43** (1996), no. 2, 291–319.

[ARR]   A. Aleman, S. Richter and W. T. Ross, *Pseudocontinuations and the backward shift*, Indiana Univ. Math. J. **47** (1998), no. 1, 223–276.

[ARS]   A. Aleman, S. Richter and C. Sundberg, *Analytic contractions, nontangential limits, and the index of invariant subspaces*, preprint, 2005.

[B]   A. Borichev, *On convolution equations with restrictions on supports*, Algebra i Analiz **14** (2002), no. 2, 1–10; St. Petersburg Math. J. **14** (2003), no. 2, 179–187.

[C]   J. B. Conway, *The theory of subnormal operators*, Amer. Math. Soc. (1991), Mathematical Surveys and Monographs, Vol. 36.

[DSS]   R. G. Douglas, H. S. Shapiro and A. L. Shields, *Cyclic vectors and invariant subspaces for the backward shift operator*, Ann. Inst. Fourier (Grenoble) **20** 1970 fasc. 1, 37–76.

[D]   Peter L. Duren, *Theory of H^p Spaces*, Pure and Applied Mathematics, Vol. 38, Academic Press, New York-London, 1970.

[E]   J. Esterle, *Toeplitz operators on weighted Hardy spaces*, Algebra i Analiz **14** (2002), no. 2, 92–116; St. Petersburg Math. J. **14** (2003), no. 2, 251–272.

[K]   P. Koosis, *Introduction to H_p Spaces*, London Mathematical Society Lecture Note Series, 40, Cambridge University Press, Cambridge-New York, 1980.

[MR]   S. McCullough and S. Richter, *Bergman-type reproducing kernels, contractive divisors, and dilations*, J. Funct. Anal. **190** (2002), 447–480.

[MT]   S. McCullough and T. Trent, *Invariant subspaces and Nevanlinna Pick kernels*, J. Funct. Anal. **178** (2000), 226–249.

[Ri]   S. Richter, *Invariant subspaces in Banach spaces of analytic functions*, Trans. Amer. Math. Soc. **304** (1987), no. 2, 585–616.

[RiSu] S. Richter and C. Sundberg, *Invariant subspaces of the Dirichlet shift and pseudocontinuations*, Trans. Amer. Math. Soc. **341** (1994), 863–879.

[RS1]   W. T. Ross and H. S. Shapiro, *Generalized analytic continuation*, University Lecture Series, Vol. 25, Amer. Math. Soc., 2002.

[RS2]   W. T. Ross and H. S. Shapiro, *Prolongations and cyclic vectors*, Comput. Methods Funct. Theory 3(2003), no.1-2, 453-483.

DEPARTMENT OF MATHEMATICS, UNIVERSITY OF LUND, BOX 118, SE-221 00 LUND, SWEDEN
E-mail address: `Aleman@maths.lth.se`

DEPARTMENT OF MATHEMATICS, UNIVERSITY OF TENNESSEE, KNOXVILLE, TN 37996-1300, USA
E-mail address: `Richter@math.utk.edu`

DEPARTMENT OF MATHEMATICS, UNIVERSITY OF TENNESSEE, KNOXVILLE, TN 37996-1300, USA
E-mail address: `Sundberg@math.utk.edu`

Contemporary Mathematics
Volume **404**, 2006

Surjectivity and Invariant Subspaces of Differential Operators on Weighted Bergman Spaces of Entire Functions

Aharon Atzmon and Bruno Brive

Dedicated to Boris Korenblum on his 80-th birthday

1. Introduction and main results

Consider the functional equation

$$(1.1) \qquad \sum_{n \geq 0} a_n \frac{d^n}{dz^n} f(z) = g(z), \quad z \in \mathbf{C},$$

where f and g are entire functions, the a_n are complex numbers, and the series converges uniformly on compact subsets of $\mathbf{C}$. This is an inhomogeneous linear differential equation of (possibly) infinite order with constant coefficients. This class of equations includes also linear difference-differential equations with constant coefficients since, for every complex number σ and every entire function f,

$$\sum_{n \geq 0} \frac{\sigma^n}{n!} \frac{d^n}{dz^n} f(z) = f(z + \sigma), \quad z \in \mathbf{C},$$

and the series converges uniformly on compact subsets of $\mathbf{C}$. We refer to [**6**] and [**9**] for results and references on such functional equations. In this paper we are concerned with equation (1.1) where f and g belong to a weighted L^p-space of entire functions, which is a Banach space on which the differentiation operator $D = d/dz$ is bounded. If $r(D)$ is the spectral radius of D, and ϕ is an analytic function on some neighbourhood of the disc $\{z \in \mathbf{C} : |z| \leq r(D)\}$, then the series $\sum_{n \geq 0} \frac{\phi^{(n)}(0)}{n!} D^n$ converges in the operator norm to a bounded linear operator on the space, which is denoted by $\phi(D)$. In this case, the left-hand side of (1.1) coincides with $\phi(D)f(z)$. We call the operator $\phi(D)$ a *linear differential operator with constant coefficients*. If ϕ is not a polynomial, then $\phi(D)$ is said to be of *infinite order*. For example, for every $\sigma \in \mathbf{C}$, the translation operator T_σ which sends a function f in the space to the function $z \mapsto f(z + \sigma)$, $z \in \mathbf{C}$, is a linear differential operator of infinite order since $T_\sigma = \exp(\sigma D)$.

This paper is a contribution to the research project HPRN-CT-2000-00116, founded by the European Commission.

We define a *weight function* as a positive continuous function w on $\mathbf{C}$. Given such a function w, and $1 \leqslant p \leqslant \infty$, we denote by $L_w^p(\mathbf{C})$ the Banach space of all complex functions f on $\mathbf{C}$ such that $fw \in L^p(\mathbf{C}, d\lambda)$, where λ is the Lebesgue measure on $\mathbf{R}^2$, endowed with the norm

$$\|f\|_{p,w} = \|fw\|_{L^p(\mathbf{R}^2)}.$$

For $p = 2$ this is a Hilbert space.

We denote by B_w^p the subspace of entire functions in $L_w^p(\mathbf{C})$. The mean-value formula and Hölder's inequality imply that for every closed disc K there exists a positive constant $C(K)$ such that for every $f \in B_w^p$

$$(1.2) \qquad\qquad |f(z)| \leqslant C(K)\|f\|_{p,w}, \quad z \in K.$$

It follows from this that convergence in B_w^p implies uniform convergence on compact subsets of $\mathbf{C}$. This shows that B_w^p is a closed subspace of $L_w^p(\mathbf{C})$, and therefore is a Banach space. Such spaces are called *weighted Bergman spaces of entire functions*.

In the sequel, we assume that w is a radial weight function on $\mathbf{C}$ of the form

$$(1.3) \qquad\qquad w(z) = \exp(-\varphi(|z|)), \quad z \in \mathbf{C},$$

where φ is a non-negative concave function on $\mathbf{R}_+ = [0, +\infty[$ such that $\varphi(0) = 0$ and

$$(1.4) \qquad\qquad \lim_{t \to +\infty} \frac{\varphi(t)}{\log t} = +\infty.$$

It follows from [7], Chap I.4, that the function $t \mapsto \varphi(t)/t$ is decreasing on $]0, +\infty[$ and we denote

$$(1.5) \qquad\qquad a = \lim_{t \to +\infty} \frac{\varphi(t)}{t}.$$

Condition (1.4) implies that B_w^p contains all polynomials.

Our first main result is

THEOREM 1. *Let $1 \leqslant p \leqslant \infty$. Then:*

(1) *The differentiation operator $D = d/dz$ is bounded on B_w^p and its spectrum $\sigma(D)$ is the closed disc*

$$\Delta_a = \{z \in \mathbf{C} : |z| \leqslant a\},$$

where a is the constant defined by (1.5).

(2) *If ϕ is an analytic function on a neighbourhood of Δ_a, then the operator $\phi(D)$ is surjective if and only if $\phi \neq 0$ on the circle*

$$C_a = \{z \in \mathbf{C} : |z| = a\}.$$

Moreover, in this case, $\phi(D)$ admits a bounded linear right inverse.

Our other results are concerned with differentiation and translation invariant subspaces of the Hilbert space B_w^2 when $w(z) = \exp(-|z|^\alpha)$, $z \in \mathbf{C}$, for some $0 < \alpha \leqslant 1$. Before stating them we recall some definitions and introduce some notations. If A is a bounded linear operator on a Banach space X, then an *invariant subspace* of A is a closed subspace of X which is mapped by A into itself; if it is different from the zero space and from the whole space, it is called a *non-trivial invariant subspace*. The collection of all invariant subspaces of A will be denoted by Lat A. We shall say that Lat A *has no proper gaps* if for every M_1 and M_2 in Lat A such that $M_1 \subset M_2$ and $\dim M_2/M_1 > 1$, there exists M in Lat A such that $M_1 \subsetneqq M \subsetneqq M_2$. An element x of X is called a *cyclic* vector for the operator A if the only element of Lat A that contains x is X, or equivalently, if the linear span of the vectors $A^n x$, $n = 0, 1, \ldots$, is dense in X.

The invariant subspace problem on Hilbert space, which is still open, asks whether every bounded linear operator on an infinite dimensional separable Hilbert space has a non-trivial invariant subspace.

In what follows we denote, for every positive integer n, by P_n the $(n+1)$-th dimensional vector space of analytic polynomials of degree at most n. Clearly they are non-trivial invariant subspaces of the operator D on B_w^p.

For every $0 < \alpha \leqslant 1$, we denote by H_α the Hilbert space B_w^2 where $w(z) = \exp(-|z|^\alpha)$, $z \in \mathbf{C}$.

Our second main result is

THEOREM 2. (a) *If $0 < \alpha < 1$, then the non-trivial invariant subspaces of the operator D on H_α are precisely the spaces P_n, $n = 0, 1, \ldots$.*

(b) *For the operator D on H_1, the problem whether* Lat D *has no proper gaps is equivalent to the invariant subspace problem on Hilbert space.*

It follows from [**2**, Th. 3.2] that a closed subspace of B_w^p is in Lat D if and only if it is *translation invariant*, that is, belongs to Lat T_σ for all $\sigma \in \mathbf{C}$. Thus Theorem 2 can be formulated in the following harmonic analysis setting.

THEOREM 3. (a) *If $0 < \alpha < 1$, then the non-trivial translation invariant subspaces of H_α are precisely the spaces P_n, $n = 0, 1, \ldots$.*

(b) *The problem whether for any two translation invariant subspaces M_1, M_2 of H_1 such that $M_1 \subset M_2$ and* $\dim M_2/M_1 > 1$, *there exists a translation invariant subspace M of H_1 such that $M_1 \subsetneq M \subsetneq M_2$, is equivalent to the invariant subspace problem on Hilbert space.*

An immediate consequence of Theorem 2 (a) and Theorem 3 (a) is

COROLLARY 1. *If $0 < \alpha < 1$, then:*

(a) *A function in H_α is a cyclic vector for the operator D if and only if it is not a polynomial.*

(b) *If f is a function in H_α then the linear span of its translates $T_\sigma f$, $\sigma \in \mathbf{C}$, is dense in H_α if and only if f is not a polynomial.*

The rest of the paper is organized as follows. In Section 2 we establish some general properties of the operator D on our spaces. In Section 3 we prove Theorem 1, and give an application to the difference equation

$$f(z+1) - f(z) = g(z), \quad z \in \mathbf{C}.$$

In section 4, we prove Theorem 2.

2. The differentiation operator

PROPOSITION 1. *Let $1 \leqslant p \leqslant \infty$, and assume that φ be a non-negative, concave function on $\mathbf{R}_+$, with $\varphi(0) = 0$, and $w(z) = e^{-\varphi(|z|)}$, $z \in \mathbf{C}$. Then:*

(a) *The differentiation operator D is bounded on B_w^p.*

(b) $\|D^n\| \leqslant n! \, r^{-n} e^{\varphi(r)}$, $r > 0$, $n = 1, 2, \ldots$.

(c) $\sigma(D) = \Delta_a$, *where a is defined by (1.5).*

PROOF. Note that (a) follows from the particular case $n = 1$ of (b). In order to prove (b), assume that $f \in B_w^p$ and $r > 0$. Cauchy's formula for the n-th derivative of f reads

$$(2.1) \qquad D^n f(z) = \frac{n!}{2i\pi} \int_{|\zeta|=r} \frac{f(z+\zeta)}{\zeta^{n+1}} d\zeta, \quad z \in \mathbf{C}.$$

Assume first that $p = \infty$. Let z and ζ in $\mathbf{C}$, with $|\zeta| = r$. By definition of B_w^∞ and the inequality

$$(2.2) \qquad \varphi(|z+\zeta|) \leqslant \varphi(|z|) + \varphi(|\zeta|),$$

which follows from the subadditivity and the monotonicity of φ (cf. [**7**, Chap. I.4]), we have

$$|f(z+\zeta)| \leqslant \|f\| e^{\varphi(|z+\zeta|)} \leqslant \|f\| e^{\varphi(|z|)} e^{\varphi(|\zeta|)}.$$

It follows then from (2.1) that

$$|D^n f(z)| \leqslant n!\, r^{-n} \sup_{|\zeta|=r} |f(z+\zeta)| \leqslant n!\, r^{-n} \|f\| e^{\varphi(|z|)} e^{\varphi(r)}$$

for all $z \in \mathbf{C}$. Hence $\|D^n\| \leqslant n!\, r^{-n} e^{\varphi(r)}$, this proves (b) for $p = \infty$.

Assume now that $1 \leqslant p < \infty$. Let $z \in \mathbf{C}$. Applying Hölder's inequality in formula (2.1), we get

$$(2.3) \qquad |D^n f(z)| \leqslant \frac{n!}{(2\pi)^{1/p} r^n} \left(\int_0^{2\pi} |f(z + re^{i\theta}|^p \, d\theta \right)^{1/p},$$

and therefore

$$(2.4) \qquad \int_{\mathbf{C}} |D^n f(z)|^p e^{-p\varphi(|z|)} \, d\lambda(z)$$

$$\leqslant \frac{n!^p}{2\pi r^{pn}} \int_0^{2\pi} \left(\int_{\mathbf{C}} |f(z + re^{i\theta})|^p e^{-p\varphi(|z|)} \, d\lambda(z) \right) d\theta.$$

By a change of variable, the inner integral can also be written

$$\int_0^{2\pi} \left(\int_{\mathbf{C}} |f(\zeta)|^p e^{-p\varphi(|\zeta - re^{i\theta}|)} \, d\lambda(\zeta) \right) d\theta.$$

Now, by the inequality

$$|\varphi(|z|) - \varphi(|z - \zeta|)| \leqslant \varphi(|\zeta|),$$

which follows from (2.2), the above integral is dominated by $2\pi \|f\|^p e^{p\varphi(r)}$. Substituting this in (2.4), we obtain the desired inequality.

In order to prove (c), consider the exponential function $e_\lambda \colon z \mapsto e^{\lambda z}$, $\lambda \in \mathbf{C}$. The equality $De_\lambda = \lambda e_\lambda$ shows that e_λ is an eigenvector of the operator D with eigenvalue λ. Since for $|\lambda| < a$ the function e_λ belongs to B_w^p, it follows that

$$(2.5) \qquad \Delta_a \subset \sigma(D).$$

We prove now the inequality $r(D) \leqslant a$. By (2.5), this will prove (c). It suffices to show that $r(D) \leqslant a + \varepsilon$ for all $\varepsilon > 0$. Given such ε, we get from (1.5) that there exists $t_0 > 0$ such that $\varphi(t) \leqslant (a + \varepsilon)t$ for $t \geqslant t_0$. Thus using (b), we see that there exists a positive constant C (which depends only on ε) such that

$$(2.6) \qquad \|D^n\| \leqslant Cn!\, r^{-n} e^{(a+\varepsilon)r}, \quad r > 0, \quad n = 1, 2 \ldots.$$

Minimizing the right-hand side of (2.6) with respect to r and applying Stirling's formula, we get $r(D) = \lim_{n \to +\infty} \|D^n\|^{1/n} \leqslant a + \varepsilon$. □

If $a = 0$, then the previous proposition asserts, in particular, that the operator D is *quasinilpotent*.

3. Surjectivity

Let $\mathcal{E}$ denote the vector space of all entire functions. Following [3], we define for every $\lambda \in \mathbf{C}$ an integral operator K_λ on $\mathcal{E}$ by

$$(3.1) \qquad K_\lambda f(z) = e^{\lambda z} \int_0^z e^{-\lambda \zeta} f(\zeta)\, d\zeta, \quad f \in \mathcal{E}, \quad z \in \mathbf{C}.$$

Then K_λ maps $\mathcal{E}$ into itself and for every function f in that space the function $K_\lambda f$ satisfies the differential equation

$$(3.2) \qquad \frac{dw}{dz} - \lambda w = f.$$

Integrating along the segment that joins 0 to z, we obtain for every $f \in \mathcal{E}$

$$(3.3) \qquad K_\lambda f(z) = z \int_0^1 e^{\lambda z(1-t)} f(zt)\, dt, \quad z \in \mathbf{C}.$$

Note that by Fubini's theorem this last formula is defined for almost all $z \in \mathbf{C}$ whenever f is a locally integrable function on $\mathbf{C}$ such that

$$\int_0^{2\pi} \int_0^1 |f(re^{i\theta})|\, dr d\theta < \infty.$$

The main ingredient in the proof of Theorem 2 is

THEOREM 4. *Assume that w is a weight function of the form* (1.3), *let a be defined by* (1.5), *and $1 \leqslant p \leqslant \infty$. Then for $|\lambda| < a$, the operator K_λ is bounded on B_w^p and is a right inverse of $D - \lambda$.*

The second assertion of the theorem follows from (3.2). The proof of the first assertion requires a preliminary result. To state it we need some notations.

If ρ is a positive continuous function on $\mathbf{R}_+$ and $1 \leqslant p \leqslant \infty$, we shall denote by $L_\rho^p(\mathbf{R}_+)$ the Banach space of all complex functions f on $\mathbf{R}_+$, such that the function $f\rho$ is in $L^p(\mathbf{R}_+)$, equipped with the norm

$$\|f\|_{p,\rho} = \|f\rho\|_{L^p(\mathbf{R}_+)}.$$

For a real number γ, we shall denote by M_γ and V_γ the linear transformations defined on the vector space $L_{loc}^1(\mathbf{R}_+)$ (of locally integrable functions on $\mathbf{R}_+$) by

$$M_\gamma f(x) = e^{\gamma x} f(x), \quad x \in \mathbf{R}_+,$$

$$V_\gamma f(x) = e^{\gamma x} \int_0^x e^{-\gamma t} f(t)\, dt, \quad x \in \mathbf{R}_+.$$

It follows that

$$(3.4) \qquad V_\gamma = M_\gamma V_0 M_{-\gamma}.$$

PROPOSITION 2. *Let ρ be a positive continuous function on $\mathbf{R}_+$ and γ a real number. Then:*

(1) *If $1 \leqslant p < \infty$ and*

(3.5)
$$\sup_{x>0} (M_\gamma \rho(x))^{-1} \|M_\gamma \rho\|_{L^p([x,+\infty[)} < \infty,$$

then V_γ is a bounded linear operator on $L^p_\rho(\mathbf{R}_+)$.

(2) *If*

(3.6)
$$\sup_{x>0} M_\gamma \rho(x) \| (M_\gamma \rho)^{-1} \|_{L^1([0,x])} < \infty,$$

then V_γ is a bounded linear operator on $L^\infty_\rho(\mathbf{R}_+)$.

PROOF. Let $\rho_\gamma = M_\gamma \rho$. Since M_γ maps $L^p_{\rho_\gamma}(\mathbf{R}_+)$ isometrically onto $L^p_\rho(\mathbf{R}_+)$, it follows from (3.4) that if the operator V_0 is bounded on $L^p_{\rho_\gamma}(\mathbf{R}_+)$, then the operator V_γ is bounded on $L^p_\rho(\mathbf{R}_+)$. This shows that it suffices to prove the theorem for $\gamma = 0$. Thus we make this assumption. Assume that $1 \leqslant p < \infty$. Then (3.5) means that there exists a positive constant C such that

(3.7)
$$\int_x^{+\infty} (\rho(t))^p \, dt \leqslant C(\rho(x))^p, \quad x \in \mathbf{R}_+.$$

Since the continuous functions on $\mathbf{R}_+$ with compact support are dense in $L^p_\rho(\mathbf{R}_+)$ and for every $f \in L^1_{loc}(\mathbf{R}_+)$,

$$|V_0 f(x)| \leqslant (V_0|f|)(x), \quad x \in \mathbf{R}_+,$$

it suffices to show that there exists a constant $b > 0$ such that if f is a non-negative continuous function on $\mathbf{R}_+$ with compact support then

(3.8)
$$\|V_0 f\|_{p,\rho} \leqslant b\|f\|_{p,\rho}.$$

If f is such a function, and h is the function on $\mathbf{R}_+$ defined by

$$h(x) = \int_x^{+\infty} (\rho(t))^p \, dt, \quad x \in \mathbf{R}_+,$$

then $(V_0 f(0))^p h(0) = 0$, and since $\lim_{t \to +\infty} h(t) = 0$ and $V_0 f$ is bounded, we have that $\lim_{t \to +\infty} (V_0 f(t))^p h(t) = 0$ and therefore, integrating by parts, we get

$$\|V_0 f\|_{p,\rho}^p = p \int_0^\infty (V_0 f(t))^{p-1} f(t) h(t) \, dt.$$

Thus using (3.7) we obtain that

$$\|V_0 f\|_{p,\rho}^p \leqslant Cp \int_0^\infty (V_0 f(t))^{p-1} f(t) (\rho(t))^p \, dt.$$

Applying to the above integral Hölder's inequality (with respect to the measure $(\rho(t))^p \, dt$), we get

$$\|V_0 f\|_{p,\rho}^p \leqslant Cp\|V_0 f\|_{p,\rho}^{p-1} \|f\|_{p,\rho},$$

which proves (3.8) with $b = Cp$. The proof of the second assertion of the theorem is elementary, and we omit the details.   □

The following gives a sufficient condition for ρ to satisfy the assumptions of Proposition 2.

PROPOSITION 3. *Let ρ be a positive continuous function on $\mathbf{R}_+$ which has the right derivative ρ'_+ in a neighbourhood of infinity, let $\gamma \in \mathbf{R}$, and assume that*

$$(3.9) \qquad \lim_{x \to +\infty} \frac{\rho'_+(x)}{\rho(x)} < -\gamma.$$

Then ρ satisfies conditions (3.5) and (3.6).

PROOF. Since the function $\rho_\gamma = M_\gamma \rho$ satisfies condition (3.9) with $\gamma = 0$, it suffices to prove the proposition for this case. Thus we assume that $\gamma = 0$. Then applying the mean value theorem to the function $\log \rho$, we get from (3.9) that there exist positive constants b and x_1 such that

$$\rho(t) \leqslant \rho(x)e^{-b(t-x)}, \quad x_1 < x < t,$$

and therefore

$$(\rho(x))^{-p} \int_x^\infty (\rho(t))^p \, dt \leqslant (pb)^{-1}, \quad x > x_1.$$

This shows that the function ρ is in $L^p(\mathbf{R}_+)$ and therefore the function $x \mapsto (\rho(x))^{-p} \int_x^\infty (\rho(t))^p \, dt$ is continuous on $\mathbf{R}_+$ hence bounded also on the interval $[0, x_1]$. So ρ satisfies (3.5) with $\gamma = 0$. The proof of (3.6) is similar, and we omit the details. $\qquad\square$

PROOF OF THEOREM 4. Since the operator K_λ maps $\mathcal{E}$ into itself, the assertion of the theorem would follow if for $|\lambda| < a$, that operator would be bounded on the space $L^p_w(\mathbf{C})$. But for $1 \leqslant p < \infty$ this is not the case. In fact, simple examples show that the operator is even not defined on this space for $1 \leqslant p \leqslant 2$. However, as we shall see, for $1 \leqslant p < \infty$, the measure $w^p(z) \, d\lambda(z)$ can be replaced by another positive Borel measure μ on $\mathbf{C}$, such that the space $L^p(\mathbf{C}, d\mu)$ includes the space B^p_w, the restriction of its norm to that subspace is equivalent to the $L^p_w(\mathbf{C})$ norm, and for $|\lambda| < a$, the operator K_λ maps $L^p(\mathbf{C}, d\mu)$ continuously into itself. We turn now to the details.

Fix $1 \leqslant p < \infty$, denote by χ the characteristic function of the unit disc $\{z \in \mathbf{C} : |z| < 1\}$, and consider the positive Borel measure μ on $\mathbf{C}$ defined by

$$(3.10) \qquad d\mu(z) = (w(z))^p[|z|^{-1}\chi(z) + (1 - \chi(z))] \, d\lambda(z).$$

Let N_p denote the norm on the Banach space $L^p(\mathbf{C}, d\mu)$. Note that

$$N_p^p(f) = \int_0^{2\pi} \int_0^1 |f(re^{i\theta})|^p w^p(r) \, dr d\theta + \int_0^{2\pi} \int_1^{+\infty} |f(re^{i\theta})|^p w^p(r) \, r dr d\theta.$$

It is clear that $L^p(\mathbf{C}, d\mu) \subset L^p_w(\mathbf{C})$, and for f in $L^p(\mathbf{C}, d\mu)$

$$\|f\|_{p,w} \leqslant N_p(f).$$

Now by (1.2), there exists a positive constant C such that for all $f \in B^p_w$

$$|f(z)| \leqslant C\|f\|_{p,w}, \quad |z| \leqslant 1.$$

Thus setting

$$C_1 = \left(2\pi C^p \int_0^1 w^p(r) \, dr + 1\right)^{1/p}$$

we get that for all f in B^p_w

$$\|f\|_{p,w} \leqslant N_p(f) \leqslant C_1\|f\|_{p,w}.$$

So on B^p_w the $L^p_w(\mathbf{C})$ and $L^p(\mathbf{C}, d\mu)$ norms are equivalent.

We show now that for $|\lambda| < a$, the operator K_λ maps $L^p(\mathbf{C}, d\mu)$ continuously into itself. To this end, consider the function ρ on $\mathbf{R}_+$ defined by $\rho(r) = w(r)$, for $0 \leqslant r \leqslant 1$ and $\rho(r) = w(r)r^{1/p}$ for $r > 1$. Recall that $w(r) = \exp(-\varphi(r))$, $r \geqslant 0$. Since the function φ is concave, it has a right derivative on $]0, +\infty[$, and

$$\lim_{t \to +\infty} \varphi'_+(t) = \lim_{t \to +\infty} \frac{\varphi(t)}{t} = a.$$

Therefore ρ has a right derivative on $]1, +\infty[$ and

$$\lim_{t \to +\infty} \frac{\rho'_+(t)}{\rho(t)} = -a.$$

Thus, since $|\lambda| < a$, we get from Propositions 2 and 3 that the operator $V_{|\lambda|}$ is bounded on $L_\rho^p(\mathbf{R}_+)$.

We show next that for all f in $L^p(\mathbf{C}, d\mu)$

$$N_p(K_\lambda f) \leqslant \|V_{|\lambda|}\| N_p(f),$$

where $\|V_{|\lambda|}\|$ denotes the norm of $V_{|\lambda|}$ as an operator on $L_\rho^p(\mathbf{R}_+)$. Since the continuous functions on $\mathbf{C}$ with compact support are dense in $L^p(\mathbf{C}, d\mu)$, it suffices to establish the inequality for such functions. Let f be a continuous function on $\mathbf{C}$ with compact support. For every $\theta \in [0, 2\pi]$, denote by f_θ the continuous function on $\mathbf{R}_+$ with compact support defined by

$$f_\theta(r) = f(re^{i\theta}), \quad r \in \mathbf{R}_+.$$

It follows from (3.3) that

$$|K_\lambda f(re^{i\theta})| \leqslant V_{|\lambda|}|f_\theta|(r), \quad r \in \mathbf{R}_+, \quad \theta \in [0, 2\pi],$$

and therefore

$$
\begin{aligned}
N_p^p(K_\lambda f) &= \int_0^{2\pi} \int_0^{+\infty} |K_\lambda f(re^{i\theta})|^p \rho^p(r) \, dr d\theta \\
&\leqslant \int_0^{2\pi} \|V_{|\lambda|}|f_\theta|\|_{p,\rho}^p \, d\theta \\
&\leqslant \|V_{|\lambda|}\|^p \int_0^{2\pi} \|f_\theta\|_{p,\rho}^p \, d\theta \\
&= \|V_{|\lambda|}\|^p N_p^p(f).
\end{aligned}
$$

This completes the proof of the theorem for $1 \leqslant p < \infty$. The assertion of the theorem for $p = \infty$ follows easily from Propositions 2 and 3. Actually, for $|\lambda| < a$, the operator K_λ is bounded on $L_w^\infty(\mathbf{C})$. We omit the details. $\square$

For the proof of the second assertion of Theorem 1, we need a simple result concerning general operators. It is probably known, but since we are unable to find it in the literature, we give its statement and proof.

PROPOSITION 4. *Let A be a bounded operator on a Banach space X. If $\lambda \in \partial\sigma(A)$ (the boundary of the spectrum of A) then $A - \lambda$ is not surjective.*

PROOF. Let $\lambda \in \partial\sigma(A)$. The operator A and its adjoint A^* have the same spectrum, so $\lambda \in \partial\sigma(A^*)$. It follows that λ belongs to the approximate point spectrum of A^* ([**15**], Theorem 0.7). This means that

$$\inf_{\|\xi\|=1} \|(A^* - \lambda)\xi\| = 0.$$

Since $A^* - \lambda = (A - \lambda)^*$, it follows then from a classical criterion of surjectivity ([**12**] Cor. 9.5) that $A - \lambda$ is not surjective. □

We turn now to the proof of Theorem 1 (2). If $a = 0$, then by the first part of the theorem and the spectral mapping theorem

$$\sigma(\phi(D)) = \{\phi(0)\}.$$

So if $\phi(0) \neq 0$, the operator $\phi(D)$ is invertible and therefore surjective, and if $\phi(0) = 0$, this operator is quasinilpotent and therefore, by Proposition 4, it is not surjective.

Assume now that $a > 0$, and denote by $\lambda_1, \ldots, \lambda_n$ the zeros (if any) of ϕ in Δ_a, counting each zero as many times as its multiplicity. Then $\phi(z) = (z - \lambda_1) \ldots (z - \lambda_n)\phi_1(z)$ where ϕ_1 is an analytic function on a neighbourhood of Δ_a which has no zeros in this disc. It follows that

(3.11) $$\phi(D) = (D - \lambda_1) \ldots (D - \lambda_n)\phi_1(D).$$

Again, by the first part of the theorem and the spectral mapping theorem

$$\sigma(\phi_1(D)) = \phi_1(\Delta_a)$$

and therefore $0 \notin \sigma(\phi_1(D))$, so the operator $\phi_1(D)$ is invertible. If $\phi \neq 0$ on the circle C_a, then $|\lambda_j| < a$, $j = 1, \ldots, n$, so by Theorem 4, each of the operators $D - \lambda_j$, $j = 1, \ldots, n$, admits a bounded linear right inverse, and therefore since $\phi_1(D)$ is invertible, it follows from (3.11) that the operator $\phi(D)$ admits a bounded linear right inverse, hence is surjective. On the other hand, if ϕ has a zero on the circle C_a, then $|\lambda_j| = a$ for some $j \in \{1, \ldots, n\}$, so by the first part of the theorem and Proposition 4, the operator $D - \lambda_j$ is not surjective. Therefore, since the operators appearing in the product on the right-hand side of (3.11) commute, it follows that the operator $\phi(D)$ is not surjective. This completes the proof of Theorem 1. □

EXAMPLE. Consider the functional equation

(3.12) $$f(z + \sigma) - f(z) = g(z), \quad z \in \mathbf{C},$$

where f and g are entire functions and σ is a fixed non-zero complex number. According to the terminology of Nörlund [**14**], when $\sigma = 1$, the function f is called the *sum* of the function g. Equation (3.12) has been intensively studied and has numerous applications (see [**6**] and [**9**]). At the end of the XIXth century, Guichard [**10**] and Hurwitz [**11**] proved that every entire function g admits an entire sum f. Around fifty years later, Whittaker [**19**] proved that every entire function g of finite order admits an entire sum f of the same order. By explicit computations and estimates, the second author obtained in [**8**] sufficient conditions for the resolution of equation (3.12) in the spaces B_w^p, with the weights $w(z) = \exp(-s|z|^\alpha)$, $s > 0$ and $0 < \alpha \leqslant 1$.

Equation (3.12) can be written in the form

$$\phi(D)f = g,$$

where $\phi(z) = e^{\sigma z} - 1$. Thus we deduce from Theorem 1:

PROPOSITION 5. *For every g in B_w^p, the equation (3.12) has a solution f in B_w^p if and only if $a \notin \{\frac{2\pi k}{|\sigma|} : k = 0, 1, \ldots\}$.*

4. Invariant subspaces

In this section, we give the proof of Theorem 2. It is based on the fact that the differentiation operator on the spaces B_w^2 is a weighted backward shift, and we begin by recalling some definitions and results concerning this class of operators.

In the sequel, we assume that H is a complex Hilbert space with orthonormal basis $(e_n)_{n \geqslant 0}$. For a bounded sequence of positive numbers $\gamma = (\gamma_n)_{n \geqslant 0}$, there is a unique bounded linear operator B_γ on H which satisfies $B_\gamma e_0 = 0$ and $B_\gamma e_n = \gamma_{n-1} e_{n-1}$ for $n \geqslant 1$. This operator is called the *weighted backward shift* with weight sequence γ.

If $\gamma = (\gamma_n)_{n \geqslant 0}$ and $\delta = (\delta_n)_{n \geqslant 0}$ are bounded sequences of positive numbers and there exist constants $0 < C_1 < C_2$ such that

$$C_1 \leqslant \frac{\delta_0 \dots \delta_n}{\gamma_0 \dots \gamma_n} \leqslant C_2, \quad n = 0, 1, \dots,$$

then the operators B_γ and B_δ are similar. This follows from [16, Proposition 3 and Theorem 2]. Thus in particular, if the sequences γ and δ are eventually equal, then the operators B_γ and B_δ are similar.

For every $n = 0, 1, \dots$, let M_n denote the $(n+1)$-th dimensional subspace of H spanned by the vectors e_j, $j = 0, 1 \dots, n$. It is clear that these spaces are non-trivial invariant subspaces for every weighted backward shift.

We shall use the following result of Yakubovich [20].

THEOREM (YAKUBOVICH). *If $\gamma = (\gamma_n)_{n \geqslant 0}$ is a sequence of positive numbers which is decreasing to zero, then the non-trivial invariant subspaces of the operator B_γ are precisely the spaces M_n, $n = 0, 1, \dots$.*

N.K. Nikolskii [13] proved earlier that the above conlusion holds under the stronger assumption that the sequence γ is decreasing and belongs to l^p for some $0 < p < \infty$. Actually, for our application, this result suffices.

We recall that a bounded linear operator T on H is called *unicellular* if $\operatorname{Lat} T$ is totally ordered by inclusion. A simple argument shows that if $M_n \in \operatorname{Lat} T$, for all $n \geqslant 0$, then T is unicellular if and only if the non-trivial invariant subspaces of T are precisely the spaces M_n, $n = 0, 1, \dots$. Therefore a backward weighted shift is unicellular if and only if its non-trivial invariant subspaces are precisely these spaces. Thus, since unicellularity is preserved under similarity, it follows from the previous observations that the conclusion in the theorem of Yakubovich remains true if the sequence γ is only assumed to be eventually decreasing to zero.

Consider now the Hilbert space B_w^2 with $w(z) = \exp(-\varphi(|z|))$, $z \in \mathbf{C}$, where φ is a non-negative concave function on $\mathbf{R}_+$ such that $\varphi(0) = 0$, which satisfies (1.4). This condition implies that the monomials $u_n(z) = z^n$, $n = 0, 1, \dots$, are in the space and since the weight w is radial they form an orthogonal family, and for every $f \in B_w^2$ we have the Parseval formula

$$(4.1) \qquad \qquad \|f\|_{2,w}^2 = \sum_{n=0}^{\infty} \left| \frac{f^{(n)}(0)}{n!} \right|^2 \|u_n\|_{2,w}^2.$$

Therefore the normalized family $e_n = \frac{u_n}{\|u_n\|_{2,w}}$, $n = 0, 1, \dots$, is an orthonormal basis for B_w^2.

Let $\gamma = (\gamma_n)_{n \geqslant 0}$ be the sequence defined by

(4.2)
$$\gamma_n = \frac{(n+1)\|u_n\|_{2,w}}{\|u_{n+1}\|_{2,w}}, \quad n = 0, 1, \ldots.$$

Then $De_0 = 0$ and $De_n = \gamma_{n-1}e_{n-1}$ for $n \geqslant 1$, and therefore, since by Theorem 2 the operator D is bounded on B_w^2, the sequence γ is bounded and $D = B_\gamma$. Thus from the theorem of Yakubovich and the comments following its statement we get:

PROPOSITION 6. *If the sequence* $\gamma = (\gamma_n)_{n \geqslant 0}$ *defined by* (4.2) *is eventually decreasing to zero, then the non-trivial invariant subspaces of* D *are precisely the polynomial spaces* P_n, $n = 0, 1, \ldots$.

A straighforward computation shows that if $\varphi(t) = t^\alpha$ for some $0 < \alpha \leqslant 1$, then

(4.3)
$$\|u_n\|_{2,w}^2 = \frac{2\pi}{\alpha} 2^{-2(n+1)/\alpha} \Gamma\left(\frac{2}{\alpha}(n+1)\right), \quad n = 0, 1, \ldots,$$

where Γ is the Gamma function. It follows that in this case

(4.4)
$$\gamma_n^2 = 2^{2/\alpha} \frac{(n+1)^2 \Gamma\left(\frac{2}{\alpha}(n+1)\right)}{\Gamma\left(\frac{2}{\alpha}(n+2)\right)}, \quad n = 0, 1, \ldots.$$

Thus, by [**17**, section 1.87], we have the asymptotic relation

(4.5)
$$\gamma_n \sim c \cdot n^{1-\frac{1}{\alpha}}, \quad n \to \infty,$$

where c is a positive constant. This shows that $\lim_{n \to \infty} \gamma_n = 0$ for $\alpha < 1$.

Thus from Proposition 6 we get that the first assertion of Theorem 2 follows from

PROPOSITION 7. *If* $\beta > 2$, *then the function* f *on* $\mathbf{R}_+$ *defined by*
$$f(x) = \frac{x^2 \Gamma(x)}{\Gamma(x+\beta)}, \quad x > 0,$$
is eventually decreasing.

PROOF. Using the fact that (see [**18**, p. 241])
$$\frac{\Gamma'(x)}{\Gamma(x)} = -\gamma - \frac{1}{x} + \sum_{n=1}^{\infty} \frac{x}{n(x+n)}, \quad x > 0,$$

(where γ is the Euler constant), we get
$$\frac{f'(x)}{f(x)} = \frac{2}{x} - \beta \sum_{n=0}^{\infty} \frac{1}{(x+n)(x+n+\beta)} = \frac{2-\beta}{x} + O\left(\frac{1}{x^2}\right), \quad x \to +\infty,$$
and this implies the claim.   □

REMARK. We used the estimate, as $x \to +\infty$,
$$\sum_{n=1}^{\infty} \frac{1}{(x+n)(x+n+\beta)} = \int_0^{\infty} \frac{dt}{(x+t)(x+t+\beta)} + O\left(\frac{1}{x^2}\right)$$
$$= \frac{1}{\beta} \log\left(1 + \frac{\beta}{x}\right) + O\left(\frac{1}{x^2}\right)$$
$$= \frac{1}{x} + O\left(\frac{1}{x^2}\right).$$

We turn now to the proof of the second assertion of Theorem 2. First we introduce some notations. For every $\beta \geqslant 0$, we denote by D_β the Hilbert space of all holomorphic functions f on the unit disc $\{z \in \mathbf{C} : |z| < 1\}$, such that the norm

$$\|f\| = \left(\sum_{n \geqslant 0} |\hat{f}(n)|^2 (n+1)^\beta \right)^{1/2}$$

(where $\hat{f}(n)$ is the n-th Taylor coefficient of f) is finite. Consider the bounded linear operator B which sends the function f in the space to the function whose Taylor expansion is $\sum_{n \geqslant 0}^{\infty} \hat{f}(n+1)z^n$. As observed in [4, p. 199], it follows from the results in [1] and [5] that for $\beta > 0$ and for the operator B on D_β, the problem whether $\operatorname{Lat} B$ has no proper gaps is equivalent to the invariant subspace problem on Hilbert space. Thus Theorem 2 (b) will be established if we show that the operator D on H_1 is similar to the operator B on $D_{1/2}$. From (4.1) and (4.3) we get by an application of Stirling's formula, that for every function f in H_1, the series $\sum_{n \geqslant 0}^{\infty} f^{(n)}(0)z^n$ is the Taylor expansion of a function g in $D_{1/2}$, and the mapping L which sends f to g is a bounded invertible operator from H_1 onto $D_{1/2}$. Since $LD = BL$, the operators D and B are similar, and the proof is complete. $\qquad \square$

References

[1]   C. Apostol, H. Bercovici, C. Foiaş, and C. Pearcy, *Invariant subspaces, dilation theory, and the structure of the predual of a dual algebra*, J. Funct. Anal. **63** (1985) 369–404.

[2]   A. Atzmon, *A model for operators with cyclic adjoint*, Integral Equations Operator Theory **10** (1987) 153–163.

[3]   A. Atzmon, *Nuclear Fréchet spaces of entire functions with transitive differentiation*, J. Anal. Math. **60** (1993) 1–19.

[4]   A. Atzmon, *Maximal, minimal, and primary invariant subspaces*, J. Funct. Anal. **185** (2001) 155–213.

[5]   H. Bercovici, C. Foiaş, and C. Pearcy, *Dual algebras with applications to invariant subspaces and dilation theory*, CBMS, 56, Amer. Math. Soc., Providence, 1985.

[6]   C. A. Berenstein, R. Gay, *Complex Analysis and Special Topics in Harmonic Analysis*, Springer, 1995.

[7]   N. Bourbaki, *Fonctions d'une Variable Réelle*, CCLS, 1976.

[8]   B. Brive, *Sums of an entire function in Certain weighted L^2-spaces*, Publicacions Matemàtiques (Barcelona) **47** (2003), 211–236.

[9]   A. O. Guelfond, *Calcul des Différences Finies*, Dunod, 1963.

[10] C. Guichard, *Sur la résolution de l'équation aux différences finies $G(x+1) - G(x) = H(x)$*, Annales Sci. de l'École Normale **4** (1887), 361–380.

[11] A. Hurwitz, *Sur l'intégrale finie d'une fonction entière*, Acta Math. **20** (1897), 285–312.

[12] R. Meise and D. Vogt, *Einführung in die Funktionalanalysis*, Vieweg, 1992.

[13] N. K. Nikol'skii, *Invariant subspaces in the theory of operators and theory of functions*, J. Soviet Math. **5** (1976), no. 2, 120–249.

[14] N.-E. Nörlund, *Sur la somme d'une Fonction*, Mémorial des Sciences Mathématiques, 24, Gauthiers Villars, 1927.

[15] H. Radjavi and P. Rosenthal, *Invariant Subspaces*, 2nd edition, Dover, 2003.

[16] A. L. Shields, *Weighted shift operators and analytic function theory*, Topics in Operator Theory, AMS, (C. Pearcy, ed.), 13, Mathematical Surveys 1974.

[17] E. C. Titchmarsh, *The Theory of Functions*, 2nd edition, Oxford Univ. Press, 1978.

[18] E. T. Whittaker and G. N. Watson, *A Course of Modern Analysis*, 4th edition, Cambridge Univ. Press, 1973.

[19] J. M. Whittaker, *Interpolatory Function Theory*, Cambridge University Press, 1935.

[20] D. V. Yakubovich, *Invariant subspaces of weighted shift operators*, J. Soviet Math. **37** (1987) no. 5, 1323–1349.

SCHOOL OF MATHEMATICAL SCIENCES, TEL AVIV UNIVERSITY, TEL AVIV 69978, ISRAEL
E-mail address: aatzmon@post.tau.ac.il

E-mail address: brive@math.univ-lille1.fr

Contemporary Mathematics
Volume **404**, 2006

Exceptional Values and the MacLane Class $\mathcal{A}$

Karl F. Barth and Philip J. Rippon

To Boris Korenblum, on the occasion of his 80th birthday.

ABSTRACT. It is known that various conditions involving exceptional values of functions holomorphic or meromorphic in $\mathbb{C}$ force these functions to be constant, or of some other particularly simple form. We show that several of these conditions also force a function holomorphic or meromorphic in the unit disc D to have asymptotic values at a dense set of points of ∂D. For example, this is true for functions f holomorphic in D such that $f f'' \neq 0$.

1. Introduction

Let D denote the unit disk $\{z : |z| < 1\}$, C denote the unit circle $\{z : |z| = 1\}$, and $\hat{\mathbb{C}}$ the extended complex plane. We say that a path $\gamma : z(t)$, $0 \leq t < 1$, in D is a *boundary path* if $|z(t)| \to 1$ as $t \to 1$. The set $\bar{\gamma} \cap C$ is called the *end* of γ. We say that a function f defined in D has the *asymptotic value* $a \in \hat{\mathbb{C}}$ if there is a boundary path $\gamma : z(t)$, $0 \leq t < 1$, such that

$$f(z(t)) \to a \quad \text{as } t \to 1.$$

If the end of γ is a singleton $\{\zeta\}$, then we say that f has the asymptotic value a at ζ. The *MacLane class* $\mathcal{A}$ consists of those functions which are holomorphic and non-constant in D, and which have asymptotic values at points of a dense subset of C; see [**20**] for a discussion of the class $\mathcal{A}$. The analogous class for meromorphic functions in D is denoted by $\mathcal{A}_m$; see [**1**] for properties of $\mathcal{A}_m$. We also write $A^*[f]$ to denote the set of points of C at which f has a *finite* asymptotic value.

Let $d(a, b)$ denote the chordal metric on $\hat{\mathbb{C}}$ and put $\mathcal{U}(a, \varepsilon) = \{z : d(a, z) < \varepsilon\}$, where $\varepsilon > 0$. Suppose that to each $\varepsilon > 0$ there corresponds a component $D(\varepsilon)$ of $f^{-1}(\mathcal{U}(a, \varepsilon))$ such that $D(\varepsilon_1) \subset D(\varepsilon_2)$, for $0 < \varepsilon_1 < \varepsilon_2$, and $\bigcap_{\varepsilon > 0} D(\varepsilon) = \emptyset$. Then we say that $\{D(\varepsilon), a\}$ is an *asymptotic tract* (or simply *tract*) of f for the value a. The set $K = \bigcap_{\varepsilon > 0} \overline{D(\varepsilon)}$ is called the *end* of the tract. It is easy to prove that K is either a point (*point tract*) or an arc (*arc tract*).

Bloch's principle (see, for example, [**25**, p. 101]) states that a condition on a holomorphic or meromorphic function in $\mathbb{C}$, which forces the function to be constant, may also force the corresponding family of functions defined in a region of $\mathbb{C}$ to be

---

2000 *Mathematics Subject Classification*. Primary 30D40.

normal. For example, the simplest version of Picard's theorem states that a non-constant entire function cannot omit two values in $\mathbb{C}$. Correspondingly, the family of functions f holomorphic in D that omit two given values is normal [25, p. 54]. In particular, each function f in this family is normal. It follows that if f is a non-constant function holomorphic in D which omits two values, then $f \in \mathcal{A}$; see [25, p. 54, p. 178], and [20, p. 43]. There are many examples of functions that omit one value but are not in $\mathcal{A}$, for instance, Example 1 below.

The object of this paper is to investigate other conditions on the exceptional values of f and its derivatives which are known to force functions analytic or meromorphic in $\mathbb{C}$ to be constant, or to take some other particularly simple form. We ask whether such conditions, when applied to functions in D, also lead to the conclusion that f has asymptotic values at a dense set of points of C. We start with the results in Theorem A below, which provides another illustration of Bloch's principle. Part (a) of Theorem A is a special case of a result of Hayman [13, Theorem 3], and part (b) is a result of Gu; see [10], [6, p. 86] or [25, p. 144]. The holomorphic case of part (b) was proved by Miranda [22].

THEOREM A. *Let $k \in \mathbb{N}$.*

(a) *If f is a function meromorphic in $\mathbb{C}$ such that the equations*

$$(1.1) \qquad\qquad f(z) = 0 \quad and \quad f^{(k)}(z) = 1,$$

have no solutions, then f is constant.

(b) *Let Ω be a region and let $\mathcal{F}$ be the family of functions meromorphic in Ω such that the equations (1.1) have no solutions in Ω. Then $\mathcal{F}$ is normal in Ω.*

By a normalization, the values 0 and 1 in (1.1) can be replaced by $a \in \mathbb{C}$ and $b \in \mathbb{C} \setminus \{0\}$, respectively. We remark that Zalcman [27, p. 225] has shown that part (b) of Theorem A can be deduced from part (a) by an elementary argument.

Corresponding to Theorem A, we have the following result on boundary behaviour in D.

THEOREM 1. *Let $k \in \mathbb{N}$. If f is meromorphic and non-constant in D and the equations (1.1) have no solutions, then $f \in \mathcal{A}_m$.*

Parts (a) and (b) of Theorem A are both clearly false if the equation $f^{(k)}(z) = 1$ is replaced by $f^{(k)}(z) = 0$. We show in Example 1 that Theorem 1 is also false, in the case $k = 1$, if the equation $f'(z) = 1$ is replaced by $f'(z) = 0$. However, the condition $ff^{(k)} \neq 0$, for $k \geq 2$, has been much studied and it leads to some remarkable conclusions.

THEOREM B. *Let $k \geq 2$.*

(a) *Let f be a function meromorphic in $\mathbb{C}$ such that $ff^{(k)} \neq 0$ in $\mathbb{C}$. Then f is of the form $f(z) = e^{az+b}$ or $f(z) = (az + b)^{-n}$, where $a, b \in \mathbb{C}, a \neq 0$ and $n \in \mathbb{N}$.*

(b) *Let Ω be a region in $\mathbb{C}$ and let $\mathcal{F}$ be the family of functions holomorphic in Ω such that $ff^{(k)} \neq 0$ in Ω. Then $\{f'/f : f \in \mathcal{F}\}$ is a normal family in Ω.*

(c) *Let Ω be a region in $\mathbb{C}$ and let $\mathcal{F}$ be the family of functions meromorphic in Ω such that $ff'' \neq 0$ in Ω. Then $\{f'/f : f \in \mathcal{F}\}$ is a normal family in Ω.*

Theorem B, part (a), was proved by Frank [8] for the case $k \geq 3$ and by Langley [19] for the case $k = 2$; see also [5]. This part had been conjectured by Hayman [13, p. 23], who obtained partial results in this direction. Part (b) was proved by Schwick [26] and part (c) by Bergweiler [5].

The following result on the boundary behaviour of functions in D involves the above condition $ff^{(k)} \neq 0$ in the case $k = 2$. It is again a special case of a result of Hayman [13, Theorem 6 and its proof]. Here and later we use the standard notations from Nevanlinna theory; see [14].

THEOREM C. *Let f be holomorphic in D and suppose that $ff'' \neq 0$ in D. Then the functions $\psi = f'/f$ and f satisfy:*

(a) $m(r, \psi) = O\left(\log \dfrac{1}{1-r}\right)$;

(b) $\log M(r, \psi) = O\left(\dfrac{1}{1-r}\right)$;

(c) $\log\log M(r, f) = O\left(\dfrac{1}{1-r}\right)$.

Hayman also showed that Theorem C, part (c), is best possible by constructing f holomorphic in D with $ff'f'' \neq 0$ but

$$\liminf_{r \to 1}(1 - r)\log\log M(r, f)\, dr > 0;$$

see [13, p. 27]. The proof of [13, Theorem 6] does not cover the case of the condition $ff^{(k)} \neq 0$, where $k > 2$, and we are not aware that Theorem C has been established in this case.

The growth estimates in Theorem C, parts (a) and (b), both imply that $\psi = f'/f \in \mathcal{A}$, as we shall see. But the estimate in part (c) is not quite sufficient to imply that $f \in \mathcal{A}$. Indeed, it is known [18, Theorem 1] that

$$\int_0^1 \log^+\log^+ M(r, f)\, dr < +\infty$$

implies that $f \in \mathcal{A}$, and that this condition is essentially best possible; see [15, Theorem 10.21]. Instead, we can show that $f \in \mathcal{A}$ by proving that $\log\log|f(z)|$ is much smaller than $1/(1 - |z|)$ for most $z \in D$.

Some evidence for this assertion can be obtained as follows. By Theorem C, part (a), (see [13, proof of Theorem 6]) there exists r_0, $0 < r_0 < 1$, such that the restriction of ψ to any disc of the form

$$D_\theta = \{z : |z - r_0 e^{i\theta}| < 1 - r_0\}, \quad \theta \in [0, 2\pi],$$

is of bounded characteristic, uniformly in θ. In particular, ψ is of locally bounded characteristic, as defined by Hayman and Korenblum in [17], and the zeros of ψ, and hence of f', satisfy the Blaschke condition in any Stolz angle for D. Now ψ is the ratio of bounded analytic functions in each disc D_θ so

$$\log|\psi(z)| = h(z) + u(z), \quad \text{for } z \in D_\theta,$$

where h is positive harmonic and u is negative subharmonic in D_θ. It follows from [24, Theorem 2] that ψ and hence f can achieve their maximal growth, as specified in Theorem C, parts (b) and (c), on at most a countable set of radii. This suggests that $\log\log|f(z)|$ is indeed much smaller than $1/(1 - |z|)$ for most $z \in D$. By making this idea more precise, we find that the estimate in Theorem C, part (a), enables us to apply to f a criterion for $\mathcal{A}$ which is based on the convergence of a certain area integral in D. Moreover, if we also know that $f' \neq 0$ in D, then we can deduce that f has *finite* asymptotic values at a dense set of points of C.

THEOREM 2. *Let f be holomorphic and non-constant in D.*

(a) *If $ff'' \neq 0$ in D, then $f'/f \in \mathcal{A}$, $\log f \in \mathcal{A}$ and $f \in \mathcal{A}$.*

(b) *If $ff'f'' \neq 0$ in D, then the sets $A^*[f'/f]$, $A^*[\log f]$, $A^*[f]$, $A^*[e^f]$, ..., are each dense in C.*

It is plausible to conjecture that Theorem 2 is true with f'' replaced by $f^{(k)}$, for some $k \geq 2$, in each of the hypotheses. The proof shows that all parts of Theorem 2 follow from Theorem C, part (b), except for the assertion that $ff'' \neq 0$ implies that $f \in \mathcal{A}$. As noted above, the proof of this assertion uses Theorem C, part (a), though it is clear from the proof that a slightly weaker estimate for $m(r, \psi)$ such as $O\left((\log 1/(1-r))^K\right)$, for some $K > 0$, would be sufficient.

Finally we give the example mentioned above, which shows that a holomorphic function and its derivative can both omit the value 0 in D and yet the function need not be in $\mathcal{A}$.

EXAMPLE 1. *There exists a function f holomorphic in D, with $ff' \neq 0$ in D, such that $f'/f \notin \mathcal{A}$, $\log f \notin \mathcal{A}$ and $f \notin \mathcal{A}$.*

The authors are grateful to Milne Anderson, Walter Bergweiler and Larry Zalcman for helpful discussions.

2. Properties of the class $\mathcal{A}$

Here we recall several properties of the MacLane class $\mathcal{A}$. Most of these are well known, but Lemma 4 is not, so we include a proof of this.

MacLane defined the set $\mathcal{L}$ to consist of functions holomorphic in D whose level sets $\Lambda_R = \{z : |f(z)| = R\}, R > 0$, 'end at points'. To be precise, for each $R > 0$ and $0 < r < 1$, we let $d(r, R)$ be the supremum of the diameters of the components of the set

$$\Lambda_R \cap \{z : r < |z| < 1\},$$

and we say that $f \in \mathcal{L}$ if and only if, for each fixed $R > 0$,

$$d(r, R) \to 0 \quad \text{as } r \to 1.$$

LEMMA 1. *We have $\mathcal{A} = \mathcal{L}$.*

This is MacLane's fundamental criterion for the class $\mathcal{A}$; see [20, Theorem 1]. Using this criterion MacLane obtained various sufficient conditions for $f \in \mathcal{A}$, including the following conditions, which can be found in [20, p. 37] and [21, proof of Theorem 13].

LEMMA 2. *Let f be holomorphic and non-constant in D.*

(a) *If*

(2.1)
$$\int_0^1 (1-r)m(r, f)\, dr < \infty,$$

then $f \in \mathcal{A}$.

(b) *If*

(2.2)
$$\int_0^1 (1-r)\log^+ M(r,f)\,dr < \infty,$$

then $f \in \mathcal{A}$; moreover, f' and $\int f(z)dz$ also satisfy (2.2), so these functions belong to $\mathcal{A}$.

We also need several results about the existence of *finite* asymptotic values, taken from [**21**, Theorem 5 and Theorem 11]; for related results see [**4**, Theorem 2], [**11**, Corollary 1.2] and [**2**, Theorem 1].

LEMMA 3. *Let f be holomorphic in D.*
(a) *If $f \in \mathcal{A}$ and $f \neq 0$, then $A^*[f]$ is dense in C.*
(b) *If $ff' \neq 0$ and $f'/f \in \mathcal{A}$, then the sets $A^*[f'/f]$, $A^*[\log f]$, $A^*[f]$, $A^*[e^f],\ldots$, are each dense in C, so the functions $\log f$, f, e^f, $\ldots$, are each in class $\mathcal{A}$.*

Finally, we give a less familiar criterion for the MacLane class, which is based on the convergence of a certain area integral. This criterion, which follows from a subharmonic majorant result that goes back to Beurling, was mentioned in [**23**] in the context of the MacLane class for subharmonic functions.

LEMMA 4. *Let $\varepsilon > 0$. If f is holomorphic and non-constant in D and*

(2.3)
$$\iint_D \left(\log^+ \log^+ |f(z)|\right)^{1+\varepsilon} dx\,dy < \infty,$$

then $f \in \mathcal{A}$.

PROOF. If $f \notin \mathcal{A}$, then by Lemma 1 the function f has a level set that does not 'end at a point'. So there exist a non-trivial compact arc γ of C, a constant $R > 0$ and compact arcs $\Gamma_n \subset D$ such that $|f(z)| = R$ for $z \in \bigcup_n \Gamma_n$ and $\Gamma_n \to \gamma$ with respect to the Hausdorff metric. Then, by making γ smaller if necessary, we can choose an open square S, with two sides perpendicular to the radius of D which passes through the midpoint of γ, such that $S \cap C = \operatorname{int} \gamma$ and such that the arcs Γ_n form a sequence of cross-cuts of S.

Now let
$$F(z) = \begin{cases} \max\{\log^+ |f(z)|, R\}, & \text{for } z \in S \cap D, \\ R, & \text{for } z \in S \setminus D. \end{cases}$$

Then F is non-negative and measurable on S and, by (2.3),
$$\iint_S \left(\log^+ |F(z)|\right)^{1+\varepsilon} dx\,dy < \infty.$$

Also, put
$$U_F(z) = \sup\{u(z) : u \text{ is subharmonic in } S, u(z) \leq F(z), \text{ for } z \in S\}.$$

Then the above integral condition on F implies, by a theorem of Beurling (see [**7**, Theorem 2]) that U_F is locally bounded above in S. Now let S_n denote that component of $S \setminus \Gamma_n$ which does not meet C. Then each of the functions
$$u_n(z) = \begin{cases} \max\{\log^+ |f(z|, R\}, & \text{for } z \in S_n, \\ R, & \text{for } z \in S \setminus S_n, \end{cases}$$

is subharmonic in S and dominated by F there. Hence the functions u_n are uniformly bounded near any interior point of γ, and so therefore is f. Since f is non-constant, the existence of the level curves Γ_n which tend to γ leads to a contradiction. This completes the proof of Lemma 4. □

Note that a slightly more general sufficient condition for $\mathcal{A}$ than (2.3) is possible; see [**23**, Theorem 8].

3. Proof of Theorem 1

Theorem 1 follows from two results in [**13**], which were used there to prove Theorem A, part (a). As pointed out in [**13**, p. 24], the hypotheses of Theorem 1 imply that

$$T(r, f) = O\left(\log \frac{1}{1-r}\right);$$

see also [**12**, pp. 200 and 204], where results of this type were anticipated. Indeed, we may assume that f is not a polynomial so, by [**13**, Theorem 1],

$$(3.1) \quad T(r, f) < \left(2 + \frac{1}{k}\right) N\left(r, \frac{1}{f}\right) + \left(2 + \frac{2}{k}\right) \overline{N}\left(r, \frac{1}{f^{(k)} - 1}\right) + S_{(k)}(r, f),$$

where $S_{(k)}(r, f)$ is a relatively small error term with the property [**13**, Theorem 2] that

$$(3.2) \quad \limsup_{r \to 1} \frac{T(r, f)}{\log \frac{1}{1-r}} = \infty \quad \text{implies} \quad \liminf_{r \to 1} \frac{S_{(k)}(r, f)}{T(r, f)} = 0.$$

However, by the hypotheses of Theorem 1, we know that

$$(3.3) \quad N\left(r, \frac{1}{f}\right) = \overline{N}\left(r, \frac{1}{f^{(k)} - 1}\right) = 0.$$

Now (3.1), (3.2) and (3.3) imply that

$$m\left(r, \frac{1}{f}\right) = T\left(r, \frac{1}{f}\right) = T(r, f) + O(1) = O\left(\log \frac{1}{1-r}\right).$$

It follows that

$$\int_0^1 (1 - r)m(r, 1/f)\, dr < \infty,$$

so $1/f \in \mathcal{A}$ by Lemma 2, part (a). Hence $f \in \mathcal{A}_m$.

4. Proof of Theorem 2

In the proof of Theorem 2, part (a), we assume that f is holomorphic in D and $ff'' \neq 0$. We can then deduce immediately from Theorem C, part (b), and Lemma 2, part (b), that $\psi = f'/f \in \mathcal{A}$ and also that $\log f \in \mathcal{A}$.

We remark that, since $\psi = f'/f$ and ψ' satisfy (2.2), the function

$$f''/f = (f'/f)^2 + \psi'$$

also satisfies (2.2). Thus $f''/f \in \mathcal{A}$ and, since $f''/f \neq 0$ in D, we can also deduce that $A^*[f''/f]$ is dense in C, by Lemma 3, part (a).

The proof that $f \in \mathcal{A}$ is more involved. It is not true in general that $g \in \mathcal{A}$ implies that $e^g \in \mathcal{A}$ (see [**3**] for a counter-example), though this implication does hold if g has no arc tracts [**21**, Theorem 10].

The idea of the proof that $f \in \mathcal{A}$ is as follows. Starting from the condition $m(r, \psi) = O\left(\log(1/(1 - r))\right)$ from Theorem C, part (a), we deduce that for values of r close to 1, there is only a small set of θ for which $|(f'/f)(re^{i\theta})|$ is close to its maximal possible value, as specified in Theorem C, part (b). From this it follows that for most θ, the value of $|(f'/f)(re^{i\theta})|$ is much less than this maximal value when r is close to 1. We can then deduce that $|f|$ is usually so much smaller than its maximal growth rate in Theorem C, part (c), that we can apply Lemma 4.

We shall need the following lemma, proved using a normal families argument from a recent paper of Hayman and Hinkkanen [16]. In this lemma $\mathcal{F}_2$ denotes the family of functions f holomorphic in D such that $ff'' \neq 0$ in D, and it is shown that for $f \in \mathcal{F}_2$ the function f'/f does not vary too much over discs of the form

$$\Delta_{z_0} = \{z : |z - z_0| \leq \delta(1 - |z_0|)\}, \quad z \in D,$$

for some suitable constant δ, $0 < \delta < 1$.

LEMMA 5. *There are positive constants λ and δ, $0 < \delta < 1$, such that if $f \in \mathcal{F}_2$, then there do not exist $z_0, z_1, z_2 \in D$ such that*

$$\left| \frac{f'(z_1)}{f(z_1)} \right| \leq \frac{e^{-\lambda}}{1 - |z_0|}, \quad \left| \frac{f'(z_2)}{f(z_2)} \right| \geq \frac{e^{\lambda}}{1 - |z_0|} \quad and \quad z_1, z_2 \in \Delta_{z_0}.$$

PROOF. By Theorem B, part (b), the family $\{f'/f : f \in \mathcal{F}_2\}$ is normal in D. Thus there are positive constants λ and δ, $0 < \delta < 1$, such that if $f \in \mathcal{F}_2$, then we cannot have

$$\left| \frac{f'(z_1)}{f(z_1)} \right| \leq e^{-\lambda}, \quad \left| \frac{f'(z_2)}{f(z_2)} \right| \geq e^{\lambda} \quad and \quad \max\{|z_1|, |z_2|\} \leq \delta.$$

For if such λ and δ do not exist, then we can obtain a contradiction to the equicontinuity property of $\{f'/f : f \in \mathcal{F}_2\}$. The required result then follows by applying this fact to the function

$$F(Z) = f(z_0 + (1 - |z_0|)Z), \quad Z \in D,$$

which belongs to $\mathcal{F}_2$. □

Now we prove that $f \in \mathcal{A}$ by showing that if f is holomorphic in D and $ff'' \neq 0$ there, then f satisfies (2.3). We may assume that $f(0) = 1$. By Theorem C, parts (a) and (b), we can choose positive constants C_1 and C_2 such that the function $\psi = f'/f$ satisfies

(4.1)
$$m(r, \psi) \leq C_1 \log \frac{1}{1 - r}, \quad 0 \leq r < 1,$$

and

(4.2)
$$\log M(r, \psi) \leq \frac{C_2}{1 - r}, \quad 0 \leq r < 1.$$

First we construct a partition of D into approximately square boxes $B_{n,k}$ with disjoint interiors, the side length of each box being less than $\delta/2$ times the distance to the boundary C, where δ is the constant in Lemma 5. To be precise we put, for $n = 1, 2, \ldots,$

(4.3) $\quad t = 1 - \delta/8, \quad r_n = 1 - t^n, \quad k_n = \left[\dfrac{16\pi r_n}{\delta t^n} \right] + 1, \quad \theta_n = \dfrac{2\pi}{k_n},$

and, for $k = 0, 1, \ldots, k_n - 1,$

(4.4) $\quad B_{n,k} = \{re^{i\theta} : r_n \leq r \leq r_{n+1}, k\theta_n \leq \theta \leq (k+1)\theta_n\}.$

Then we have, for all n and k,

$$
\begin{aligned}
\operatorname{diam} B_{n,k} &\leq (r_{n+1} - r_n) + r_n \theta_n \\
&= \frac{1}{8} \delta t^n + \frac{2\pi r_n}{k_n} \\
&\leq \frac{1}{4} \delta t^n \\
&\leq \frac{1}{2} \delta t^{n+1} \\
&= \frac{1}{2} \delta (1 - r_{n+1}).
\end{aligned}
$$

Thus

$$(4.5) \qquad z_0 \in B_{n,k} \quad \text{implies} \quad B_{n,k} \subset \left\{ z : |z - z_0| \leq \frac{1}{2} \delta (1 - |z_0|) \right\} \subset \Delta_{z_0},$$

where Δ_{z_0} was defined just before Lemma 5.

Now take a fixed ε with $0 < \varepsilon < \frac{1}{2}$, and choose a positive integer $N = N(\varepsilon, \lambda)$ such that

$$(4.6) \qquad \log \frac{1}{1 - r_n} > \lambda \quad \text{and} \quad \frac{1}{(1 - r_n)^{1-\varepsilon}} > \lambda + \log \frac{1}{1 - r_{n+1}}, \quad \text{for } n \geq N,$$

where λ is the constant in Lemma 5. We say that $B_{n,k}$ is a *bad* box if $n \geq N$ and

$$(4.7) \qquad \log |\psi(z)| > \frac{1}{(1 - |z|)^{1-\varepsilon}}, \quad \text{for some } z \in B_{n,k}.$$

Otherwise we say that $B_{n,k}$ is a *good* box.

For any bad box $B_{n,k}$ there exists, by (4.6) and (4.7), a point $z_0 \in B_{n,k}$ such that

$$(4.8) \qquad \log |\psi(z_0)| > \frac{1}{(1 - |z_0|)^{1-\varepsilon}} > \lambda + \log \frac{1}{1 - |z_0|},$$

since $r_n \leq |z_0| \leq r_{n+1}$, so

$$\left| \frac{f'(z_0)}{f(z_0)} \right| \geq \frac{e^\lambda}{1 - |z_0|}.$$

Thus, by Lemma 5, (4.5) and (4.6), we have

$$\log |\psi(z)| > \log \frac{1}{1 - |z_0|} - \lambda > 0, \quad \text{for } z \in \Delta_{z_0}.$$

Hence $\log |\psi(z)|$ is positive harmonic in Δ_{z_0}, so by (4.3), (4.5), (4.8) and Harnack's inequality, we have

$$(4.9) \qquad \log |\psi(z)| > \left(\frac{1 - \frac{1}{2}}{1 + \frac{1}{2}} \right) \frac{1}{(1 - |z_0|)^{1-\varepsilon}} \geq \frac{t^{1-\varepsilon}}{3(1 - |z|)^{1-\varepsilon}}, \quad \text{for } z \in B_{n,k}.$$

For $n \geq N$, let E_n denote the union of the bad boxes of the form $B_{n,k}$, and let $F_n = \{ \theta \in [0, 2\pi] : r_n e^{i\theta} \in E_n \}$. For $r_n \leq r \leq r_{n+1}$, $n \geq N$, we then have

$$C_1 \log \frac{1}{1 - r} \geq m(r, \psi) \geq \frac{1}{2\pi} \int_{F_n} \log^+ |\psi(r e^{i\theta})| \, d\theta \geq \frac{t^{1-\varepsilon}}{6\pi (1 - r)^{1-\varepsilon}} |F_n|,$$

by (4.1) and (4.9), where $|.|$ denotes length. Hence, for $n \geq N$,

$$
\begin{aligned}
|F_n| &\leq \left(\frac{6\pi C_1}{t^{1-\varepsilon}}\right)(1-r)^{1-\varepsilon}\log\frac{1}{1-r} \\
&\leq \left(\frac{6\pi C_1}{t^{1-\varepsilon}}\right)(1-r_n)^{1-\varepsilon}\log\frac{1}{1-r_{n+1}} \\
&= \left(\frac{6\pi C_1 \log(1/t)}{t^{1-\varepsilon}}\right)(n+1)\left(t^{1-\varepsilon}\right)^n,
\end{aligned}
$$

by (4.3). Thus

(4.10) $$\qquad |F_n| \leq C_3(n+1)(t^{1-\varepsilon})^n, \quad \text{for } n \geq N,$$

where $C_3 = (6\pi C_1 \log(1/t))/t^{1-\varepsilon}$, so

$$
\sum_{n=N}^{\infty} |F_n| < \infty.
$$

Let $R_\theta = \{re^{i\theta} : 0 \leq r < 1\}$. Then it follows that the set

$$
F = \{\theta \in [0, 2\pi] : R_\theta \text{ meets infinitely many } E_n\} = \bigcap_{m=N}^{\infty} \bigcup_{n=m}^{\infty} F_n
$$

has length 0.

Now put

$$
G_N = \{\theta \in [0, 2\pi] : R_\theta \cap E_n = \emptyset, \text{ for } n \geq N\},
$$

and

$$
G_n = \{\theta \in [0, 2\pi] : R_\theta \cap E_{n-1} \neq \emptyset, R_\theta \cap E_m = \emptyset, \text{ for } m \geq n\}, \quad \text{for } n > N.
$$

Then G_n, $n \geq N$, is a disjoint sequence of sets whose union is $[0, 2\pi] \setminus F$. Also, we have $G_n \subset F_{n-1}$, for $n > N$. Suppose that $\theta \in G_n$, where $n \geq N$. Then, by (4.2) and (4.7), we have

$$
\log^+ |\psi(re^{i\theta})| \leq
\begin{cases}
\dfrac{C_2}{1-r}, & \text{for } 0 \leq r \leq r_n, \\[2mm]
\dfrac{1}{(1-r)^{1-\varepsilon}}, & \text{for } r_n < r < 1,
\end{cases}
$$

since $R_\theta \cap \{z : r_n < |z| < 1\}$ meets only good boxes. Thus, since $f(0) = 1$, we have, for $n \geq N$ and $0 \leq r < 1$,

$$
\begin{aligned}
\log^+ |f(re^{i\theta})| &\leq \int_0^r |\psi(se^{i\theta})| \, ds \\
&\leq
\begin{cases}
\exp\left(\dfrac{C_2}{1-r}\right), & \text{for } 0 \leq r \leq r_n, \\[3mm]
\exp\left(\dfrac{C_2}{1-r_n}\right) + \exp\left(\dfrac{1}{(1-r)^{1-\varepsilon}}\right), & \text{for } r_n < r < 1,
\end{cases}
\end{aligned}
$$

since, for $r_n < r < 1$,

$$
\int_{r_n}^r |\psi(se^{i\theta})| \, ds \leq \int_{r_n}^r \exp\left(\frac{1}{(1-s)^{1-\varepsilon}}\right) ds \leq \exp\left(\frac{1}{(1-r)^{1-\varepsilon}}\right).
$$

Next put $\rho_n = 1 - (1 - r_n)^{1/(1-\varepsilon)}$, so $r_n < \rho_n < 1$ and also $(1-r)^{1-\varepsilon} < 1 - r_n$ if and only if $1 > r > \rho_n$. Then, since $x + y \le xy$, for $x, y \ge 2$, and without loss of generality $C_2 > 1$, we have for $\theta \in G_n$, $n \ge N$,

$$\int_0^1 \left(\log^+ \log^+ |f(re^{i\theta})|\right)^{1+\varepsilon} dr$$

$$\le \int_0^{r_n} \left(\frac{C_2}{1-r}\right)^{1+\varepsilon} dr + \int_{r_n}^1 \left(\frac{C_2}{1-r_n} + \frac{1}{(1-r)^{1-\varepsilon}}\right)^{1+\varepsilon} dr$$

$$\le C_2^{1+\varepsilon} \int_0^{r_n} \left(\frac{1}{1-r}\right)^{1+\varepsilon} dr + C_2^{1+\varepsilon} \int_{r_n}^1 \left(\frac{1}{1-r_n} + \frac{1}{(1-r)^{1-\varepsilon}}\right)^{1+\varepsilon} dr$$

$$\le \frac{C_2^{1+\varepsilon}/\varepsilon}{(1-r_n)^\varepsilon} + C_2^{1+\varepsilon} \int_{r_n}^{\rho_n} \left(\frac{2}{1-r_n}\right)^{1+\varepsilon} dr + C_2^{1+\varepsilon} \int_{\rho_n}^1 \left(\frac{2}{(1-r)^{1-\varepsilon}}\right)^{1+\varepsilon} dr$$

$$\le \frac{C_2^{1+\varepsilon}\left(1/\varepsilon + 2^{1+\varepsilon}\right)}{(1-r_n)^\varepsilon} + \frac{(2C_2)^{1+\varepsilon}}{\varepsilon^2}(1-r_n)^{\varepsilon^2}$$

$$\le \frac{C_4}{t^{n\varepsilon}},$$

by (4.3), where C_4 depends on C_2 and ε.

Now $G_n \subset F_{n-1}$, for $n > N$, so

$$(4.11) \qquad |G_n| \le |F_{n-1}| \le C_3 n (t^{1-\varepsilon})^{n-1}, \quad \text{for } n > N,$$

by (4.10). Also, $G_N \le 2\pi$, so we may assume that (4.11) holds for $n \ge N$ by increasing C_3 if necessary. Thus, since $|F| = 0$ and $1 - 2\varepsilon > 0$, we have

$$\iint_D \left(\log^+ \log^+ |f(z)|\right)^{1+\varepsilon} dx\,dy$$

$$\le \sum_{n=N}^\infty \int_{G_n} d\theta \int_0^1 \left(\log^+ \log^+ |f(re^{i\theta})|\right)^{1+\varepsilon} dr$$

$$\le \sum_{n=N}^\infty C_3 n (t^{1-\varepsilon})^{n-1} \left(\frac{C_4}{t^{n\varepsilon}}\right)$$

$$= C_3 C_4 t^{-1+\varepsilon} \sum_{n=N}^\infty n (t^{1-2\varepsilon})^n < \infty.$$

Thus f satisfies (2.3), so the proof of Theorem 2, part (a), is complete.

In Theorem 2, part (b), we add the hypothesis that $f' \neq 0$, so we know that $f'/f \in \mathcal{A}$, from part (a), and also that $ff' \neq 0$. Thus, by Lemma 3, part (b), we deduce that $A^*[f'/f]$, $A^*[\log f]$, $A^*[f]$, $A^*[e^f]$, ..., are dense in C. This completes the proof of Theorem 2.

5. Proof of Example 1

The construction of Example 1 is similar to that of [**2**, Example 1]. Let $\gamma_1(t) = r_1(t)e^{it}$, $\gamma_2(t) = r_2(t)e^{it}$, $t \ge 0$, be two smooth spirals contained in D, disjoint except for $\gamma_1(0) = \gamma_2(0) = 0$, and such that, for $k = 1, 2$, we have $r_k(t) \uparrow 1$ as $t \uparrow \infty$, $\gamma_k'(t) \neq 0$, for $t \ge 0$, and $\gamma_2'(0) > 0$. For $k = 1, 2$, define functions ϕ_k to be continuous on γ_k and satisfy $\phi_k(\gamma_k(0)) = 0$,

$$(5.1) \qquad \exp \phi_1(\gamma_1(t)) = \frac{\gamma_1'(0)}{\gamma_1'(t)} \frac{1}{1+t^2}, \text{ for } t \geq 0,$$

and

$$(5.2) \qquad \exp \phi_2(\gamma_2(t)) = \frac{\gamma_2'(0)}{\gamma_2'(t)}, \text{ for } t \geq 0.$$

Then use Arakelyan's Theorem [9] to choose a function h, holomorphic in D, such that

$$(5.3) \quad |h(z) - \phi_1(z)| < \frac{1}{2}, \text{ for } z \in \gamma_1, \quad \text{and} \quad |h(z) - \phi_2(z)| < \frac{1}{2}, \text{ for } z \in \gamma_2,$$

and put

$$g(z) = \int_0^z \exp h(\zeta) \, d\zeta \quad \text{and} \quad f(z) = \exp g(z).$$

Clearly $f f' \neq 0$.

Now for $z = \gamma_1(t)$, $t > 0$, we have

$$g(z) = \int_0^t \exp(h(\gamma_1(s))) \gamma_1'(s) \, ds = \gamma_1'(0) \int_0^t \frac{\exp(h(\gamma_1(s)) - \phi_1(\gamma_1(s)))}{1+s^2} \, ds,$$

by (5.1). Since this last integral is uniformly bounded for $0 < t < \infty$, by (5.3), we deduce that both $g = \log f$ and f are bounded on the spiral γ_1.

On the other hand, for $z = \gamma_2(t)$, $t > 0$, we have

$$g(z) = \int_0^t \exp(h(\gamma_2(s))) \gamma_2'(s) \, ds = \gamma_2'(0) \int_0^t \exp(h(\gamma_2(s)) - \phi_2(\gamma_2(s))) \, ds,$$

by (5.2). By (5.3) the values of this last integrand lie in the set

$$K = \{w : |w - 1| \leq \frac{1}{2} e^{1/2}\},$$

which is a compact convex subset of $\{w : \Re(w) > 0\}$. It follows from the mean value theorem that, for $t > 0$, $g(z) = g(\gamma_2(t))$ lies in the set $\gamma_2'(0) t K$. Since $\gamma_2'(0) > 0$, both $g = \log f$ and f have asymptotic value ∞ along γ_2. Thus $\log f$ and f have asymptotic values at no points of C, so neither function belongs to the class $\mathcal{A}$.

Finally we show that $f'/f \notin \mathcal{A}$. For if $f'/f = g' \in \mathcal{A}$, then because $g' \neq 0$ we deduce from Lemma 3, part (b), that $g = \log f \in \mathcal{A}$, which is false.

References

[1] K. F. Barth, *Asymptotic values of meromorphic functions*, Michigan Math. J. **13** (1966), 321–340.

[2] K. F. Barth and P. J. Rippon, *Asymptotic tracts of locally univalent functions*, Bull. London Math. Soc. **34** (2002), 200–204.

[3] K. F. Barth and W. J. Schneider, *Exponentiation of functions in MacLane's class A*, J. Reine Angew. Math. **236** (1969), 120–130.

[4] K. F. Barth and W. J. Schneider, *Integrals and derivatives of functions in MacLane's class A and of normal functions*, Proc. Camb. Phil. Soc. **71** (1972), 111–121.

[5] W. Bergweiler, *Normality and exceptional values of derivatives*, Proc. Amer. Math. Soc. **129** (2000), 121–129.

[6] C. T. Chuang, *Normal Families of Meromorphic Functions*, World Scientific, 1993.

[7] Y. Domar, *On the existence of a largest subharmonic minorant for a given function*, Ark. Mat. **3** (1954–58), 429–440.

[8] G. Frank, *Eine Vermutung von Hayman über Nullstellen meromorpher Funktionen*, Math. Z. **149** (1976), 29–36.

[9] D. Gaier, *Lectures on Complex Approximation*, Birkhäuser, 1985.

[10] Y. X. Gu, *A criterion for normality of families of meromorphic functions* (Chinese), Sci. Sinica (special issue) **1** (1979), 267–274.

[11] R. L. Hall, *Line sets and asymptotic behavior of functions holomorphic in the unit disc*, Ann. Acad. Sci. Fenn. **519** (1972), 2–11.

[12] W. K. Hayman, *Uniformly normal families*, Lectures on functions of a complex variable, University of Michigan Press, Ann Arbor (1955), 199–212.

[13] W. K. Hayman, *Picard values of meromorphic functions and their derivatives*, Annals of Math. **70** (1959), 9–42.

[14] W. K. Hayman, *Meromorphic Functions*, Oxford University Press, 1964.

[15] W. K. Hayman, *Subharmonic Functions*, Volume 2, Academic Press, 1989.

[16] W. K. Hayman and A. Hinkkanen, *Generalisations of uniformly normal families*, preprint.

[17] W. K. Hayman and B. Korenblum, *An extension of the Riesz–Herglotz formula*, Ann. Acad. Sci. Fenn. Ser. AI Math. **2** (1976), 175–201.

[18] R. J. M. Hornblower, *A growth condition for the MacLane class $\mathcal{A}$*, Proc. London Math. Soc.(3) **23** (1971), 371–384.

[19] J. K. Langley, *Proof of a conjecture of Hayman concerning f and f''*, J. London Math. Soc. **48** (1993), 500–514.

[20] G. R. MacLane, *Asymptotic values of holomorphic functions*, Rice Univ. Studies, 49 (1963), 83 pp.

[21] G. R. MacLane, *Exceptional values of $f^{(n)}(z)$, asymptotic values of $f(z)$ and linearly accessible asymptotic values*, Mathematical Essays Dedicated to A.J. Macintyre, Ohio University Press, 1970, pp. 271–288.

[22] C. Miranda, *Sur un nouveau critère de normalité pour les familles de fonctions holomorphes*, Bull. Soc. Math. France, 63 (1935), 185–196.

[23] P. J. Rippon, *On subharmonic functions which satisfy a growth restriction*, Proc. NATO ASI Durham, Academic Press, 1980, pp. 485–492.

[24] P. J. Rippon, *The fine boundary behaviour of certain delta-subharmonic functions*, J. London Math. Soc. **26** (1982), 487–503.

[25] J. L. Schiff, *Normal Families*, Springer-Verlag, 1993.

[26] W. Schwick, *Normality criteria for families of meromorphic functions*, J. Analyse Math. **52** (1989), 241–289.

[27] L. Zalcman, *Normal families: new perspectives*, Bull. Amer. Math. Soc. **35** (1998), 215–230.

DEPARTMENT OF MATHEMATICS, SYRACUSE UNIVERSITY, SYRACUSE, NY 13244, USA
E-mail address: `kfbarth@mailbox.syr.edu`

THE OPEN UNIVERSITY, DEPARTMENT OF PURE MATHEMATICS, WALTON HALL, MILTON KEYNES MK7 6AA, UK
E-mail address: `p.j.rippon@open.ac.uk`

Contemporary Mathematics
Volume 404, 2006

Operators on Weighted Bergman Spaces

Oscar Blasco

ABSTRACT. Let $\rho : (0,1] \to \mathbb{R}^+$ be a weight function and let X be a complex Banach space. We denote by $A_{1,\rho}(\mathbb{D})$ the space of analytic functions in the disc $\mathbb{D}$ such that $\int_{\mathbb{D}} |f(z)|\rho(1-|z|)dA(z) < \infty$ and by $\text{Bloch}_\rho(X)$ the space of analytic functions in the disc $\mathbb{D}$ with values in X such that $\sup_{|z|<1} \frac{1-|z|}{\rho(1-|z|)}\|F'(z)\| < \infty$. We prove that, under certain assumptions on the weight, the space of bounded operators $L(A_{1,\rho}(\mathbb{D}), X)$ is isomorphic to $\text{Bloch}_\rho(X)$ and some applications of this result are presented. Several properties of generalized vector-valued Bloch functions are also considered.

1. Introduction and Preliminaries

Weighted Bergman spaces appeared, denoted by B^p, in [**15**], when looking at the Banach envelop of the Hardy spaces H^p for $0 < p < 1$, although they had been implicitly used in the work of Hardy and Littlewood (see [**16**], or [**14**, p. 87]) who established that for $0 < p < 1$

$$(1) \qquad \int_0^1 (1-r)^{1/p-2} M_1(F,r)dr \leq C\|F\|_{H^p}$$

for any $F \in H^p$, where $M_q(F,r) = (\int_0^{2\pi} |F(re^{it})|^q \frac{dt}{2\pi})^{1/q}$, $0 < q < \infty$ and $\|F\|_{H^p} = \sup_{0<r<1} M_p(F,r)$.

This inequality was a crucial point in proving the duality $(H^p)^* = \Lambda_\alpha$ for $0 < p < 1$ and $\alpha = 1/p - 1$ between the Hardy classes H^p and the Lipschitz classes Λ_α (see [**15**]). In that paper, they denoted by B^p the space of analytic functions in the disc such that

$$\|F\|_{B^p} = \int_0^1 (1-r)^{1/p-2} M_1(F,r) < \infty.$$

It is known that for $0 < p < 1$, $F \in B^p$ if and only if

$$\int_0^1 (1-r)^{1/p-1} M_1(F',r)dr < \infty.$$

2000 *Mathematics Subject Classification*. Primary 46E40; Secondary 46E15, 47B38, 30D45.
Partially supported by Proyectos BMF2002-04013 and MTM2005-08350.

Making $p = 1$ in the last formula, one can denote by B^1 the space of analytic functions such that

$$\|F\|_{B^1} = |F(0)| + \int_0^1 M_1(F', r)dr < \infty.$$

This space was later shown to be the predual of the Bloch space (see [1]).

Of course, using polar coordinates, one can realize the spaces as particular weighted Bergman spaces, or weighted Besov spaces, given by analytic functions in the unit disk such that

$$(2) \qquad \int_{\mathbb{D}} |F(z)|(1 - |z|)^{1/p-2}dA(z) < \infty,$$

or

$$(3) \qquad \int_{\mathbb{D}} |F'(z)|(1 - |z|)^{1/p-1}dA(z) < \infty,$$

where $dA(z)$ stands for the normalized Lebesgue measure. The reader is referred to [20] for a proof of the duality between $A_1(\mathbb{D})$ and Bloch.

There are some natural conditions on a weight function $\rho : (0, 1] \to \mathbb{R}^+$ which allow to extend several results that hold for $\rho(t) = t^\alpha$ to a more general context.

DEFINITION 1.1. Let $0 \leq p, q < \infty$ and let $\rho : (0, 1] \to \mathbb{R}^+$ be a continuous function. It is said to be a (b_q)-weight if there exists a constant $C > 0$ such that

$$(4) \qquad \int_s^1 \frac{\rho(t)}{t^{1+q}}dt \leq C\frac{\rho(s)}{s^q}, 0 < s < 1.$$

It is said to be a (d_p)-weight (or to satisfy Dini condition of order p) if there exists a constant $C > 0$ such that

$$(5) \qquad \int_0^s t^p\rho(t)\frac{dt}{t} \leq Cs^p\rho(s), 0 < s < 1.$$

These notions turned out to be relevant for different purposes (see [2, 12, 13, 18, 19]). We refer the reader to those papers for examples and properties of these classes of weights.

In this paper, we consider the weighted Bergman spaces $A_{1,\rho}(\mathbb{D})$. The just mentioned B^p-spaces correspond to the case $\rho(t) = t^{1/p-2}$ for $0 < p < 1$.

DEFINITION 1.2. Let $\rho : (0, 1] \to \mathbb{R}^+$ be an integrable function. An analytic function f in the unit disc $\mathbb{D}$ is said to belong to $A_{1,\rho}(\mathbb{D})$ if

$$\|f\|_{A_{1,\rho}} = \int_{\mathbb{D}} |f(z)|\rho(1 - |z|)dA(z) < \infty.$$

REMARK 1.1. (1) If $\rho \in L^1((0, 1])$, then $H^\infty(\mathbb{D}) \subset A_{1,\rho}(\mathbb{D})$.
(2) $A_{1,\rho}(\mathbb{D})$ is a closed subspace of the space $L^1(\mathbb{D}, \rho(1 - |z|)dA(z))$.
(3) The polynomials are dense in $A_{1,\rho}(\mathbb{D})$.

Let us denote by P_α and P_α^*, $\alpha > -1$, the operators

$$P_\alpha(f)(z) = (\alpha + 1) \int_{\mathbb{D}} \frac{(1 - |w|^2)^\alpha}{(1 - \bar{w}z)^{2+\alpha}} f(w)dA(w)$$

$$P_\alpha^*(f)(z) = (\alpha + 1) \int_{\mathbb{D}} \frac{(1 - |w|^2)^\alpha}{|1 - \bar{w}z|^{2+\alpha}} f(w)dA(w)$$

for $f \in L^1(\mathbb{D}, (1 - |w|^2)^\alpha dA(w))$.

It is well-known (see [20]) that P_α^* in bounded on $L^p(\mathbb{D})$ and that P_α is a projection on the Bergman spaces $A^p(\mathbb{D})$ if and only if $p > 1 + \alpha$.

The Bergman projection corresponds to $\alpha = 0$ and it will be denoted by P. Since P is not bounded on $L^1(\mathbb{D})$, we now study its boundedness in $L^1(\mathbb{D}, \rho(1 - |z|)dA)$.

PROPOSITION 1.3. *Let $\alpha \geq 0$ and $\rho : (0, 1] \to \mathbb{R}^+$ be a continuous function. If ρ is a (d_1) and (b_α)-weight, then P_α^* is a bounded on $L^1(\mathbb{D}, \rho(1 - |z|)dA)$.*

In particular, P_α defines a projection from $L^1(\mathbb{D}, \rho(1 - |z|)dA)$ onto $A_{1,\rho}(\mathbb{D})$.

PROOF. Let f belong to $L^1(\mathbb{D}, \rho(1 - |z|)dA)$. We have

$$
\begin{aligned}
\|P_\alpha^* f\|_{A_{1,\rho}(\mathbb{D})} &= \int_{\mathbb{D}} \rho(1 - |z|) |P_\alpha^* f(z)| dA(z) \\
&= C \int_{\mathbb{D}} \rho(1 - |z|) \left(\int_{\mathbb{D}} \frac{(1 - |w|^2)^\alpha}{|1 - z\bar{w}|^{2+\alpha}} |f(w)| dA(w) \right) dA(z) \\
&\leq C \int_{\mathbb{D}} |f(w)|(1 - |w|^2)^\alpha \left(\int_{\mathbb{D}} \frac{\rho(1 - |z|)}{|1 - z\bar{w}|^{2+\alpha}} dA(z) \right) dA(w) \\
&\approx C \int_{\mathbb{D}} (1 - |w|^2)^\alpha |f(w)| \left(\int_0^1 \frac{\rho(1 - r)}{(1 - r|w|)^{1+\alpha}} dr \right) dA(w) \\
&\approx C \int_{\mathbb{D}} |f(w)|(1 - |w|^2)^\alpha \left(\int_0^1 \frac{\rho(t)}{t + (1 - |w|)^{1+\alpha}} dt \right) dA(w) \\
&\approx C \int_{\mathbb{D}} \frac{|f(w)|}{1 - |w|} \left(\int_0^{1-|w|} \rho(t) dt \right) dA(w) \\
&\quad + C \int_{\mathbb{D}} |f(w)|(1 - |w|)^\alpha \left(\int_{1-|w|}^1 \frac{\rho(t)}{t^{1+\alpha}} dt \right) dA(w) \\
&\leq C \int_{\mathbb{D}} |f(w)| \rho(1 - |w|) dA(w).
\end{aligned}
$$

$\square$

PROPOSITION 1.4. *Let $\rho : (0, 1] \to \mathbb{R}^+$ be non-increasing, $\rho(1) > 0$ and (d_1)-weight. Set $\rho_1(t) = t\rho(t)$. Then $f \in A_{1,\rho}(\mathbb{D})$ if and only if $f' \in A_{1,\rho_1}(\mathbb{D})$. Moreover, $\|f'\|_{A_{1,\rho_1}(\mathbb{D})} + |f(0)| \approx \|f\|_{A_{1,\rho}(\mathbb{D})}$.*

PROOF. Since $M_1(f', r^2) \leq C \frac{M_1(f,r)}{1-r}$,

$$
\begin{aligned}
\|f'\|_{A_{1,\rho_1}(\mathbb{D})} &\approx \int_0^1 (1 - r^2)\rho(1 - r^2) M_1(f', r^2) r dr \\
&\leq C \int_0^1 \rho(1 - r^2) M_1(f, r) dr \leq C \|f\|_{A_{1,\rho}(\mathbb{D})}.
\end{aligned}
$$

On the other hand,

$$
|f(0)| \leq \int_{\mathbb{D}} |f(w)| dA(w) \leq \rho(1)^{-1} \int_{\mathbb{D}} |f(w)| \rho(1 - |w|) dA(w).
$$

To prove the other inequality, we use $M_1(f, r) \leq \int_0^r M_1(f', s) ds + |f(0)|$.

Hence

$$
\begin{aligned}
\int_0^1 \rho(1-r)M_1(f,r)dr &\leq \int_0^1 \rho(1-r)\Big(\int_0^r M_1(f',s)ds\Big)dr + |f(0)|\int_0^1 \rho(t)dt \\
&\leq \int_0^1 \Big(\int_s^1 \rho(1-r)dr\Big)M_1(f',s)ds + |f(0)|\int_0^1 \rho(t)dt \\
&\leq \int_0^1 \Big(\int_0^{1-s}\rho(u)du\Big)M_1(f',s)ds + |f(0)|\int_0^1 \rho(t)dt \\
&\leq C\int_0^1 (1-s)\rho(1-s)M_1(f',s)ds + |f(0)|\int_0^1 \rho(t)dt \\
&\leq C(\|f'\|_{A_{1,\rho_1}(\mathbb{D})} + |f(0)|).
\end{aligned}
$$

$\square$

Let us now introduce the generalized Bloch classes, extending the notion of Bloch functions:

$$
\mathrm{Bloch}(X) = \Big\{ F : \mathbb{D} \to X \text{ analytic} : \sup_{|z|<1}(1-|z|^2)\|F'(z)\| < \infty \Big\}.
$$

The reader is referred to [**3, 4, 5, 6, 8, 9, 10**] for different results concerning vector-valued Bloch functions.

DEFINITION 1.5. Let $\rho : (0,1] \to \mathbb{R}^+$ be a continuous function such that $\frac{\rho(t)}{t}$ is non-increasing and $\rho(1) > 0$ and let X be a complex Banach space. An analytic function from the disc $\mathbb{D}$ into X, $F(z) = \sum_{n=0}^\infty x_n z^n$ where $x_n \in X$, is said to belong to $\mathrm{Bloch}_\rho(X)$ if there exists a constant $C > 0$ such that

$$
\|F'(z)\| \leq C\frac{\rho(1-|z|)}{1-|z|}, \qquad z \in \mathbb{D}.
$$

It is easy to see that $\mathrm{Bloch}_\rho(X)$ becomes a Banach space under the norm

$$
\|F\|_{\mathrm{Bloch}_\rho(X)} = \|F(0)\| + \sup_{|z|<1}\left\{ \frac{1-|z|}{\rho(1-|z|)}\|F'(z)\| \right\}.
$$

DEFINITION 1.6. Let $\rho : (0,1] \to \mathbb{R}^+$ be a continuous function such that $\frac{\rho(t)}{t}$ is non-increasing, $\rho(1) > 0$ and $\lim_{t\to 0^+}\frac{\rho(t)}{t} = \infty$ and let X be a complex Banach space. The *little Bloch space* $\mathrm{bloch}_\rho(X)$ is the closed subspace of $\mathrm{Bloch}_\rho(X)$ given by those functions for which

$$
\lim_{|z|\to 1}\frac{1-|z|}{\rho(1-|z|)}\|F'(z)\| = 0.
$$

REMARK 1.2. (1) There is no loss of generality in assuming $\frac{\rho(t)}{t}$ non-increasing, because the function $\tilde{\rho}$ defined by

$$
\frac{\tilde{\rho}(1-t)}{1-t} = \sup\{M_\infty(F',t) : \|F\|_{\mathrm{Bloch}_\rho(X)} \leq 1\}
$$

is non-increasing and $\mathrm{Bloch}_\rho(X) = \mathrm{Bloch}_{\tilde{\rho}}(X)$.

(2) The assumptions $\frac{\rho(t)}{t} \geq \rho(1) > 0$ and $\lim_{t\to 0}\frac{\rho(t)}{t} = \infty$ are needed to have the vector-valued polynomials in the spaces $\mathrm{Bloch}_\rho(X)$ and $\mathrm{bloch}_\rho(X)$, respectively.

(3) If $\frac{\rho(t)}{t}$ non-increasing, then ρ is a (b_q)-weight for $q > 1$.

Indeed,

$$\int_s^1 \frac{\rho(t)}{t^{1+q}}\,dt \le C\frac{\rho(s)}{s}\int_s^1 \frac{dt}{t^q} \le C\frac{\rho(s)}{s^q}.$$

(4) If ρ is a (b_1)-weight and $\frac{\rho(t)}{t} \ge C > 0$, then $\lim_{t\to 0}\frac{\rho(t)}{t} = \infty$.

Indeed,

$$C\log(\frac{1}{s}) = C\int_s^1 \frac{dt}{t}\,dt \le \int_s^1 \frac{\rho(t)}{t^2}\,dt \le C'\frac{\rho(s)}{s}.$$

In [5], it was shown that the boundedness of operators between weighted Bergman spaces and a general Banach space X can be characterized by the fact that a fixed associated vector-valued function belongs to certain Lipschitz space. In the papers [6] and [9], similar results were extended to weighted spaces $B_p(\rho)$ for $0 < p \le 1$ and certain generalized Lipschitz classes for weights introduced by Janson (see [18]). Some applications of these results to multipliers, Carleson measures and composition operators were achieved.

In this paper, we present an independent proof of some of those results, where we shall addecuate the duality between the Bergman space $A_1(\mathbb{D})$ and Bloch (see [20]) to our vector-valued and weighted situation.

Let us now give a natural correspondence between operators and vector-valued analytic functions, that will allow us to identify the bounded operators from $A_{1,\rho}(\mathbb{D})$ into X with $\mathrm{Bloch}_\rho(X)$ with equivalent norms. This idea has been used by the author several times with slight modifications (see [4], [5], [6] or [9]).

Given an analytic function $F(z) = \sum_{n=0}^\infty x_n z^n$ where $x_n \in X$, we can define a linear operator T_F which acts on polynomials as follows:

$$(6) \qquad T_F\left(\sum_{k=0}^n \alpha_k z^k\right) = \sum_{k=0}^n \frac{\alpha_k x_k}{k+1}.$$

Conversely, given a linear operator T defined on some space of analytic functions on the unit disc containing the polynomials and with range in a Banach space X, one can define the vector-valued analytic function F_T given by

$$(7) \qquad F_T(z) = \sum_{n=0}^\infty (n+1)T(u_n)z^n,$$

where $u_n(z) = z^n$.

Now we are ready to state the main theorem of the paper:

THEOREM 1.7. Let $\rho : (0,1] \to \mathbb{R}^+$ be a continuous function such that $\frac{\rho(t)}{t}$ is non-increasing and $\rho(1) > 0$ and let X be a complex Banach space. Assume ρ is a (d_1) and (b_1) weight.

(i) If $F(z) = \sum_{n=0}^\infty T_n z^n \in \mathrm{Bloch}_\rho(X)$, then T_F extends to a bounded operator in $L(A_{1,\rho}(\mathbb{D}), X)$.

(ii) If T extends to a bounded operator in $L(A_{1,\rho}(\mathbb{D}), X)$, then F_T belongs to $\mathrm{Bloch}_\rho(X)$.

In particular, $\mathrm{Bloch}_\rho(X) = L(A_{1,\rho}(\mathbb{D}), X)$ with equivalent norms.

It is known that $\mathrm{Bloch}_\rho(\mathbb{C})$ for $\rho(t) = t^\alpha$, $0 < \alpha < 1$ and $X = \mathbb{C}$ coincides with the Lipschitz class defined in terms of the modulus of continuity Λ_α (see

Theorem 5.1 in [**14**]). The proof works also in the vector-valued case, and then we can say that the space $\text{Bloch}_\rho(X)$ for $\rho(t) = t^\alpha$ and $0 < \alpha < 1$ coincides with

$$(8) \qquad \Lambda_\alpha(X) = \{f \in C_X(\mathbb{T}) : w(f,t) = \sup_{s \in \mathbb{T}} ||f(e^{i(t+s)}) - f(e^{is})|| = O(t^\alpha)\}.$$

COROLLARY 1.8. *Let* $1/3 < p < 1/2$, $\alpha = 1/p - 2$ *and let* X *be a Banach space. Then the following are equivalent.*

(i) $T : H_p \to X$ *is bounded.*
(ii) $F_T \in \Lambda_\alpha(X)$.

PROOF. (i) $\implies$ (ii). We use that $F_T'(z) = \sum_{n=1}^\infty (n+1)nT(u_n)z^{n-1}$. Hence if $G_z(w) = \sum_{n=1}^\infty (n+1)nw^n z^{n-1} = \frac{2w}{(1-wz)^3}$, one has $F_T'(z) = T(G_z)$. Hence one has to estimate $||G_z||_{H^p}$. Observe now that

$$||G_z||_{H^p} \le \left(\int_0^{2\pi} \frac{dt}{|1 - ze^{it}|^{3p}}\right)^{1/p} \le C\frac{1}{(1-|z|)^{3-1/p}} = C\frac{1}{(1-|z|)^{1-\alpha}}.$$

(ii) $\implies$ (i) Since $\rho(t) = t^{1/p-2}$ satisfies the assumptions in Theorem 1.7, one gets T is bounded from $A_{1,\rho}(\mathbb{D})$ into X. Now, by (1), $H^p \subset A_{1,\rho}(\mathbb{D})$ and then $T = T_{F_T}$ is also bounded from H^p. $\square$

The reader is referred to [**3, 6, 9**] for applications to multipliers, Carleson measures and composition operators of similar nature.

2. Proof of the main theorem

We need some lemmas before starting the proof.

LEMMA 2.1. *Let* X *be complex Banach spaces,* $F(z) = \sum_{n=0}^\infty x_n z^n$ *for* $x_n \in X$ *and* $g(z) = \sum_{n=0}^m \alpha_n z^n$ *for* $\alpha_n \in \mathbb{C}$. *Then*

(i) $T_F(g) = \int_{\mathbb{D}} F(w)g(\bar{w})dA(w)$.
(ii) $\int_{\mathbb{D}}(1 - |w|^2)F'(w)g_1(\bar{w})dA(w) = \sum_{n=1}^m \frac{\alpha_n x_n}{n+1} = T_F(g) - F(0)g(0)$, *where*
$g_1(z) = \sum_{n=1}^m x_n z^{n-1} = \frac{g(z)-g(0)}{z}$.

PROOF. (i)

$$\int_{\mathbb{D}} F(w)g(\bar{w})dA(w) = \sum_{n,k\ge 0} \int_{\mathbb{D}} x_n\alpha_k w^n \bar{w}^k dA(w)$$
$$= \sum_{n=0}^m \int_{\mathbb{D}} x_n\alpha_n|w|^{2n}dA(w) = \sum_{n=0}^m \frac{\alpha_n x_n}{n+1}.$$

(ii)

$$\int_{\mathbb{D}}(1-|w|^2)F'(w)g_1(\bar{w})dA(w) = \sum_{n\ge 1,k\ge 1} \int_{\mathbb{D}}(1-|w|^2)nx_n\alpha_k w^{n-1}\bar{w}^{k-1}dA(w)$$
$$= \sum_{n=1}^m n\alpha_n x_n \int_{\mathbb{D}}(1-|w|^2)|w|^{2n-2}dA(w)$$
$$= \sum_{n=1}^m \frac{\alpha_n x_n}{n+1}.$$

$\square$

LEMMA 2.2. *Let $\rho : (0,1] \to \mathbb{R}^+$ belong to $L^1((0,1])$ and $\rho(t) \geq Ct$ for $0 < t < 1$. Let $f(z) = \sum_{n=0}^{\infty} \alpha_n z^n$ and $f_1(z) = \sum_{n=1}^{\infty} \alpha_n z^{n-1} = \frac{f(z)-f(0)}{z}$. Then $f \in A_{1,\rho}(\mathbb{D})$ if and only if $f_1 \in A_{1,\rho}(\mathbb{D})$.*
Moreover, $\|f_1\|_{A_{1,\rho}(\mathbb{D})}) + |f(0)| \approx \|f\|_{A_{1,\rho}(\mathbb{D})}$.

PROOF. Using $|\alpha_n| r^n \leq M_1(f, r)$ we obtain for $n \geq 0$

$$\|f\|_{A_{1,\rho}} \geq \int_0^1 |\alpha_n| r^n \rho(1-r)dr$$

$$\geq C \int_0^1 |\alpha_n| r^n (1-r)dr$$

$$= C\frac{|\alpha_n|}{(n+1)^2}.$$

Hence

(9) $$|\alpha_n| \leq C(n+1)^2 \|f\|_{A_{1,\rho}}.$$

Assume first that $f_1 \in A_{1,\rho}(\mathbb{D})$. Since $f(z) = f(0) + zf_1(z)$, we have

$$\|f\|_{A_{1,\rho}(\mathbb{D})} \leq |f(0)| \int_0^1 \rho(t)dt + \|f_1\|_{A_{1,\rho}(\mathbb{D})}.$$

Conversely, assume that $f \in A_{1,\rho}(\mathbb{D})$. By (9)

$$\int_{\mathbb{D}} |f_1(z)| \rho(1-|z|) dA(z) \leq C \left(\sum_{n=0}^{\infty} \frac{(n+1)^2}{2^n} \right) \|f\|_{A_{1,\rho}(\mathbb{D})}$$

$$+ 2 \int_{1/2}^1 (|f(0)| + M_1(f, r)) \rho(1-r)dr.$$

This gives $\|f_1\|_{A_{1,\rho}(\mathbb{D})} \leq C\|f\|_{A_{1,\rho}(\mathbb{D})}$. $\qquad\square$

PROOF OF THEOREM 1.7. **(i)** $\implies$ **(ii)**. From (ii) in Lemma 2.1, we have

$$T_F(g) - F(0)g(0) = \int_{\mathbb{D}} (1-|z|^2) F'(z) g_1(\bar{z}) dA(z)$$

for each polynomial g. Hence

$$\|T_F(g)\| \leq \|F(0)\| \|g(0)\| + \int_{\mathbb{D}} (1-|z|^2) \|F'(z)\| |g_1(\bar{z})| dA(z)$$

$$\leq \|F(0)\| \|g(0)\| + \sup_{|z|<1} \frac{1-|z|}{\rho(1-|z|)} \|F'(z)\| \int_{\mathbb{D}} \rho(1-|z|) |g_1(\bar{z})| dA(z)$$

$$\leq \|F\|_{\text{Bloch}_\rho(X)} (|g(0)| + \|g_1\|_{A_{1,\rho}}).$$

Now we apply Lemma 2.2 to get $\|T_F(g)\| \leq C\|F\|_{\text{Bloch}_\rho(X)} \|g\|_{A_{1,\rho}}$. Hence we can extend now to $A_{1,\rho}$ using the density of polynomials.

(ii) $\implies$ **(i)**. Assume T extends to a bounded operator from $A_{1,\rho}(\mathbb{D})$ into X. We write $u_n(z) = z^n$ and $K_z(w) = \frac{1}{(1-wz)^2}$ for the Bergman kernel.

Clearly, for $n \in \mathbb{N}$,

$$\|u_n\|_{A_{1,\rho}} = \int_0^1 r^n \rho(1-r)dr \leq C.$$

This shows that $K_z = \sum_{n=0}^{\infty}(n+1)u_n z^n$ is an absolutely convergent series in $A_{1,\rho}(\mathbb{D})$. Therefore

$$F_T(z) = \sum_{n=0}^{\infty}(n+1)T(u_n)z^n = T(K_z).$$

The same argument gives

$$F_T'(z) = \sum_{n=1}^{\infty}(n+1)nT(u_n)z^{n-1} = T\left(\sum_{n=1}^{\infty}(n+1)nu_n z^{n-1}\right).$$

Write $G_z(w) = \sum_{n=1}^{\infty}(n+1)nw^n z^{n-1} = \frac{2w}{(1-wz)^3}$.

Hence

$$\|F_T'(z)\| \leq \|T\| . \|G_z\|_{A_{1,\rho}}.$$

Let us now estimate $\|G_z\|_{A_{1,\rho}}$

$$
\begin{aligned}
\|G_z\|_{A_{1,\rho}} &\leq C\int_0^1 \rho(1-r)\left(\int_0^{2\pi}\frac{d\theta}{|1-rze^{i\theta}|^3}\right)dr \\
&\leq C\int_0^1 \frac{\rho(1-r)}{(1-r|z|)^2}dr \\
&= \int_0^{|z|}\frac{\rho(1-r)}{(1-r|z|)^2}dr + \int_{|z|}^1\frac{\rho(1-r)}{(1-r|z|)^2}dr \\
&\leq \int_0^{|z|}\frac{\rho(1-r)}{(1-r)^2}dr + \frac{1}{(1-|z|)^2}\int_{|z|}^1\rho(1-r)dr \\
&\leq \int_{1-|z|}^1\frac{\rho(t)}{t^2}dt + \frac{1}{(1-|z|)^2}\int_0^{1-|z|}\rho(r)dr.
\end{aligned}
$$

Using now the (d_1) and (b_1) assumptions on ρ, one gets

$$\|F_T'(z)\| \leq \|T\| . \|G_z\|_{A_{1,\rho}} \leq C\frac{\rho(1-|z|)}{1-|z|}.$$

$\square$

3. Vector-valued generalized Bloch spaces.

Let us now indicate how to construct examples of functions in the generalized Bloch classes in the vector-valued case using Theorem 1.7. See [5, 8, 9, 10] for more examples.

PROPOSITION 3.1. *Let ρ be a (d_1)-weight and a (b_1)-weight on $(0,1]$ and set $\rho_n = \int_0^1 r^n\rho(1-r)dr$. Then*

(10) $$F(z) = (\rho_n z^n)_n = \sum_{n=0}^{\infty}\rho_n e_n z^n \in \text{Bloch}_\rho(\ell^1),$$

where $\{e_n\}$ stands for the canonical basis of ℓ^1.

PROOF. It suffices to see that $F(z) = T(K_z)$ for a bounded operator $T \in L(A_{1,\rho}(\mathbb{D}), \ell^1)$.

Consider $T(g) = (\frac{\alpha_n \rho_n}{n+1})_{n\geq 0}$ for $g(z) = \sum_{n=0}^{\infty}\alpha_n z^n \in A_{1,\rho}(\mathbb{D})$.

Note that Hardy inequality (see [**14**]) gives

$$\sum_{n=0}^{\infty} \frac{|\alpha_n| r^n}{n+1} \leq CM_1(g, r),$$

and then

$$\sum_{n=0}^{\infty} \frac{|\alpha_n| \rho_n}{n+1} = \int_0^1 \rho(1-r) \sum_{n=0}^{\infty} \frac{|\alpha_n| r^n}{n+1} dr \leq C \|g\|_{A_{1,\rho}}.$$

This shows that T is bounded from $A_{1,\rho}(\mathbb{D})$ into ℓ^1. Note now that

$$T(K_z) = \left(\frac{(n+1) z^n \rho_n}{n+1} \right)_{n \geq 0} = \sum_{n=0}^{\infty} \rho_n e_n z^n.$$

$\square$

PROPOSITION 3.2. *Let $\gamma_n \geq 0$ such that $\gamma_0 > 0$ and*

$$\gamma_n \leq C \frac{1}{n+1} \sum_{k=0}^{n} \gamma_k, \qquad n \geq 0.$$

Then

(11)
$$\sum_{n=0}^{\infty} \frac{\gamma_n |\alpha_n|}{n+1} \leq C \|g\|_{A_{1,\rho}(\mathbb{D})},$$

where $g(z) = \sum_{n=0}^{\infty} \alpha_n z^n$ and $\rho(1-t) = \sum_{n=0}^{\infty} \gamma_n t^n$.

PROOF. Observe first that

$$\frac{\rho(1-t)}{1-t} = \sum_{n=0}^{\infty} \left(\sum_{k=0}^{n} \gamma_k \right) t^n \geq C^{-1} \sum_{n=0}^{\infty} (n+1) \gamma_n t^n \geq C^{-1} \gamma_0.$$

Define $T(g) = (\frac{\gamma_n \alpha_n}{n+1})_{n \geq 0}$ for $g(z) = \sum_{n=0}^{\infty} \alpha_n z^n$. We have to show that T is bounded from $A_{1,\rho}(\mathbb{D})$ into ℓ^1. It suffices to show that $F(z) = (\gamma_n z^n)_n = \sum_{n=1}^{\infty} \gamma_n e_n z^n \in \text{Bloch}_\rho(\ell^1)$ where $\{e_n\}$ stands for the canonical basis of ℓ^1. (The reader should observe that the assumptions (b_1) and (d_1) on the weight are only used in the other implication.)

Note that $F'(z) = \sum_{n=1}^{\infty} n \gamma_n e_n z^{n-1}$ and $\|F'(z)\| = (\sum_{n=1}^{\infty} n \gamma_n |z|^{n-1})$. Hence

$$\|F'(z)\| \leq C \left(\sum_{n=1}^{\infty} \left(\sum_{k=1}^{n} \gamma_k \right) |z|^n \right) \leq C \frac{\rho(1-|z|)}{1-|z|}.$$

$\square$

We now introduce certain related spaces which can be used to produce more examples.

DEFINITION 3.3. Let X be a Banach space and let $\sigma : (0, 1] \to \mathbb{R}^+$ be a continuous function, we define by $L^\sigma(X)$ (respectively, $A^\sigma(X)$) the space of measurable (respectively, analytic) functions in the unit disc $\mathbb{D}$ such that there exists $C > 0$ for which

$$\|F(z)\| \leq C \sigma(1 - |z|)$$

for almost all $z \in \mathbb{D}$ with respect to the normalized Lebesgue measure $dA(z)$ (respectively, for all $z \in \mathbb{D}$).

It becomes a Banach space under the norm $\|F\|_{L^\sigma(X)}$ (respectively, $\|F\|_{A^\sigma(X)}$) given by $\sup\limits_{z\in\mathbb{D}} \frac{1}{\sigma(1-|z|)}\|F(z)\|$.

REMARK 3.1. $A^\sigma(X) = H^\infty(X)$ for $\sigma(t) = 1$, $A^\sigma(X) = A^\alpha(X)$ (see [17]) for $\sigma(t) = t^{-\alpha}$ for $\alpha > 0$ and $A^\sigma(X) = A^0(X)$ for $\sigma(t) = \log(\frac{1}{t})$ (see [17]). We use the notation $L^0(X)$ for the space $L^\sigma(X)$ for $\sigma(t) = \log(\frac{1}{t})$ (see [11]).

PROPOSITION 3.4. Let X be a Banach space and let $\sigma : (0,1] \to \mathbb{R}^+$ be non-increasing and denote $\sigma_1(t) = \int_t^1 \frac{\sigma(s)}{s}ds$ for $0 < t < 1$. Then
  (i) $A^\sigma(X) \subset \mathrm{Bloch}_\sigma(X)$.
  (ii) $\mathrm{Bloch}_\sigma(X) \subset A^{\sigma_1}(X)$.
  If $\rho(t) = t\sigma(t)$ is a (b_1)-weight, then $A^\sigma(X) = \mathrm{Bloch}_\sigma(X)$. In particular, $A^\alpha(X) = \mathrm{Bloch}_\rho(X)$ for $\rho(t) = t^{-\alpha}$.

PROOF. (i) Since
$$F'(z) = \frac{1}{2\pi i}\int_\Gamma \frac{F(\xi)}{(z-\xi)^2}d\xi$$
for $\Gamma = \{re^{it} : t \in [0,2\pi]\}$ where $r = \lambda|z|$ for some $1 < \lambda < \frac{1}{|z|}$, we get

$$
\begin{aligned}
\|F'(z)\| &\leq \frac{1}{2\pi}\int_\Gamma \frac{\|F(\xi)\|}{|z-\xi|^2}|d\xi| \\
&\leq C\int_\Gamma \frac{\sigma(1-|\xi|)}{|z-\xi|^2}|d\xi| \\
&\leq C\sigma(1-r)\int_0^{2\pi} \frac{1}{|z-re^{it}|^2}r\,dt \\
&\leq C\sigma(1-\lambda|z|)\int_0^{2\pi} \frac{1}{r|1-\frac{|z|}{r}e^{-it}|^2}dt \\
&\leq C\frac{\sigma(1-\lambda|z|)}{(r-|z|)} \\
&\leq C\frac{\sigma(1-|z|)}{(\lambda-1)|z|}.
\end{aligned}
$$

Hence, taking limits as λ goes to $1/|z|$, one gets
$$\|F'(z)\| \leq C\frac{\sigma(1-|z|)}{1-|z|}.$$

(ii) Let $F \in \mathrm{Bloch}_\sigma(X)$. Simply observe that $F(z) - F(0) = \int_0^1 zF'(rz)dr$ and then
$$\|F(z) - F(0)\| \leq \int_0^1 |z|\|F'(rz)\|dr \leq C|z|\int_0^1 \frac{\sigma(1-r|z|)}{1-r|z|}dr \leq C\int_{1-|z|}^1 \frac{\sigma(s)}{s}ds.$$
This shows that $\|F(z)\| \leq \|F(0)\| + C\sigma_1(1-|z|)$.
Note that ρ in (b_1), then $\sigma_1 \leq C\sigma$ which easily gives the final observation. □

Let us now give general properties of the spaces.

PROPOSITION 3.5. Let $\rho : (0,1] \to \mathbb{R}^+$ be a continuous function such that $\frac{\rho(t)}{t}$ is non-increasing and $\rho(1) > 0$. Let X be a complex Banach space and $F : \mathbb{D} \to X$ be analytic. Then
$$\|F\|_{\mathrm{Bloch}_\rho(X)} = \lim_{r\to 1} \|F_r\|_{\mathrm{Bloch}_\rho(X)},$$

where $F_r(z) = F(rz)$ for $0 < r < 1$.

PROOF. Note that $F_r(0) = F(0)$ for all r. Observe that

$$\frac{1-|z|}{\rho(1-|z|)}\|F_r'(z)\| = r\frac{1-|z|}{\rho(1-|z|)}\|F'(rz)\| = r\frac{(1-|z|)^{1+\alpha}}{\rho(1-|z|)(1-|z|)^\alpha}\|F'(rz)\|.$$

Due to the fact that $\frac{t}{\rho(t)}$ is non-decreasing, we have

$$\frac{1-|z|}{\rho(1-|z|)}\|F_r'(z)\| \le \frac{1-|rz|}{\rho(1-|rz|)}\|F'(rz)\|$$

which implies that $\|F_r\|_{\mathrm{Bloch}_\rho(X)} \le \|F\|_{\mathrm{Bloch}_\rho(X)}$ for all $0 < r < 1$.

Now, given $\varepsilon > 0$ take $z_0 \in \mathbb{D}$ such that $\frac{1-|z_0|}{\rho(1-|z_0|)}\|F'(z_0)\| > \|F\|_{\mathrm{Bloch}_\rho(X)} - \varepsilon/2$ and take r_0 verifying that

$$r\frac{1-|z_0|}{\rho(1-|z_0|)}\|F'(rz_0)\| > \frac{1-|z_0|}{\rho(1-|z_0|)}\|F'(z_0)\| - \varepsilon/2$$

for any $r > r_0$. Hence

$$\|F_r\|_{\mathrm{Bloch}_\rho(X)} > \|F\|_{\mathrm{Bloch}_\rho(X)} - \varepsilon.$$

$\square$

THEOREM 3.6. *Let $\rho : (0,1] \to \mathbb{R}^+$ be a continuous function such that $\frac{\rho(t)}{t}$ is non-increasing, $\rho(1) > 0$ and $\lim_{t\to 0^+}\frac{\rho(t)}{t} = \infty$. Let X be a complex Banach space and $F \in \mathrm{Bloch}_\rho(X)$. The following are equivalent:*
   (i) $F \in \mathrm{bloch}_\rho(X)$.
   (ii) $\lim_{r\to 1}\|F - F_r\|_{\mathrm{Bloch}_\rho(X)} = 0$.
   (iii) *F belongs to the closure of the X-valued polynomials.*

PROOF. (i) $\Rightarrow$ (ii) Assume that $\lim_{s\to 1}\frac{1-s}{\rho(1-s)}M_\infty(F', s) = 0$. Note that for all $0 < s < 1$ we have

$$\sup_{|z|<1}\frac{1-|z|}{\rho(1-|z|)}\|F'(z) - rF'(rz)\| \le 2\sup_{|z|>s}\frac{1-|z|}{\rho(1-|z|)}M_\infty(F', |z|)$$
$$+ \ C\sup_{|z|\le s}\|F'(z) - F_r'(z)\|.$$

Hence, given $\varepsilon > 0$ choose $s_0 < 1$ such that $\sup_{|z|>s_0}\frac{1-|z|}{\rho(1-|z|)}M_\infty(F', |z|) < \frac{\varepsilon}{4}$ and then use that F_r' converges uniformly on compact sets to get $r_0 < 1$ such that $\sup_{|z|\le s_0}\|F'(z) - F_r'(z)\| < \frac{\varepsilon}{2}$ for $r > r_0$. Then $\|F - F_r\|_{\mathrm{Bloch}_\rho} < \varepsilon$ for $r > r_0$.

(ii) $\Rightarrow$ (iii) Assume now that for each $\varepsilon > 0$ there exists $r_0 < 1$ such that $\|F - F_{r_0}\|_{\mathrm{Bloch}_\rho} < \varepsilon/2$. Now we can take a Taylor polynomial of F_{r_0}, $P_N = P_N(F_{r_0})$, such that $\|F_{r_0}' - P_N'\|_{H_\infty} < \varepsilon/2A$ where $A = sup_{0<r<1}\frac{1-r}{\rho(1-r)}$. Therefore

$$\|F - P_N(F_{r_0})\|_{\mathrm{Bloch}_\rho} \le \|F - F_{r_0}\|_{\mathrm{Bloch}_\rho} + \left(\sup_{0<r<1}\frac{1-r}{\rho(1-r)}\right)\|F_{r_0}' - P_N'\|_{H_\infty} < \varepsilon.$$

(iii) $\Rightarrow$ (i) Let P be an X-valued polynomial. Using that

$$\frac{1-r}{\rho(1-r)}M_\infty(P', r) \le \frac{1-r}{\rho(1-r)}\max_{|z|\le 1}\|P'(z)\|,$$

one has that $P \in \text{bloch}_\rho(X)$. The result follows because $\text{bloch}_\rho(X)$ is closed in $\text{Bloch}_\rho(X)$. $\qquad\square$

PROPOSITION 3.7. *Let X be a complex Banach space and let $\rho : (0,1] \to \mathbb{R}^+$ be a continuous function such that $\frac{\rho(t)}{t}$ is non-increasing and $\rho(1) > 0$. Assume that ρ is a weight in (d_1) and (b_1).*

If $F \in \text{bloch}_\rho(X)$, then T_F is a compact operator from $A_{1,\rho}(\mathbb{D})$ into X.

PROOF. Using (4) in Remark 1.2 and Theorem 3.6, we have a sequence of polynomials P_n with values in X which approaches F in $\text{bloch}_\rho(X)$. Note that the associated operators T_{P_n} are finite rank operators. Due to Theorem 1.7, one gets that T_{P_n} converges to T in norm. Therefore T is compact. $\qquad\square$

REMARK 3.2. The converse of Proposition 3.7 is not true.

Take $F \in \text{Bloch}_\rho(\mathbb{C}) \setminus \text{bloch}_\rho(\mathbb{C})$ and $T = T_F$ the corresponding operator for $X = \mathbb{C}$. Now T is compact but $F_T = F \notin \text{bloch}_\rho(\mathbb{C})$.

We observe now that the Bergman projection is also well-defined for X-valued integrable functions f in $L^1(\mathbb{D}, dA, X)$:

$$P(f)(z) = \int_{\mathbb{D}} \frac{f(w)}{(1 - \bar{w}z)^2} dA(w).$$

THEOREM 3.8. *Let X be a Banach space and let σ be a (d_1)-weight and (b_1)-weight in $(0,1]$. Then the Bergman projection P defines a bounded operator from $L^\sigma(X)$ onto $\text{Bloch}_\sigma(X)$.*

PROOF. Let f belong to $L^\sigma(X)$. Since σ is a (d_1)-weight, in particular, $\sigma \in L^1((0,1])$, and therefore $f \in L^1(\mathbb{D}, dA, X)$.

Since $(Pf)'(z) = \int_{\mathbb{D}} \frac{2\bar{w}}{(1 - \bar{w}z)^3} f(w) dA(w)$, we have

$$
\begin{aligned}
\|(Pf)'(z)\| &\leq \|f\|_{L^\sigma} \int_{\mathbb{D}} \frac{2\sigma(1 - |w|)}{|1 - z\bar{w}|^3} dA(w) \\
&\leq 2\|f\|_{L^\sigma} \int_0^1 \sigma(1 - r) \left(\int_0^{2\pi} \frac{1}{|1 - \bar{z}re^{it}|^3} dt \right) dr \\
&\leq C\|f\|_{L^\sigma} \int_0^1 \frac{\sigma(1 - r)}{(1 - |z|r)^2} dr \\
&\approx C\|f\|_{L^\sigma} \int_0^1 \frac{\sigma(1 - r)}{((1 - |z|) + (1 - r))^2} dr \\
&\approx C\|f\|_{L^\sigma} \left(\int_0^{1 - |z|} \frac{\sigma(r)}{(1 - |z|)^2} dr + \int_{1 - |z|}^1 \frac{\sigma(r)}{r^2} dr \right) \\
&\leq C\|f\|_{L^\sigma} \left(\frac{1}{(1 - |z|)^2} \int_0^{1 - |z|} \sigma(t) dt + \frac{\sigma(1 - |z|)}{1 - |z|} \right) \\
&\leq C\|f\|_{L^\sigma} \frac{\sigma(1 - |z|)}{1 - |z|}
\end{aligned}
$$

Let us prove the surjectivity. Let $f \in \text{Bloch}_\sigma(X)$ with $f(0) = f'(0) = 0$.

If $f(z) = \sum_{n=2}^\infty x_n z^n$, let g be given by

$$g(z) = \frac{(1 - |z|^2) f'(z)}{\bar{z}}.$$

We have that $g \in L^\sigma(X)$ since $f \in \mathrm{Bloch}_\sigma(X)$ and $f'(0) = 0$. Now write $Pg(z) = \sum_{n=0}^\infty y_n z^n$ and take $n \geq 1$

$$
\begin{aligned}
y_n &= (n+1) \int_{\mathbb{D}} g(z) \bar{z}^n dA(z) = (n+1) \int_{\mathbb{D}} (1 - |z|^2) f'(z) \bar{z}^{n-1} dA(z) \\
&= (n+1) \int_{\mathbb{D}} f'(z) \bar{z}^{n-1} dA(z) - (n+1) \int_{\mathbb{D}} z f'(z) \bar{z}^n dm(z) = x_n.
\end{aligned}
$$

Also $y_0 = 0$. That is, $Pg = f$.

The general case follows by writing $f(z) = f(0) + f'(0)z + f_1(z)$ where f_1 is as above. So if $Pg_1 = f_1$ then $P(f(0) + f'(0)z + g_1) = f$. $\qquad\square$

COROLLARY 3.9. *The Bergman projection maps $L^0(X)$ to $A^0(X)$.*

PROOF. Note that $\rho(t) = \log(\frac{e}{t})$ is a (d_1)-weight and a (b_1)-weight. Indeed,

$$
\int_0^s \log\left(\frac{e}{t}\right) dt = s(\log(\frac{e}{s}) + 1) \leq Cs \log\left(\frac{e}{s}\right),
$$

$$
\int_s^1 \frac{\log\left(\frac{e}{t}\right)}{t^2} dt \leq \log\left(\frac{e}{s}\right) \int_s^1 \frac{dt}{t^2} \leq C \frac{\log\left(\frac{e}{s}\right)}{s}.
$$

$\qquad\square$

REMARK 3.3. Denote by R the adjoint operator of P_2^*, that is,

$$
Rf(z) = (1 - |z|^2)^2 \int_{\mathbb{D}} \frac{f(w)}{|1 - \bar{z}w|^4} dA(w),
$$

which corresponds to the Berezin transform. The reader is referred to the recent paper [11] for the results on the Berezin transform of functions on L^σ.

References

[1] J. M. Anderson, J. Clunie and C. Pommerenke, *On Bloch functions and normal functions*, J. Reine Angew. Math. **270** (1974), 12–37.

[2] J. L. Ansorena and O. Blasco, *Characterization of weighted Besov spaces*, Math. Nachr. **171**, (1995), 5–17.

[3] J. L. Arregui and O. Blasco, *Bergman and Bloch spaces of vector-valued functions*, Math. Nachr. **261–262** (2003), 3–22.

[4] J. L. Arregui and O. Blasco, *Multipliers on vector-valued Bergman spaces*, Canad. J. Math. **54** (2002), 1165–1186.

[5] O. Blasco, *Spaces of vector valued analytic functions and applications*, London Math. Soc. Lecture Notes Series 158, (1990), 33–48.

[6] O. Blasco, *Operators on weighted Bergman spaces and applications*, Duke Math. J. **66** (1992), 443–467.

[7] O. Blasco, *Multipliers on weighted Besov spaces of analytic functions*, Contemporary Math. **144** (1993), 23–33.

[8] O. Blasco, *Remarks on vector valued BMO and vector-valued multipliers*, Positivity **4** (2000), 339–356.

[9] O. Blasco, *A note on the boundedness of operators on weighted Bergman spaces*, Margarita Mathematica, (2001), 477–486.

[10] O. Blasco, *On Taylor coefficients of vector-valued Bloch functions*, preprint.

[11] O. Blasco, A. Kukuryka and M. Nowak, *Luecking's condition for zeros of analytic functions*, Ann. Univ. Marie Courie-Sklodowska **58** (2004), 1-15.

[12] O. Blasco and G. S. de Souza, *Spaces of analytic functions on the disc where the growth of $M_p(F,r)$ depends on a weight*, J. Math. Ann. Appl. **147** (1990), 580–598.

[13] S. Bloom and G. S. de Souza, *Atomic decomposition of generalized Lipschitz spaces*, Illinois J. Math. **33** (1989), 189–209.

[14] P. Duren, *Theory of H^p-Spaces*, Adademic Press, New York, 1970.

[15] P. L. Duren, B. W. Romberg and A. L. Shields, *Linear functionals on H^p spaces, $0 < p < 1$*, J. Reine Angew. Math. **238** (1969), 32–60.

[16] G. H. Hardy and J. E. Littlewood, *Some properties of fractional integrals*, II, Math. Z. **34** (1932), 403–439.

[17] H. Hendenmalm, B. Korenblum and K. Zhu, *Theory of Bergman Spaces*, Springer Verlag, New York, 2000.

[18] S. Janson, *Generalization of Lipschitz and application to Hardy spaces and bounded mean oscillation*, Duke Math. J. **47** (1980), 959–982.

[19] A. L. Shields and D. L. Williams, *Bounded projections, duality and multipliers in spaces of analytic functions*, Trans. Amer. Math. Soc. **162** (1971), 287–302.

[20] K. Zhu, *Operator Theory in Function Spaces*, Marcel Dekker, New York, 1990.

DEPARTAMENTO DE MATEMÁTICAS, UNIVERSIDAD DE VALENCIA, 46100 BURJASSOT, VALENCIA, SPAIN

E-mail address: oblasco@uv.es

Contemporary Mathematics
Volume **404**, 2006

A Wiener Tauberian Theorem for Weighted Convolution Algebras of Zonal Functions on the Automorphism Group of the Unit Disc

Anders Dahlner

Dedicated to Boris Korenblum on the occasion of his 80th birthday

ABSTRACT. Our main result gives necessary and sufficient conditions, in terms of Fourier transforms, for an ideal in the algebra $L^1(G//K, \omega)$, the convolution algebra of zonal functions on the automorphism group on the unit disc which are integrable with respect to the weight function ω, to be dense in the algebra, or to have as closure an ideal of functions whose set of common zeros of the Fourier transforms is a finite set on the boundary of the maximal ideal space of the algebra. The weights considered behave like Legendre functions of the first kind.

Introduction

Given a locally compact group G, there is a positive measure on the group which is invariant under left translations, and the L^1-space induced by that measure becomes a Banach algebra under convolution. By the convolution of two locally integrable functions $f, g : G \to \mathbb{C}$, we mean the function $f * g$ given by

$$f * g(x) = \int_G f(xy)g(y^{-1}) \, dy = \int_G f(y)g(y^{-1}x) \, dy,$$

defined wherever the right-hand side exists.

If G is a locally compact abelian group, Wiener's Tauberian Theorem (WTT) asserts that if the Fourier transforms of the elements of a closed ideal I of the convolution Banach algebra $L^1(G)$ have no common zero, then $I = L^1(G)$.

In the non-abelian case, the analog of WTT fails for all non-compact connected Lie groups, [8], [19]. In the 1950's, Ehrenpreis and Mautner studied harmonic analysis on the Lie group $G = SU(1,1) \cong SL(2, \mathbb{R})$ (see [7]). They used the ideal structure on the disc algebra $A(\mathbb{D})$ to show that the analog of WTT fails even for the commutative sub-algebra $L^1(G//K)$ of zonal functions. They realized that in addition to the non-vanishing of the Fourier transforms, a condition on the rate of

---

2000 *Mathematics Subject Classification.* Primary 43A20; Secondary 43A80.

Key words and phrases. Wiener Tauberian Theorem, estimates of Legendre functions, resolvent transform.

decay of the Fourier transforms at infinity is needed as well. However, for technical reasons, they had to impose smoothness conditions on the Fourier transforms in their version of WTT.

It is known that smoothness conditions make WTT much easier. However, in 1996, Ben Natan, Benyamini, Hedenmalm, and Weit proved a genuine analog of WTT for $L^1(G//K)$ $(G = SU(1,1))$ [3], without any superfluous smoothness conditions on the Fourier transforms. The main ingredient in their proof is the resolvent transform method, as developed by Carleman, Gelfand and Beurling (most probably Carleman was the first who used this method in his seminar on Wiener's theorem held at the Mittag-Leffler institute in 1935; however, Gelfand seems to be the first to publish a paper containing the method). Gelfand's point of view was later rediscovered by Domar [6], and applied and extended by Hedenmalm and Borichev in the study of harmonic analysis on the real line, half-line, and the first quadrant in the plane [12], [5], [13].

Now, let G be a locally compact group (abelian or not). Given a weight function $\omega : G \to (0,\infty)$, one considers the weighted space $L^1(G,\omega)$ with norm

$$\|f\|_{L^1(G,\omega)} = \int_G |f(x)|\omega(x)dx, \quad f \in L^1(G,\omega).$$

It is easy to see that $L^1(G,\omega)$ is a Banach algebra under convolution if and only if ω is sub-multiplicative, i.e. $\omega(a \cdot b) \leq C\omega(a)\omega(b)$ holds for all $a, b \in G$, and some constant $C > 0$ not depending on a or b. It is known that there are sub-multiplicative weights ω on $\mathbb{R}$, the real line, such that $L^1(\mathbb{R},\omega)$ fails to satisfy WTT.

In this paper, we study the analog of WTT for the algebra $L^1(G//K,\omega)$, with $G = SU(1,1)$, to give a qualitative generalization of the WTT contained in [3]. Since the underlying space, $G//K$, is not a group, the sub-multiplicative condition is replaced with another condition – ω should satisfy a "sub-translative" condition. We find a natural class of such weight functions, and prove that WTT holds for $L^1(G//K,\omega)$ with these weights. As in [3], we use the resolvent transform method in our proof. Another necessity in the proof are estimates of certain combinations of Legendre functions, (in fact, the characters on $G//K$ are Legendre functions of the first kind), sharp estimates of those are obtained using relations and asymptotics; this method simplifies the proof of the main theorem of [3], and improves several of the estimates contained therein.

For further references on WTT, we refer to the book [22], and the articles [3] and [5].

1. Preliminaries

The standard references for the topic of this section are [23], [14], [17], [3] and [7].

Let G be the multiplicative group $SU(1,1)$, consisting of all complex matrices of the form

$$M(\alpha, \beta) = \begin{pmatrix} \alpha & \beta \\ \bar{\beta} & \bar{\alpha} \end{pmatrix},$$

with determinant 1. The group G acts on the unit disc $\mathbb{D} = \{z \in \mathbb{C} : |z| < 1\}$ via conformal automorphisms

$$M(\alpha, \beta)(z) = \frac{\alpha z + \beta}{\bar{\beta} z + \bar{\alpha}}, \quad z \in \mathbb{D}, \, M(\alpha, \beta) \in SU(1,1);$$

in fact, $G/\{\pm 1\} = \text{Aut}(\mathbb{D})$. Let K be the subgroup of all "rotations" $k_\theta = M(e^{i\theta}, 0)$, $\theta \in \mathbb{R}$, and let A be the subgroup of all matrices of the form $a_\zeta = M(\cosh \zeta, \sinh \zeta)$, $\zeta \in \mathbb{R}$. The group K is a maximal compact subgroup of G, and we have *the polar decomposition* $G = KA_+K$, where $A_+ = \{a_\zeta \in A : \zeta \geq 0\}$. Moreover, the factor $a \in A_+$ in the polar decomposition $g = k_1 a k_2$ of an element $g \in G$ is unique.

The *double coset space* is the set $G//K$ of all K-double cosets $\bar{g} = KgK = \{k_1 g k_2 : k_1, k_2 \in K\}$, $g \in G$. After considering the polar decomposition, we see that to each $\bar{g} \in G//K$, there is a unique $\zeta \geq 0$ such that $\bar{g} = \overline{a_\zeta}$; this gives an identification of $G//K$ with $\mathbb{R}_+ = [0, +\infty)$. The topology on $G//K$ is the quotient topology; to be precise, the relation $\sim$ defined by $g \sim g' \Leftrightarrow g \in Kg'K$ is an equivalence relation on G, and $G//K = G/\sim$. The projection $\pi : G \to G//K$ is closed and proper (i.e., the inverse image of each compact set in $G//K$ is compact); hence the topology on $G//K$ is locally compact.

A function $f : G \to \mathbb{C}$ is called *zonal* if $f(k_1 g k_2) = f(g)$ for all $g \in G$, $k_1, k_2 \in K$. If X is a locally compact Hausdorff topological space, then we write $C_0(X)$ for the space of all continuous functions $f : X \to \mathbb{C}$, such that the support of f, $\text{supp}(f) = \text{clos}\{x : f(x) \neq 0\}$, is compact. It is easy to check that there is a one-to-one correspondence between $C_0(G//K)$ and $C_0^\natural(G) = \{f \in C_0(G) : f$ is zonal$\}$.

As mentioned in the introduction, it is well-known that for any locally compact group G, there is a (non-trivial) positive linear functional μ on $C_0(G)$ such that $\mu(L_x f) = \mu(f)$, where $L_x f(y) = f(xy)$, and that μ is unique up to multiplication by a positive constant. This measure μ is known as *the left invariant Haar measure on G*. Locally compact groups where the Haar measure is both left and right invariant are called *unimodular*.

Note that $G//K$ is *not* a group under the induced multiplication. In fact, if $\zeta, \xi \geq 0$, then by multiplication of matrices, we see that

$$Ka_\zeta K \cdot Ka_\xi K = \bigcup_{|\zeta - \xi| \leq \eta \leq \zeta + \xi} Ka_\eta K.$$

However, due to the correspondence between $C_0(G//K)$ and $C_0^\natural(G)$, the Haar measure on G induces a measure on $G//K$. To be explicit, if μ is the left invariant Haar measure on G and if $\pi : G \to G//K$ is the projection, we define $\bar{\mu}$ – thought of as a linear functional in view of the Riesz representation theorem – on $G//K$ by $\bar{\mu}(f) = \mu(f \circ \pi)$ for f in $C_0(G//K)$. With this identification, it is clear that the integration theory on $G//K$ is the same as the integration theory for zonal functions on G. We may therefore write $L^1(G//K)$ and think of the elements as zonal functions. Moreover, the translation operator on $L^1(G)$ induces a translation operator on the space $L^1(G//K)$, and in this way $L^1(G//K)$ becomes a convolution sub-algebra of $L^1(G)$. The convolution structure on $L^1(G//K)$ will be explained below.

To the polar decomposition of G corresponds a normalization of the Haar measure such that

$$(1.1) \qquad \mu(f) = \frac{1}{2\pi^2} \int_0^{2\pi} \int_0^\infty \int_0^{2\pi} f(k_\theta a_\zeta k_{\theta'}) \sinh 2\zeta \, d\theta \, d\zeta \, d\theta', \quad \text{for } f \in L^1(G).$$

Using the symmetry of this formula, it is easy to see that G is unimodular; namely, the identity $a_\zeta = k_{-\pi/2} a_{-\zeta} k_{\pi/2}$ implies

$$\begin{aligned} \mu(f) &= \frac{1}{2\pi^2} \int_0^{2\pi} \int_0^\infty \int_0^{2\pi} f(k_\theta a_\zeta k_{\theta'}) \sinh 2\zeta \, d\theta \, d\zeta \, d\theta' \\ &= \frac{1}{2\pi^2} \int_0^{2\pi} \int_0^\infty \int_0^{2\pi} f(k_{-\theta} a_{-\zeta} k_{-\theta'}) \sinh 2\zeta \, d\theta \, d\zeta \, d\theta' = \mu(\check{f}), \end{aligned}$$

where $\check{f}(x) = f(x^{-1})$. Now, if μ is a left invariant Haar measure on a locally compact group G, then the condition $\mu(f) = \mu(\check{f})$ for all $f \in L^1(G)$ is equivalent to unimodularity (see [20]). For a proof of (1.1), we refer to the book [23] by Sugiura.

Using (1.1), we get a convolution formula for $f, g \in L^1(G//K)$:

$$(1.2) \quad f * g(x) = \mu(L_x(f)\check{g}) = \mu(L_x(f)g) = \frac{1}{\pi} \int_0^\infty \int_0^{2\pi} f(x k_\theta a_\zeta) g(a_\zeta) \sinh 2\zeta \, d\theta \, d\zeta,$$

where we again have used the identity $a_{-\zeta} = k_{\pi/2} a_\zeta k_{-\pi/2}$.

When f is a zonal function, it will be convenient for us to write $\tilde{f}$ for the function $\tilde{f}(x) = f(a_\zeta)$, where $x \in [1, +\infty)$ corresponds to a_ζ via $x = \cosh 2\zeta$. Under this convention, we have

$$(1.3) \qquad \tilde{f} * \tilde{g}(x) = \frac{1}{2\pi} \int_1^\infty \int_0^{2\pi} \tilde{f}\left(xy + \sqrt{x^2 - 1}\sqrt{y^2 - 1} \cos\theta\right) \tilde{g}(y) \, d\theta \, dy,$$

for $f, g \in L^1(G//K)$. This is easily seen from (1.2) and the following simple lemma.

LEMMA 1.1. *In $G//K$, we have*

$$K a_\xi k_\theta a_\zeta K = K a_\eta K,$$

where $\cosh 2\eta = \cosh 2\xi \cosh 2\zeta + \sinh 2\xi \sinh 2\zeta \cos 2\theta$.

PROOF. For each $\xi, \zeta \geq 0$ and $\theta \in \mathbb{R}$, we have

$$(1.4) \qquad a_\xi k_\theta a_\zeta = k_{\theta_1} a_\eta k_{\theta_2} \sim a_\eta$$

for some $\eta \geq 0$ and some $\theta_1, \theta_2 \in \mathbb{R}$. Multiplying the matrices to the left of the equality in (1.4), we get

$$(1.5) \qquad a_\xi k_\theta a_\zeta = M(\cosh_\theta(\xi, \zeta), \sinh_\theta(\xi, \zeta))$$

where we agree that

$$\begin{aligned} \cosh_\theta(\xi, \zeta) &= e^{i\theta} \cosh\xi \cosh\zeta + e^{-i\theta} \sinh\xi \sinh\zeta, \\ \sinh_\theta(\xi, \zeta) &= e^{i\theta} \cosh\xi \sinh\zeta + e^{-i\theta} \sinh\xi \cosh\zeta. \end{aligned}$$

On the other hand,

$$(1.6) \qquad k_{\theta_1} a_\eta k_{\theta_2} = M(e^{i(\theta_1 + \theta_2)} \cosh\eta, e^{i(\theta_1 - \theta_2)} \sinh\eta).$$

Using (1.4), (1.5) and (1.6), we get

$$\cosh^2 \eta = |\cosh_\theta(\xi, \zeta)|^2$$
$$= \cosh^2 \xi \cosh^2 \zeta + \sinh^2 \xi \sinh^2 \zeta + 2^{-1} \sinh 2\xi \sinh 2\zeta \cos 2\theta;$$

in the same way, we get

$$\sinh^2 \eta = \cosh^2 \xi \sinh^2 \zeta + \sinh^2 \xi \cosh^2 \zeta + 2^{-1} \sinh 2\xi \sinh 2\zeta \cos 2\theta.$$

Summing up and simplifying, we obtain

$$\cosh 2\eta = \cosh^2 \eta + \sinh^2 \eta = \cosh 2\xi \cosh 2\zeta + \sinh 2\xi \sinh 2\zeta \cos 2\theta,$$

which completes the proof of our lemma. $\qquad\square$

It is a classical theorem that if $f, g \in L^1(G//K)$, then $f * g \in L^1(G//K)$ and $f * g = g * f$ – this is in sharp contrast to group algebras $L^1(G)$ (here G may be any locally compact group), which are commutative if and only if G is abelian. The commutativity of convolution in $L^1(G//K)$ may be deduced from (1.3) using the change of variables $z = xy + \sqrt{x^2 - 1}\sqrt{y^2 - 1} \cos \theta$.

By a weight function, we mean a positive continuous zonal function ω. When ω is a weight function, then the space $L^1(G//K, \omega)$ of all functions f on $G//K$ such that $f\omega \in L^1(G//K)$ is a Banach space under the norm given by $\|f\|_{L^1(\omega)} = \|f\omega\|_{L^1}$. For $L^1(G//K, \omega)$ to be a commutative Banach algebra under convolution, it is necessary and sufficient that $\|f * g\|_{L^1(\omega)} \leq C\|f\|_{L^1(\omega)}\|g\|_{L^1(\omega)}$, for all $f, g \in L^1(G//K, \omega)$, and some constant $C > 0$ not depending on f or g. Such convolution inequalities are easily solved, and the condition on ω is

$$\frac{1}{2\pi} \int_0^{2\pi} \omega(a_\xi k_\theta a_\zeta) \, d\theta \leq C\omega(a_\xi)\omega(a_\zeta),$$

for all $\xi, \zeta \in \mathbb{R}_+$, or equivalently,

$$(1.7) \qquad \frac{1}{2\pi} \int_0^{2\pi} \tilde{\omega}\left(xy + \sqrt{x^2 - 1}\sqrt{y^2 - 1} \cos \theta\right) d\theta \leq C\tilde{\omega}(x)\tilde{\omega}(y),$$

for all $x, y \geq 1$. Weight functions satisfying (1.7) will be called *Banach algebra weights*.

OBSERVATION 1.2. Banach algebra weights need not be sub-multiplicative, as is the case for group algebras $L^1(G, \omega)$.

This is *not* a deep result, we just want to emphasize that the algebras are different. The explanation for this is of course that the translation operator on groups is an algebraic operation, while in our space, the translation operator is a mean value.

Next, we want to determine the Gelfand transform on $L^1(G//K, \omega)$, when ω is a Banach algebra weight. To do this, we need to know the algebra homomorphisms $\Lambda : L^1(G//K, \omega) \to \mathbb{C}$ ($\Lambda \not\equiv 0$). First of all, it is clear that if Λ is such an algebra homomorphism, then, by the Riesz representation theorem, there is a function $g_\Lambda \in L^\infty(G//K, 1/\omega)$, the dual space of $L^1(G//K, \omega)$, such that

$$\Lambda(f) = \int_1^\infty \tilde{f}(x)\tilde{g}_\Lambda(x) \, dx,$$

for all $f \in L^1(G//K, \omega)$. Moreover, if Λ is an algebra homomorphism, then $\Lambda(f * g) = \Lambda(f)\Lambda(g)$, for all $f, g \in L^1(G//K, \omega)$ – this can only happen if the function g_Λ is a *character* on $G//K$; by a character on $G//K$, we mean a function χ ($\chi \not\equiv 0$) such that

$$(1.8) \qquad \frac{1}{2\pi} \int_0^{2\pi} \tilde\chi \left(xy + \sqrt{x^2 - 1}\sqrt{y^2 - 1}\cos\theta \right) d\theta = \tilde\chi(x)\tilde\chi(y), \quad x, y \geq 1.$$

Characters on $G//K$ are also known as *zonal spherical functions*. It is well-known (see [**14**]) that the only functions χ which satisfy (1.8) are given by $\tilde\chi(x) = P_s(x)$ ($s \in \mathbb{C}$), where P_s is the Legendre functions of the first kind, as given by

$$(1.9) \qquad P_s(x) = \frac{1}{2\pi} \int_0^{2\pi} \left(x + \sqrt{x^2 - 1}\cos\theta \right)^s d\theta$$

$$= \frac{1}{\pi} \int_0^1 \left(x + \sqrt{x^2 - 1}(2t - 1) \right)^s \frac{1}{\sqrt{t(1-t)}} \, dt, \quad x \geq 1.$$

By the above discussion, it follows that the Gelfand transform on $L^1(G//K, \omega)$, where ω satisfies (1.7), is given by

$$(1.10) \qquad \hat{f}(s) = \int_1^\infty \tilde{f}(x)P_s(x) \, dx, \quad s \in \Sigma_\omega,$$

where

$$\Sigma_\omega = \left\{ s \in \mathbb{C} : \sup_{x \in (1, +\infty)} |P_s(x)|/\tilde\omega(x) < \infty \right\}$$

is *the maximal ideal space corresponding to* ω. The terms Fourier transform and spherical transform are also used for this Gelfand transform. In the case $\omega = 1$, it is more common to write P_{s-1} instead of P_s in (1.10).

In the lemma below, we list some basic properties of the function P_s. In the lemma, and in the sequel, we will use the notation $f(x) \sim g(x)$ as $x \to \infty$, meaning that $f(x)/g(x) \to 1$ as $x \to \infty$.

LEMMA 1.3. *The Legendre function of the first kind* P_s *satisfies:*

(1) *For fixed* $x \in [1, +\infty)$, $P_s(x)$ *is an entire function of the complex variable* s, *and obeys the symmetry properties* $P_s(x) = P_{-s-1}(x) = \bar{P}_{-\bar{s}-1}(x)$.

(2) *We have* $|P_s(x)| \leq P_{\mathrm{Re}\,s}(x)$.

(3) *When* $\mathrm{Re}\,s > -1/2$, *we have the asymptotic identity*

$$P_s(x) \sim \frac{\Gamma(s + 1/2)}{\sqrt{\pi}\Gamma(s+1)}(2x)^s, \qquad as\ x \to \infty.$$

(4) *When* $s = -1/2$, *we have*

$$P_{-1/2}(x) \sim \frac{\sqrt{2}}{\pi} x^{-1/2} \log x, \qquad as\ x \to \infty.$$

(5) $P_s(1) = 1$.

REMARK. All these facts are well-known, but, for the convenience of the reader, we give a short proof based on (1.9).

PROOF. (2) and (5) are obvious. A calculation shows that if we make the change of variables $x + \sqrt{x^2 - 1}(2t - 1) \mapsto (x + \sqrt{x^2 - 1}(2t - 1))^{-1}$ in (1.9), we get

$P_s(x) = P_{-s-1}(x)$; the rest of (1) is then obvious. In order to prove (3), we note that

$$P_s(x)/x^s = \frac{1}{\pi} \int_0^1 \left(1 + \frac{\sqrt{x^2-1}}{x}(2t-1) \right)^s \frac{1}{\sqrt{t(1-t)}} \, dt,$$

for $x > 1$, and that

$$\min(2t, 1) \leq 1 + \frac{\sqrt{x^2-1}}{x}(2t-1) \leq \max(2t, 1).$$

It follows that the modulus of the expression under the integral sign is dominated by $\max((2t)^{\mathrm{Re}\, s}, 1)(t(1-t))^{-1/2}$, which, since $\mathrm{Re}\, s > -1/2$, is integrable on the interval $(0, 1)$. Hence, by the dominated convergence theorem, we have

$$\lim_{x \to \infty} P_s(x)/x^s \to \frac{2^s}{\pi} \int_0^1 t^{s-1/2}(1-t)^{-1/2} \, dt = \frac{2^s \Gamma(s+1/2)}{\sqrt{\pi}\,\Gamma(s+1)},$$

provided $\mathrm{Re}\, s > -1/2$. To prove (4), we again use the second representation of P_s in (1.9). First, note that

$$\int_0^1 \left(x + \sqrt{x^2-1}(2t-1) \right)^{-1/2} (t(1-t))^{-1/2} \, dt$$

$$= \int_0^{1/\log x} \left(x + \sqrt{x^2-1}(2t-1) \right)^{-1/2} (t(1-t))^{-1/2} \, dt + \varepsilon(x),$$

where $\varepsilon(x)x^{1/2}/\log x \to 0$ as $x \to \infty$, this is easily seen by estimating the integral on the intervals $(1/\log x, 1/2)$ and $(1/2, 1)$ separately. Next,

$$\int_0^{1/\log x} \left(x + \sqrt{x^2-1}(2t-1) \right)^{-1/2} (t(1-t))^{-1/2} \, dt$$

$$= x^{-1/2} b(x) \int_0^{1/\log x} \left(1 + \frac{\sqrt{x^2-1}}{x}(2t-1) \right)^{-1/2} t^{-1/2} \, dt,$$

where $b(x) \to 1$ as $x \to \infty$. By completing the expression under the integral sign into a square, the last integral is easily computed, and is asymptotically equal to $\sqrt{2} \log x$ as $x \to \infty$. Comparing with (1.9), the statement follows. $\qquad \square$

The next proposition summarizes some properties of the Gelfand transform on the algebra $L^1(G//K, \omega)$. As usual, we write $C_0^\infty(G//K)$ for the space of infinitely differentiable functions with compact support.

PROPOSITION 1.4. *Suppose ω is a Banach algebra weight. Then the following assertions are valid:*

(1) *For each $f \in L^1(G//K, \omega)$, the Gelfand transform $\hat{f}$ is continuous on Σ_ω and analytic in the interior of Σ_ω.*

(2) *The space $C_0^\infty(G//K)$ is dense in $L^1(G//K, \omega)$, and if $f \in C_0^\infty(G//K)$, then $\hat{f}$ is an entire function of finite exponential type, in the sense that for each integer $m \geq 0$, there are positive constants C_m and a_m such that*

$$(1 + |s|^2)^m |\hat{f}(s)| \leq C_m e^{a_m |\mathrm{Re}\, s|}, \quad s \in \mathbb{C}.$$

(3) *$|\hat{f}(s)| \leq C\|f\|_\omega$, for all $s \in \Sigma_\omega$ and $f \in L^1(G//K, \omega)$.*

(4) *$\hat{f}(s) = \hat{f}(-s-1)$, for all $s \in \Sigma_\omega$ and $f \in L^1(G//K, \omega)$.*

(5) $\hat{f}(s) \to 0$ *as* $|s| \to \infty$, $s \in \Sigma_\omega$.

Suppose ω is a Banach algebra weight and consider the maximal ideal space Σ_ω related to ω, i.e., the set of all s such that $P_s(x)/\tilde{\omega}(x)$ is bounded in x. Put $\alpha = \sup\{\operatorname{Re} s : s \in \Sigma_\omega\}$, note that Lemma 1.3 shows that $\alpha \in [-1/2, \infty]$. If $-1/2 < \alpha < \infty$, then Lemma 1.3 shows that Σ_ω is a strip of the form $\Sigma(\alpha) = \{s : -\alpha - 1 \le \operatorname{Re} s \le \alpha\}$. The possibility $\alpha = \infty$ can be excluded due to the regularity assumption on ω; in fact, since ω is continuous it follows that $C_0(G//K)$ is dense in $L^1(G//K, \omega)$, and if $\Sigma_\omega = \mathbb{C}$, then Proposition 1.4 together with Liouville's theorem shows that $\hat{f} = 0$ for every $f \in L^1(G//K, \omega)$ – this contradicts the semi-simplicity of, say, the algebra $L^1(G//K)$. To determine Σ_ω in the remaining case, $\alpha = -1/2$, one needs more knowledge of the asymptotic behavior of $P_{-1/2+it}$, $t \in \mathbb{R}$; we shall not get in to this problem here.

In the case $\alpha > -1/2$, we note that $s \in \Sigma(\alpha)$ if and only if that $P_s/\tilde{\omega}$ is bounded; hence $\tilde{\omega}(x) = P_\alpha(x)\rho(x)$ for some continuous function ρ such that $\inf \rho(x) > 0$ and $\liminf_{x \to \infty} \rho(x)/x^\varepsilon = 0$ as $x \to \infty$, for all $\varepsilon > 0$. However, our results need more conditions on the function ρ. In this paper we shall assume that $\omega \in W(\alpha)$, where $\alpha > -1/2$, and $W(\alpha)$ denotes the collection of all weight functions ω with $\tilde{\omega}(x) = P_\alpha(x)\rho(x)$, where $\sup \rho < \infty$ and $\inf \rho > 0$.

PROPOSITION 1.5. *If* $\omega \in W(\alpha)$*, then* $L^1(G//K, \omega) = L^1(G//K, \omega_\alpha)$*, where* $\tilde{\omega}_\alpha = P_\alpha$*, and the norms are equivalent. In particular,* ω *is a Banach algebra weight, i.e.,* ω *satisfies the condition* (1.7).

PROOF. The first statement is obvious since $\tilde{\omega}(x) = P_\alpha(x)\rho(x)$, where ρ is bounded from above by $\sup \rho < \infty$ and from below by $\inf \rho > 0$. The second statement then follows from (1.8). $\qquad\square$

2. The completeness theorem

For $f \in L^1(G//K, \omega)$, $\omega \in W(\alpha)$, we need quantitative measures of decay of the Gelfand transform at ∞ and at a boundary point $s \in \alpha + i\mathbb{R}$ of $\Sigma(\alpha)$. As in [3], we define, for $f \in L^1(G//K, \omega)$,

$$(2.1) \qquad \delta_\infty(f) = \delta_{\infty,\omega}(f) = -\limsup_{t \to +\infty} \exp\left(-\frac{\pi}{2\alpha + 1}t\right) \log\left|\hat{f}(-1/2 + it)\right|,$$

and

$$(2.2) \qquad \delta_s(f) = \delta_{s,\omega}(f) = -\limsup_{t \to \alpha^-}(\alpha - t)\log\left|\hat{f}(t + i\operatorname{Im} s)\right|.$$

Let $\Sigma(\alpha)^\circ$ denote the interior of $\Sigma(\alpha)$. Since $\hat{f} \in H^\infty(\Sigma(\alpha)^\circ)$, and since $\Sigma(\alpha)^\circ$ is conformally equivalent to the unit disc, we can factor $\hat{f}$ into an inner and an outer factor (unless $\hat{f}$ vanishes everywhere). Regarded as a function on the unit disc, the inner factor is the product of a Blaschke product and a singular inner function, the quantities $\delta_\infty(f)$ and $\delta_s(f)$ measures the atomic parts of the positive Borel measure associated with the singular inner function at the points corresponding to $-\frac{1}{2} + i\infty$ and s respectively; moreover, since $\hat{f}(z) = \hat{f}(-z - 1)$, this also applies to the points $-\frac{1}{2} - i\infty$ and $-1 - s$. For a collection $\mathfrak{S}$ of functions of $L^1(G//K, \omega)$, we define

$$\delta_\infty(\mathfrak{S}) = \inf_{f \in \mathfrak{S}} \delta_\infty(f),$$

and

$$\delta_s(\mathfrak{S}) = \inf_{f \in \mathfrak{S}} \delta_s(f).$$

Let Φ be a finite subset of $\Sigma(\alpha)$, and let $L^1_\Phi(G//K, \omega)$ be the closed convolution ideal of $L^1(G//K, \omega)$ of all functions f with $\hat{f}(s) = \hat{f}(-1-s) = 0$, for all $s \in \Phi$. We are now able to state our main theorem.

THEOREM 2.1. *Let $\mathfrak{S}$ be a family of functions of $L^1(G//K, \omega)$, where $\omega \in W(\alpha)$ and $\alpha > -1/2$. Let $I(\mathfrak{S})$ be the smallest closed ideal in $L^1(G//K, \omega)$ containing $\mathfrak{S}$, and let Φ be a finite subset of $\partial\Sigma(\alpha)$.*

Then $I(\mathfrak{S}) = L^1_\Phi(G//K, \omega)$ holds if and only if the set of common zeros of the Gelfand transform of the elements of $\mathfrak{S}$ is equal Φ, and $\delta_\infty(\mathfrak{S}) = \delta_s(\mathfrak{S}) = 0$, for all $s \in \Phi$.

In particular, if Φ is empty, then $I(\mathfrak{S}) = L^1(G//K, \omega)$ holds if and only if the Gelfand transform of the elements of $\mathfrak{S}$ has no common zero, and $\delta_\infty(\mathfrak{S}) = 0$.

The classical proof of Wiener's Tauberian Theorem for $L^1(G)$ when G is a locally compact abelian group strongly depends on the fact that $C_0(\widehat{G}) \cap \widehat{L^1(G)}$ is dense in $\widehat{L^1(G)}$, where $\widehat{G}$ is the dual group to G and $\widehat{L^1(G)} = \{\hat{f} : f \in L^1(G)\}$. However, the space $(L^1(G//K, \omega))^\wedge = \{\hat{f} : f \in L^1(G//K, \omega)\}$ ($\omega \in W(\alpha)$, $\alpha > -1/2$) contains only one function with compact support, the trivial function $\hat{f} = 0$. Hence, to prove our theorem we are forced to choose another approach than the classical one. We shall use the resolvent transform technique. Here is a sketch of this method as it applies to the algebra $L^1(G//K, \omega)$.

Let $L^1_\delta(G//K, \omega)$ be the unitization of $L^1(G//K, \omega)$; we identify the unit with δ, the Dirac measure at the unit element of G. In §4, we find to each $\lambda \in \mathbb{C} \setminus \Sigma(\alpha)$ a function $b_\lambda \in L^1(G//K, \omega)$, such that

(2.3) $$\hat{b}_\lambda(s) = \frac{1}{\lambda(\lambda+1) - s(s+1)}, \quad s \in \Sigma(\alpha),$$

and then show that the set $\{b_\lambda : \lambda \in \mathbb{C} \setminus \Sigma(\alpha)\}$ spans a dense subspace of $L^1(G//K, \omega)$. In fact, when $\operatorname{Re}\lambda > \alpha$, b_λ is given by $b_\lambda(a_\zeta) = Q_\lambda(\cosh(2\zeta))$ for $\zeta > 0$, where Q_λ is the Legendre function of the second kind (see §3 for the definition). The function Q_λ is analytic in λ when $\lambda \in \mathbb{C} \setminus \{-1, -2, \dots\}$; in particular, b_λ is analytic when $\operatorname{Re}\lambda > \alpha$. When $\operatorname{Re}\lambda < -\alpha - 1$, we put $b_\lambda = b_{-\lambda-1}$, as is consistent with (2.3). It is clear that b_λ depends analytically on λ for all $\lambda \in \mathbb{C} \setminus \Sigma(\alpha)$.

For each g in $L^\infty(G//K, 1/\omega)$, the dual Banach space of $L^1(G//K, \omega)$, we associate its *resolvent transform*

(2.4) $$\mathfrak{R}[g](\lambda) = \langle b_\lambda, g \rangle, \quad \lambda \in \mathbb{C} \setminus \Sigma(\alpha),$$

which is analytic in λ for $\lambda \in \mathbb{C} \setminus \Sigma(\alpha)$. In the sequel, we shall attempt to continue $\mathfrak{R}[g]$ analytically. For this purpose, we fix a point $\xi \in \mathbb{C} \setminus \Sigma(\alpha)$ and notice that by (2.3) we have

(2.5) $$\hat{b}_\lambda(s) = \left(1 - \hat{b}_\xi(\lambda)^{-1} \hat{b}_\xi(s)\right)^{-1} \hat{b}_\xi(s), \quad s \in \Sigma(\alpha),$$

for all $\lambda \in \mathbb{C} \setminus \Sigma(\alpha)$; here, we use the notation $\hat{b}_\xi(\lambda)$ as given by (2.3), *not* as a Gelfand transform.

The maximal ideal space of $L_\delta^1(G//K, \omega)$ is identified with $\Sigma(\alpha) \cup \{\infty\}$, the one-point compactification of $\Sigma(\alpha)$. For each closed ideal I in $L_\delta^1(G//K, \omega)$, we identify the maximal ideal space of the quotient algebra $L_\delta^1(G//K, \omega)/I$ with the hull $Z_\infty(I)$ of I, in the standard way. Here,

$$Z_\infty(I) = \left\{ z \in \Sigma(\alpha) \cup \{\infty\} : \hat{f}(z) = 0 \text{ for all } f \in I \right\}.$$

Later, we also need the notation

$$Z(f) = \left\{ z \in \Sigma(\alpha) : \hat{f}(z) = 0 \right\}.$$

Let $\mathfrak{S}$ be a collection of functions in $L^1(G//K, \omega)$, and let $I(\mathfrak{S})$ denote the closed ideal in $L^1(G//K, \omega)$ generated by $\mathfrak{S}$.

Recall that ξ is a fixed point in $\mathbb{C} \setminus \Sigma(\alpha)$. Since $\hat{b}_\xi(\infty) = 0$, and since $\hat{b}_\xi(\lambda) = \hat{b}_\xi(s)$ if and only if $\lambda = s$ or $\lambda = -s - 1$, it follows that if $\lambda \in \mathbb{C} \setminus Z_\infty(I(\mathfrak{S}))$, then $\hat{\delta} - \hat{b}_\xi(\lambda)^{-1}\hat{b}_\xi$ is non-zero on $Z_\infty(I(\mathfrak{S}))$. Hence $\delta - \hat{b}_\xi(\lambda)^{-1}b_\xi + I(\mathfrak{S})$ is invertible in the quotient algebra $L_\delta^1(G//K, \omega)/I(\mathfrak{S})$. Put

$$(2.6) \qquad B_\lambda = \left(\delta - \hat{b}_\xi(\lambda)^{-1}b_\xi + I(\mathfrak{S}) \right)^{*-1} * (b_\xi + I(\mathfrak{S}))$$

as an element in $L_\delta^1(G//K, \omega)/I(\mathfrak{S})$; here $*$ stands for convolution, which is the operation of multiplication in the algebra $L_\delta^1(G//K, \omega)/I(\mathfrak{S})$, and the inversion is taken in the convolution sense modulo the ideal. If we take the Gelfand transform of both sides of (2.6) and compare the result with (2.5), we find that

$$(2.7) \qquad B_\lambda = b_\lambda + I(\mathfrak{S}), \quad \lambda \in \mathbb{C} \setminus \Sigma(\alpha).$$

In particular, B_λ does not depend on the point ξ.

Let us return to the function $g \in L^\infty(G//K, 1/\omega)$, and suppose it annihilates $I(\mathfrak{S})$. It follows that g may be considered as a bounded linear functional on $L_\delta^1(G//K, \omega)/I(\mathfrak{S})$, and by (2.7) the resolvent transform $\mathfrak{R}[g]$ of g, defined by (2.4), may also be represented by

$$\mathfrak{R}[g](\lambda) = \langle B_\lambda, g \rangle, \quad \lambda \in \mathbb{C} \setminus \Sigma(\alpha).$$

However, according to (2.6), B_λ is defined for all $\lambda \in \mathbb{C} \setminus Z_\infty(I(\mathfrak{S}))$, and clearly depends analytically on λ. Thus, the formula

$$\mathfrak{R}[g](\lambda) = \langle B_\lambda, g \rangle, \quad \lambda \in \mathbb{C} \setminus Z_\infty(I(\mathfrak{S})),$$

gives an analytic continuation of $\mathfrak{R}[g]$ to $\mathbb{C} \setminus Z_\infty(I(\mathfrak{S}))$.

From now on, we assume that the elements of $I(\mathfrak{S})$ only have finitely many zeros in common, say,

$$Z_\infty(I(\mathfrak{S})) = \{s_1, s_2, \ldots, s_n, \infty\}.$$

To prove Theorem 2.1, later on we will show, under appropriate conditions on $\mathfrak{S}$, that

(1) The functions b_λ, $\lambda \in \mathbb{C} \setminus \Sigma(\alpha)$, span a dense subspace of $L^1(G//K, \omega)$.

(2) $\mathfrak{R}[g]$ is analytic at ∞ and it vanishes there.

(3) The singularities of $\mathfrak{R}[g]$ at $s_1, s_2, \ldots, s_n$ are simple poles.

Indeed, from (2), (3), and the fact $\mathfrak{R}[g](\lambda) = \mathfrak{R}[g](-\lambda - 1)$, it follows that $\mathfrak{R}[g]$ may be represented in the form

$$\mathfrak{R}[g](\lambda) = \sum_{j=1}^{n} \frac{a_j}{s_j(s_j + 1) - \lambda(\lambda + 1)},$$

with some complex numbers $a_1, a_2, \ldots, a_n$. Let m_j be the complex-valued algebra homomorphism of $L^1(G//K, \omega)$ corresponding to the point $s_j \in \Sigma(\alpha)$, $m_j = P_{s_j}$, and consider the functional m with complex conjugate given by

$$\bar{m} = \sum_{j=1}^{n} a_j m_j.$$

Taking the resolvent transform of m using (2.3), we see that $\mathfrak{R}[m] = \mathfrak{R}[g]$, and hence $m - g$ annihilates all the functions b_λ, with $\lambda \in \mathbb{C} \setminus \Sigma(\alpha)$. By (1) we deduce that $g = m$. This shows that if $g \in L^\infty(G//K, 1/\omega)$ annihilates $I(\mathfrak{S})$, then g annihilates each $f \in L^1(G//K, \omega)$ with $\hat{f}(s_j) = 0$, $j = 1, \ldots, n$. This completes the proof of Theorem 2.1.

To implement this sketch, and show that $\mathfrak{R}[g]$ is indeed analytic at ∞ and has simple poles at $s_1, s_2, \ldots, s_n$, the method requires estimates. To achieve the required estimates, we need an explicit formula for the function $\mathfrak{R}[g]$; this is done by finding representatives in $L^1(G//K, \omega)$ for the cosets $B_\lambda \in L^1_\delta(G//K, \omega)/I(\mathfrak{S})$. In §5, we show that for every $f \in L^1(G//K, \omega)$ and $\lambda \in \Sigma(\alpha)^\circ$, there exists $T_\lambda f \in L^1(G//K, \omega)$ such that

$$(2.8) \qquad \widehat{T_\lambda f}(s) = \frac{\hat{f}(\lambda) - \hat{f}(s)}{\lambda(\lambda + 1) - s(s + 1)}, \quad s \in \Sigma(\alpha) \setminus \{\lambda, -1 - \lambda\}.$$

Consider the identity

$$\widehat{T_\lambda f}(s) \left(1 - \hat{b}_\xi(\lambda)^{-1} \hat{b}_\xi(s) \right) = \hat{f}(\lambda)\hat{b}_\xi(s) - \hat{f}(s)\hat{b}_\xi(s).$$

Suppose $f \in I(\mathfrak{S})$, apply the inverse Gelfand transform to the above identity, and mod out $I(\mathfrak{S})$, to obtain

$$(T_\lambda f + I(\mathfrak{S})) * \left(\delta - \hat{b}_\xi(\lambda)^{-1} b_\xi + I(\mathfrak{S}) \right) = \hat{f}(\lambda) b_\xi + I(\mathfrak{S}).$$

Together with (2.6), this shows that for $f \in I(\mathfrak{S})$ and $\lambda \in \Sigma(\alpha)^\circ \setminus Z(f)$, we have

$$T_\lambda f / \hat{f}(\lambda) \in B_\lambda;$$

in other words, $T_\lambda f / \hat{f}(\lambda)$ is a representative for the coset B_λ. It follows that

$$(2.9) \qquad \mathfrak{R}[g](\lambda) = \frac{\langle T_\lambda f, g \rangle}{\hat{f}(\lambda)}, \quad \lambda \in \Sigma(\alpha)^\circ \setminus Z(f).$$

In §5, we show that $T_\lambda f$, where $f \in L^1(G//K, \omega)$ and $\lambda \in \Sigma(\alpha)^\circ$, is given by the explicit formula

$$(2.10) \quad T_\lambda f(\tau)$$
$$= \begin{cases} Q_\lambda(\tau) \int_\tau^\infty f(x) P_\lambda(x)\, dx - P_\lambda(\tau) \int_\tau^\infty f(x) Q_\lambda(x)\, dx, & \operatorname{Re} \lambda \geq -\frac{1}{2}, \\ Q_{-\lambda-1}(\tau) \int_\tau^\infty f(x) P_\lambda(x)\, dx - P_\lambda(\tau) \int_\tau^\infty f(x) Q_{-\lambda-1}(x)\, dx, & \operatorname{Re} \lambda \leq -\frac{1}{2}, \end{cases}$$

where P_λ and Q_λ are the Legendre functions of the first and second kind, respectively, and $\tau = \cosh(2\zeta) \in (1, \infty)$. The formulas (2.9) and (2.10) will then be used to derive the necessary estimates of $\mathfrak{R}[g]$.

We now indicate how the rest of this paper is organized. In §3, we gather facts on Legendre functions. In §4, we use the facts from §3 to find b_λ satisfying (2.3), and shows that they span a dense subspace of $L^1(G//K, \omega)$. In §5, we again apply the results from §3 to prove that the function $T_\lambda f$ appearing in (2.10) satisfies (2.8), and estimate its norm. In §6, we supply results from the theory of analytic functions, which are applied in §7 to prove that $\Re[g]$ satisfies (2) and (3) in the above discussion. We thus obtain the announced completeness theorem for $L^1(G//K, \omega)$.

3. Some notes on Legendre functions

In this section, we gather some facts on the Legendre functions of the first and second kind. Standard references for these are [18], [9], [21] and [24].

For $a, b, c, z \in \mathbb{C}$, $c \notin \{0, -1, -2, \ldots\}$, the hypergeometric function of Gauss is given by

$$_2F_1(a, b; c; z) = \sum_{j=0}^{\infty} \frac{(a)_j (b)_j}{j!(c)_j} z^j, \quad |z| < 1,$$

where $(d)_j$ is the Pochhammer symbol defined by

$$(d)_0 = 1, \quad (d)_j = d(d+1)\cdots(d+j-1) \quad \text{for } j = 1, 2, \ldots.$$

One has the integral representation

$$(3.1) \qquad _2F_1(a, b; c; z) = \frac{\Gamma(c)}{\Gamma(b)\Gamma(c-b)} \int_0^1 t^{b-1}(1-t)^{c-b-1}(1-tz)^{-a}\, dt,$$

valid when $\operatorname{Re} c > \operatorname{Re} b > 0$ and $|z| < 1$.

The ordinary differential equation

$$(3.2) \qquad \frac{d}{dx}\left((x^2 - 1)\frac{du}{dx}\right) = \nu(\nu+1)u$$

has two linearly independent solutions P_ν and Q_ν, which may be expressed in terms of the hypergeometric function $_2F_1$,

$$P_\nu(x) = {_2F_1}\left(-\nu, \nu+1; 1; \frac{1-x}{2}\right), \quad \text{when } |1-x| < 2$$

$$Q_\nu(x) = \frac{\sqrt{\pi}\Gamma(\nu+1)}{2^{\nu+1}\Gamma(\nu+3/2)} x^{-\nu-1}{_2F_1}\left(\frac{\nu}{2}+1, \frac{1}{2}(\nu+1); \nu+\frac{3}{2}; x^{-2}\right),$$

where in the last formula $x \in \mathbb{C} \setminus (-\infty, 1]$, $|x| > 1$. The functions P_ν and Q_ν are called *the Legendre functions of the first and second kind*, respectively, of degree ν. In the definition of Q_ν, we assumed $\nu \notin \{-1, -2, \ldots\}$. The function P_ν extends analytically to $\mathbb{C} \setminus (-\infty, -1]$, and for fixed x, the function $\nu \mapsto P_\nu(x)$ is an entire function. The function Q_ν extends analytically to $\mathbb{C} \setminus (-\infty, 1]$, and for fixed x the function $\nu \mapsto Q_\nu(x)$ is meromorphic with simple poles at the negative integers. We shall always assume $x \geq 1$.

Sometimes the following representation of Q_ν is more convenient to work with:

$$(3.3) \quad Q_\nu(x) = \frac{\sqrt{\pi}\Gamma(\nu+1)}{2^{\nu+1}\Gamma(\nu+3/2)}(1+x)^{-\nu-1}{_2F_1}\left(\nu+1, \nu+1; 2(\nu+1); 2/(1+x)\right).$$

Using (3.1) and (3.3), together with the duplication formula for the Gamma function, $2^{2z-1}\Gamma(z)\Gamma(z+1/2) = \sqrt{\pi}\Gamma(2z)$, we get the integral formula

$$(3.4) \qquad Q_\nu(x) = 2^\nu \int_0^1 t^\nu (1-t)^\nu (1+x-2t)^{-\nu-1}\, dt,$$

valid at least when $x > 1$ and $\operatorname{Re}\nu > -1$. It is clear from (3.4) that if $\nu \in \mathbb{C}$ and $\operatorname{Re}\nu > -1$, then

(3.5) $$|Q_\nu(x)| \leq Q_{\operatorname{Re}\nu}(x), \quad x \in (1, \infty).$$

The Legendre function of the first kind satisfies certain relations:

(3.6) $$(\nu + 1)P_{\nu+1}(x) + \nu P_{\nu-1}(x) = (2\nu + 1)xP_\nu(x),$$

(3.7) $$P_\nu'(x) = \frac{\nu + 1}{x^2 - 1}\left(P_{\nu+1}(x) - xP_\nu(x)\right) = \frac{\nu}{x^2 - 1}\left(xP_\nu(x) - P_{\nu-1}(x)\right),$$

(3.8) $$P_\nu(x) = P_{-\nu-1}(x).$$

The Legendre function of the second kind also satisfies these relations, with the exception that (3.8) is replaced by

(3.9) $$Q_\nu(x) - Q_{-\nu-1}(x) = \pi \cot \nu\pi P_\nu(x), \quad \nu \notin \mathbb{Z}.$$

Another relation between the Legendre functions which is sometimes handy is given by

(3.10) $$Q_\nu(x)P_{\nu+1}(x) - Q_{\nu+1}(x)P_\nu(x) = \frac{1}{\nu + 1};$$

this formula is verified by calculating the Wronskian $W(Q_\nu, P_\nu)$ (see [**18**]), and then apply the relation (3.7).

Later on, we shall need sharp estimates for the Legendre functions and their derivatives. For this purpose, we need the following lemma.

LEMMA 3.1. *Suppose $\nu \in \mathbb{R}$. As a function of $x \in [1, \infty)$, $P_\nu(x)$ decreases with x, provided $\nu \in [-1, 0]$, and increases with x, provided $\nu \geq 0$ or $\nu \leq -1$. In particular, $P_\nu(x) \leq 1$ when $\nu \in [-1, 0]$, and $P_\nu(x) \geq 1$ when $\nu \geq 0$.*

PROOF. First note that $P_\nu(x) \geq 0$ for real ν and $x \in [1, \infty)$. The differential equation (3.2) shows that $(x^2 - 1)P_\nu'(x)$ decreases (increases) with x, when $\nu \in (-1, 0)$ ($\nu \in \mathbb{R} \setminus (-1, 0)$). Since $P_\nu'(1) = \nu(\nu + 1)/2$ is finite, we conclude that $(x^2 - 1)P_\nu'(x)$ has the same sign as $P_\nu'(x)$. $\qquad\square$

PROPOSITION 3.2. *For $\nu \in \mathbb{R}$ and $x \in (1, \infty)$ we have the estimates*

$$\frac{\Gamma(\nu + 1/2)}{\sqrt{\pi}\Gamma(\nu + 1)}(2x)^\nu \leq P_\nu(x) \leq x^\nu, \quad \text{when} \quad 0 < \nu < 1,$$

$$x^\nu \leq P_\nu(x) \leq \frac{\Gamma(\nu + 1/2)}{\sqrt{\pi}\Gamma(\nu + 1)}(2x)^\nu, \quad \text{when} \quad -1/2 < \nu < 0 \text{ or } \nu > 1,$$

and

$$\max\left(\frac{1}{\sqrt{x}}, \frac{\sqrt{2x}\log x}{\pi(x - 1)}\right) \leq P_{-1/2}(x) \leq \frac{\sqrt{x}\log x}{x - 1}.$$

PROOF. Using (3.6) and (3.7), one derives the identity

$$\frac{d}{dx}\left(x^{-\nu}P_\nu(x)\right) = x^{-\nu-1}P_{\nu-1}'(x);$$

hence the first two statements follow from Lemma 3.1 together with Lemma 1.3.

We now turn to the case $\nu = -1/2$. First, note that by the above argument and Lemma 3.1, we have $x^{-1/2} \leq P_{-1/2}(x) \leq 1$. The inequality $P_{-1/2}(x) \leq \sqrt{x}\log x/(x - 1)$ will certainly follow if we can show that the function Ψ_+, given by $\Psi_+(x) = x^{-1/2}(x-1)P_{-1/2}(x) - \log x$, decreases with x, for $x \in [1, \infty)$. A calculation shows that

$$\Psi'_+(x) = \frac{2x+1}{2x^{3/2}(x+1)} P_{-1/2}(x) + \frac{1}{2\sqrt{x}(x+1)} P_{1/2}(x) - \frac{1}{x} \le -\frac{(\sqrt{x}-1)^2}{2x^2}.$$

The claimed inequality follows. Similarly, one shows that the function Ψ_-, given by $\Psi_-(x) = x^{-1/2}(x-1)P_{-1/2}(x) - \sqrt{2}\log x/\pi$, increases with x, and hence that $P_{-1/2}(x) \ge \sqrt{2x}\log x/\pi(x-1)$. □

REMARK. By the above lemma, it follows that the weight function ω, with $\tilde{\omega}(x) = x^\alpha$, belongs to $W(\alpha)$ for each $\alpha > -1/2$. Moreover, since Proposition 1.5 obviously holds in the limit case $\alpha = -1/2$, Proposition 3.2 shows that the weight function ω, with $\tilde{\omega}(x) = \frac{\sqrt{x}\log x}{x-1}$, makes $L^1(G//K, \omega)$ a Banach algebra.

We shall need the asymptotic behavior of $Q_\nu(x)$ at the end points $x = 1$ and $x = +\infty$. For $x = +\infty$, the definition of Q_ν shows that

$$(3.11) \qquad Q_\nu(x) \sim \frac{\sqrt{\pi}\Gamma(\nu+1)}{2^{\nu+1}\Gamma(\nu+3/2)} x^{-\nu-1}, \quad \text{as } x \to +\infty.$$

To achieve the asymptotic nature at $x = 1$, we apply the formula (9.7.5) in [**18**] to the hypergeometric function given in formula (3.3); then it is not difficult to prove that

$$(3.12) \qquad Q_\nu(x) = \frac{1}{2}\log\frac{2}{x-1} - \gamma - \psi(\nu+1) + o(1), \quad \text{as } x \to 1^+,$$

where $\gamma = -\psi(1)$ is Euler's constant, and ψ is the logarithmic derivative of the Gamma function $\psi(x) = (\log(\Gamma(x)))'$.

We now turn to the estimates of Q_ν and Q'_ν.

PROPOSITION 3.3. *For $\nu \in \mathbb{R}$, $\nu > -1$, and $x \in (1, \infty)$, we have the estimates*

$$-\frac{\sqrt{\pi}\Gamma(\nu+2)}{2^{\nu+1}\Gamma(\nu+3/2)} \frac{x^{-\nu}}{x^2-1} \le Q'_\nu(x) \le -\frac{x^{-\nu}}{x^2-1}, \quad \text{when } -1 < \nu \le 0,$$

and

$$-\frac{x^{-\nu}}{x^2-1} \le Q'_\nu(x) \le -\frac{\sqrt{\pi}\Gamma(\nu+2)}{2^{\nu+1}\Gamma(\nu+3/2)} \frac{x^{-\nu}}{x^2-1}, \quad \text{when } \nu \ge 0.$$

PROOF. Recall that the relation (3.7) is satisfied by Q_ν, applying this formula, together with (3.2), we get the identity

$$\frac{d}{dx}\left((x^2-1)x^\nu Q'_\nu(x)\right) = \nu(\nu+1)x^{\nu-1}Q_{\nu+1}(x).$$

Since $\nu > -1$, (3.5) shows that $Q_{\nu+1} \ge 0$, and hence it follows that $(x^2-1)x^\nu Q'_\nu(x)$ is a monotone function of x. The estimates in the statement now follows from the relation (3.7) together with (3.11) and (3.12), when we take the limits as $x \to +\infty$ and $x \to 1^+$, respectively. □

PROPOSITION 3.4. *For $\nu \in \mathbb{R}$, $\nu > -1$, and $x \in (1, \infty)$, we have the estimates*

$$Q_\nu(x) \le \min\left(q_\nu(x-1)^{-\nu-1}, x^{-\nu-1}\left(q_\nu + \frac{1}{2}\log\frac{x+1}{x-1}\right)\right),$$

and

$$Q_\nu(x) \geq x^{-\nu-1} \max\left(q_\nu, \frac{1}{2} \log \frac{x+1}{x-1} - \gamma - \psi(\nu+1)\right),$$

where $q_\nu = \dfrac{\sqrt{\pi}\Gamma(\nu+1)}{2^{\nu+1}\Gamma(\nu+3/2)}$.

PROOF. Since the relation (3.7) is satisfied by Q_ν, (3.5) shows that

$$\frac{d}{dx}\left((x-1)^{\nu+1}Q_\nu(x)\right) = (\nu+1)\frac{(x-1)^{\nu+1}}{x^2-1}(Q_{\nu+1}(x) + Q_\nu(x)) > 0,$$

which implies the first of the upper bounds for Q_ν. Similarly, the first of the lower bounds follows from

$$\frac{d}{dx}\left(x^{\nu+1}Q_\nu(x)\right) = x^\nu Q'_{\nu+1}(x) < 0.$$

For the remaining estimates, it is sufficient to show that the function Ψ_ν, given by

$$\Psi_\nu(x) = x^{\nu+1}Q_\nu(x) - \frac{1}{2}\log\frac{x+1}{x-1},$$

is increasing. By Proposition 3.3, we have

$$\Psi'_\nu(x) = x^\nu Q'_{\nu+1}(x) + \frac{1}{x^2-1} \geq \frac{1}{x(x+1)} > 0.$$

Hence

$$-\gamma - \psi(\nu+1) \leq \Psi_\nu(x) \leq \frac{\sqrt{\pi}\Gamma(\nu+1)}{2^{\nu+1}\Gamma(\nu+3/2)}.$$

This completes the proof. $\qquad\square$

Next, we introduce integration formulas from 1 to x:

$$(3.13) \qquad \int_1^x P_\lambda(t)P_v(t)\,dt = S(\lambda, v)F_{(\lambda,v)}(x), \quad \lambda \neq v, -v-1$$

and

$$(3.14) \qquad \int_1^x Q_\lambda(t)P_v(t)\,dt = S(\lambda, v)(1 + G_{(\lambda,v)}(x)), \quad \lambda \neq v, -v-1.$$

Here

$$S(\lambda, v) = \frac{1}{\lambda(\lambda+1) - v(v+1)} = \frac{1}{(\lambda+v+1)(\lambda-v)},$$

whereas

$$F_{(\lambda,v)}(x) = (v-\lambda)xP_\lambda(x)P_v(x) + (\lambda+1)P_{\lambda+1}(x)P_v(x) - (v+1)P_\lambda(x)P_{v+1}(x),$$
$$G_{(\lambda,v)}(x) = (v-\lambda)xQ_\lambda(x)P_v(x) + (\lambda+1)Q_{\lambda+1}(x)P_v(x) - (v+1)Q_\lambda(x)P_{v+1}(x).$$

These formulas stem from the differential equation (3.2). In fact, the relation (3.7) shows that

$$F_{(\lambda,v)}(x) = (x^2-1)\left(P_v(x)P'_\lambda(x) - P'_v(x)P_\lambda(x)\right).$$

On the other hand, by (3.2), we have

$$\frac{d}{dx}\left((x^2-1)(P_v(x)P'_\lambda(x) - P'_v(x)P_\lambda(x))\right) = (\lambda(\lambda+1) - v(v+1))P_\lambda(x)P_v(x);$$

hence we end up with (3.13), since both sides of (3.13) attain the same value at $x = 1$. In the same manner, one proves (3.14), except that one has to take the limit as $x \to 1^+$ using (3.12).

The next lemma follows easily from the integration formulas (3.13) and (3.14).

LEMMA 3.5. *Let $\lambda, \upsilon \in \mathbb{R}$, and suppose that $\lambda > -1$ and $\upsilon > -1/2$.*
If $\lambda > \upsilon$ or $\lambda < -\upsilon - 1$, then $S(\lambda, \upsilon) > 0$; hence $F_{(\lambda, \upsilon)}$ and $G_{(\lambda, \upsilon)}$ are increasing as
functions on $[1, +\infty)$; moreover, we then have $F_{(\lambda, \upsilon)}(x) \geq 0$ and $G_{(\lambda, \upsilon)}(x) \geq -1$.
If $-\upsilon - 1 < \lambda < \upsilon$, then $S(\lambda, \upsilon) < 0$; hence $F_{(\lambda, \upsilon)}$ and $G_{(\lambda, \upsilon)}$ are decreasing as
functions on $[1, +\infty)$; moreover, we then have $F_{(\lambda, \upsilon)}(x) \leq 0$ and $G_{(\lambda, \upsilon)}(x) \leq -1$.

COROLLARY 3.6. *If $-1/2 \leq \lambda < \alpha$, then $P_\lambda \leq P_\alpha$ on $[1, \infty)$.*

PROOF. It is sufficient to prove that P_λ / P_α is decreasing; the derivative is given
by

$$(P_\lambda(x)/P_\alpha(x))' = \frac{P'_\lambda(x)P_\alpha(x) - P_\lambda(x)P'_\alpha(x)}{P_\alpha(x)^2} = \frac{F_{(\lambda, \alpha)}(x)}{(x^2 - 1)P_\alpha(x)^2} \leq 0,$$

where the inequality holds by Lemma 3.5.   □

In §5, we need estimates of the function $H_{(\lambda, \upsilon)}$, defined by

(3.15)
$$H_{(\lambda, \upsilon)}(x) = P_\lambda(x) \int_1^x Q_\lambda(t)P_\upsilon(t)\, dt + Q_\lambda(x) \int_1^x P_\lambda(t)P_\upsilon(t)\, dt$$
$$= S(\lambda, \upsilon)\big[P_\lambda(x)(1 + G_{(\lambda, \upsilon)}(x)) + Q_\lambda(x)F_{(\lambda, \upsilon)}(x)\big].$$

A somewhat more convenient formula for $H_{(\lambda, \upsilon)}$ is given by

(3.16) $$H_{(\lambda, \upsilon)}(x) = S(\lambda, \upsilon)\big[P_\upsilon(x) + P_\lambda(x) + 2P_\lambda(x)G_{(\lambda, \upsilon)}(x)\big].$$

To prove (3.16), we apply the relation (3.10) in order to show that

(3.17) $$Q_\lambda(x)F_{(\lambda, \upsilon)}(x) = P_\upsilon(x) + P_\lambda(x)G_{(\lambda, \upsilon)}(x).$$

In the case $-1/2 < \upsilon \leq 0$, the following lemma contains the required estimates
of $H_{(\lambda, \upsilon)}$.

LEMMA 3.7. *Suppose $-1/2 < \upsilon \leq 0$ and $-1/2 < \lambda < \upsilon$. Then*

$$H_{(\lambda, \upsilon)}(x) \leq S(\upsilon, \lambda)\frac{1}{2\lambda + 1}P_\upsilon(x),$$

for $x \in (1, \infty)$.

PROOF. In this case $S(\lambda, \upsilon) = -S(\upsilon, \lambda) < 0$, hence by (3.16) it suffices to
estimate $P_\upsilon + P_\lambda + 2P_\lambda G_{(\lambda, \upsilon)}$ from below.
By Lemma 3.1, $P'_\upsilon(x) \leq 0$, whence

$$G_{(\lambda, \upsilon)}(x) = (x^2 - 1)\left(Q'_\lambda(x)P_\upsilon(x) - Q_\lambda(x)P'_\upsilon(x)\right) \geq (x^2 - 1)Q'_\lambda(x)P_\upsilon(x);$$

and, therefore, Proposition 3.2 and Proposition 3.3 imply

$$H_{(\lambda, \upsilon)}(x) \leq S(\lambda, \upsilon)\left(P_\upsilon(x) - 2\frac{\lambda + 1}{2\lambda + 1}P_\upsilon(x)\right) \leq S(\upsilon, \lambda)\frac{1}{2\lambda + 1}P_\upsilon(x),$$

for $x \in (1, \infty)$.   □

Next, we estimate $G_{(\lambda, \upsilon)}$ when $-1 < \lambda < \upsilon$ and $\upsilon \geq 0$. Observe that the
behavior of $G_{(\lambda, \upsilon)}$ at $+\infty$ in terms of asymptotic expansion can be calculated using
Lemma 1.3 and (3.11); for $\text{Re}\,\upsilon > -1/2$ and $\lambda \in \mathbb{C} \setminus \{-1, -2, -3, \ldots\}$, we have

$$G_{(\lambda, \upsilon)}(x) \sim -2^{\upsilon - \lambda - 1}(\upsilon + \lambda + 1)\frac{\Gamma(\lambda + 1)}{\Gamma(\lambda + 3/2)}\frac{\Gamma(\upsilon + 1/2)}{\Gamma(\upsilon + 1)}x^{\upsilon - \lambda}, \quad \text{as} \quad x \to +\infty.$$

LEMMA 3.8. *Fix v with $v \geq 0$. If $-1 < \lambda < v$, then $G_{(\lambda,v)}(x)x^{\lambda-v}$ decreases with x, for $x \in [1,\infty)$. In particular, we then have*

$$-2^{v-\lambda-1}(v+\lambda+1)\frac{\Gamma(\lambda+1)}{\Gamma(\lambda+3/2)}\frac{\Gamma(v+1/2)}{\Gamma(v+1)}x^{v-\lambda} \leq G_{(\lambda,v)}(x) \leq -x^{v-\lambda},$$

for $x \in [1,\infty)$.

PROOF. First note that

$$\frac{d}{dx}\left(G_{(\lambda,v)}(x)x^{\lambda-v}\right) = x^{\lambda-v-1}\left(xG'_{(\lambda,v)}(x) + (\lambda-v)G_{(\lambda,v)}(x)\right).$$

The definition of $G_{(\lambda,v)}$, together with the recurrence formula for the Legendre function of the first kind (3.6), shows that

$$xG'_{(\lambda,v)}(x) + (\lambda-v)G_{(\lambda,v)}(x) = (\lambda-v)\Big((\lambda+1)Q_{\lambda+1}(x)P_v(x)$$

$$+ Q_\lambda(x)\big[(2v+1)xP_v(x) - (v+1)P_{v+1}(x)\big]\Big)$$

$$= (\lambda-v)\left[(\lambda+1)Q_{\lambda+1}(x)P_v(x) + vQ_\lambda(x)P_{v-1}(x)\right] \leq 0.$$

Hence, the statement of the lemma follows from the asymptotic expansion of $G_{(\lambda,v)}$ together with the fact $G_{(\lambda,v)}(1) = -1$. $\square$

The needed estimate of $H_{(\lambda,v)}$ in the case $v > 0$ is contained in the next lemma.

LEMMA 3.9. *Suppose $v > 1$ and $x \in (1,\infty)$. Then*

$$H_{(\lambda,v)}(x) \leq S(v,\lambda)2^{v+1}\frac{2v+1}{2\lambda+1}\frac{\Gamma(v+1/2)}{\sqrt{\pi}\Gamma(v+1)}P_v(x),$$

provided $-1/2 < \lambda \leq 0$ or $1 \leq \lambda < v$; and

$$H_{(\lambda,v)}(x) \leq S(v,\lambda)2^{v+1}(v+1)\frac{\Gamma(v+1/2)}{\sqrt{\pi}\Gamma(v+1)}P_v(x),$$

provided $0 \leq \lambda \leq 1$.

Suppose $0 < v \leq 1$ and $x \in (1,\infty)$. Then

$$H_{(\lambda,v)}(x) \leq S(v,\lambda)(2v+1)P_v(x),$$

provided $0 \leq \lambda < v$; and

$$H_{(\lambda,v)}(x) \leq S(v,\lambda)\frac{2v+1}{2\lambda+1}P_v(x).$$

provided $-1/2 < \lambda \leq 0$.

PROOF. We first consider the case when $v > 1$. For λ with $-1/2 < \lambda \leq 0$ or $1 \leq \lambda < v$, (3.16), Proposition 3.2, and Lemma 3.8 imply

$$H_{(\lambda,v)}(x) \leq S(v,\lambda)2^v\frac{v+\lambda+1}{\lambda+1/2}\frac{\Gamma(v+1/2)}{\sqrt{\pi}\Gamma(v+1)}x^v$$

$$\leq S(v,\lambda)2^{v+1}\frac{2v+1}{2\lambda+1}\frac{\Gamma(v+1/2)}{\sqrt{\pi}\Gamma(v+1)}P_v(x).$$

In the case when $0 \leq \lambda \leq 1$, (3.16), Proposition 3.2, and Lemma 3.8 imply

$$H_{(\lambda,v)}(x) \leq S(v,\lambda)2^{v-\lambda}(v+\lambda+1)\frac{\Gamma(\lambda+1)}{\Gamma(\lambda+3/2)}\frac{\Gamma(v+1/2)}{\Gamma(v+1)}x^v$$

$$\leq S(v,\lambda)2^{v+1}(v+1)\frac{\Gamma(v+1/2)}{\sqrt{\pi}\Gamma(v+1)}P_v(x).$$

In the last inequality, we used Proposition 3.2 (or Proposition 3.3), which shows that

(3.18) $$\frac{\Gamma(\lambda+1)}{2^\lambda\Gamma(\lambda+3/2)} \le \frac{2}{\sqrt{\pi}(\lambda+1)}.$$

Finally, we consider the case $0 < \upsilon \le 1$. For $0 \le \lambda < \upsilon$, (3.16), Proposition 3.2, Lemma 3.8, and (3.18) imply

$$
\begin{aligned}
H_{(\lambda,\upsilon)}(x) &\le S(\upsilon,\lambda)\left(2^{\upsilon-\lambda}(\upsilon+\lambda+1)\frac{\Gamma(\lambda+1)}{\Gamma(\lambda+3/2)}\frac{\Gamma(\upsilon+1/2)}{\Gamma(\upsilon+1)}x^\upsilon - P_\upsilon(x)\right)\\
&\le S(\upsilon,\lambda)\left(2^{\upsilon+1}(\upsilon+1)\frac{\Gamma(\upsilon+1/2)}{\sqrt{\pi}\Gamma(\upsilon+1)}x^\upsilon - P_\upsilon(x)\right)\\
&\le S(\upsilon,\lambda)(2\upsilon+1)P_\upsilon(x).
\end{aligned}
$$

In the case when $-1/2 < \lambda \le 0$, (3.16), Proposition 3.2, and Lemma 3.8 imply

$$
\begin{aligned}
H_{(\lambda,\upsilon)}(x) &\le S(\upsilon,\lambda)\left(2^\upsilon\frac{\upsilon+\lambda+1}{\lambda+1/2}\frac{\Gamma(\upsilon+1/2)}{\sqrt{\pi}\Gamma(\upsilon+1)}x^\upsilon - P_\upsilon(x)\right)\\
&\le S(\upsilon,\lambda)\frac{2\upsilon+1}{2\lambda+1}P_\upsilon(x).
\end{aligned}
$$

The proof is complete. $\qquad\square$

Finally, we need the following two lemmas concerning Legendre functions.

LEMMA 3.10. *For each fixed $x > 1$, the function $\nu \mapsto Q_\nu(x)$ decreases as a function on $(-1,\infty)$.*

PROOF. By (3.4),

$$\frac{d}{d\nu}Q_\nu(x) = 2^\nu\int_0^1 t^\nu(1-t)^\nu(1+x-2t)^{-\nu-1}\log\frac{2t(1-t)}{1+x-2t}\,dt \le 0,$$

because $\log\dfrac{2t(1-t)}{1+x-2t} \le \log t < 0$, when $t \in (0,1)$ and $x > 1$. $\qquad\square$

LEMMA 3.11. *If $\operatorname{Re}\lambda > \operatorname{Re}\upsilon > -1/2$, then*

$$\int_1^\infty Q_\lambda(t)P_\upsilon(t)\,dt = S(\lambda,\upsilon).$$

PROOF. This follows when we let x tend to $+\infty$ in (3.14), together with the asymptotic expansion of $G_{(\lambda,\upsilon)}$. $\qquad\square$

4. The construction of b_λ

In this section, we find functions $b_\lambda \in L^1(G//K,\omega)$ such that (2.3) holds, and then we prove that the set $\{b_\lambda : \lambda \in \mathbb{C}\setminus\Sigma(\alpha)\}$ spans a dense linear subspace of $L^1(G//K,\omega)$.

LEMMA 4.1. *Suppose $\omega \in W(\alpha)$, $\alpha > -1/2$. For each λ with $\operatorname{Re}\lambda > \alpha$, define $b_\lambda(a_\zeta) = Q_\lambda(\tau)$, where $\tau = \cosh 2\zeta$ as in §1; then, for $\operatorname{Re}\lambda > \alpha$, we have*

(1) $b_\lambda \in L^1(G//K, \omega)$,

(2) $\hat{b}_\lambda(s) = \dfrac{1}{\lambda(\lambda+1) - s(s+1)}$, $for \ s \in \Sigma(\alpha)$,

(3) $\|b_\lambda\|_{L^1(\omega)} \leq C_\omega \dfrac{1}{(\operatorname{Re}\lambda + \alpha + 1)(\operatorname{Re}\lambda - \alpha)}$,

where in (3), C_ω is a constant depending only on ω.

PROOF. Since (3) implies (1), and since (2) follows from Lemma 3.11, we only have to prove (3).

By Proposition 1.5, the statement follows if we can prove it for the weight function ω_α, given by $\tilde{\omega}_\alpha(x) = P_\alpha(x)$. Hence, (3) follows from (3.5) and Lemma 3.11. □

Recall, from §2, that $b_\lambda = b_{-\lambda-1}$ when $\operatorname{Re}\lambda < -\alpha - 1$.

LEMMA 4.2. *Suppose* $\omega \in W(\alpha)$, $\alpha > -1/2$. *The linear span of the functions* b_λ, $\lambda \in \mathbb{C} \setminus \Sigma(\alpha)$, *is a dense subspace of* $L^1(G//K, \omega)$.

PROOF. Since $C_0^\infty(G//K)$ is a dense subspace of $L^1(G//K, \omega)$, it will suffice to prove that $C_0^\infty(G//K)$ is contained in the closure of the subspace spanned by the functions b_λ, $\lambda \in \mathbb{C} \setminus \Sigma(\alpha)$.

Choose $f \in C_0^\infty(G//K)$ arbitrary. By Proposition 1.4, the Gelfand transform $\hat{f}$ is an entire function of finite exponential type.

Fix $\delta > 0$ and $s \in \Sigma(\alpha)$. By Cauchy's integral formula and the rapid decay of $\hat{f}$, we have

$$\hat{f}(s) = \frac{1}{2\pi i} \int_{\gamma_1(\delta)+\gamma_2(\delta)} \frac{\hat{f}(z)}{z-s}\, dz = \frac{1}{2\pi i} \int_{\gamma_1(\delta)} \frac{\hat{f}(z)}{z-s}\, dz + \frac{1}{2\pi i} \int_{\gamma_2(\delta)} \frac{\hat{f}(z)}{z-s}\, dz,$$

where $\gamma_1(\delta)$ is the line $\alpha + \delta + i\mathbb{R}$ with upward direction and $\gamma_2(\delta)$ is the line $-1 - \alpha - \delta + i\mathbb{R}$ with downward direction. If we apply the substitution $z \mapsto -1 - z$ and the identity $\hat{f}(-1-z) = \hat{f}(z)$ in the second integral, we get

$$\int_{\gamma_2(\delta)} \frac{\hat{f}(z)}{z-s}\, dz = \int_{\gamma_1(\delta)} \frac{\hat{f}(z)}{1+z+s}\, dz.$$

Hence, since

$$\frac{1}{z-s} + \frac{1}{1+z+s} = \frac{2z+1}{z(z+1) - s(s+1)},$$

Lemma 4.1 shows that

$$\hat{f}(s) = \frac{1}{2\pi i} \int_{\gamma_1(\delta)} (2z+1)\hat{f}(z)\hat{b}_z(s)\, dz.$$

Lemma 4.1, together with the rapid decay of $\hat{f}$ on $\gamma_1(\delta)$ (Proposition 1.4), shows that the $L^1(G//K, \omega)$-valued integral

$$\frac{1}{2\pi i} \int_{\gamma_1(\delta)} (2z+1)\hat{f}(z)b_z\, dz$$

converges, and hence it must converge to f. Thus, the Riemann sums, which are linear combinations of the b_λ's, converge to f.

The statement follows. □

5. Representatives for the cosets $B_\lambda + I(\mathfrak{S})$

In this section, we find for $f \in L^1(G//K, \omega)$ and $\lambda \in \Sigma(\alpha)^\circ$, the function $T_\lambda f \in L^1(G//K, \omega)$ satisfying (2.8) whose existence was claimed in §2.

LEMMA 5.1. *Suppose* $\omega \in W(\alpha)$, $\alpha > -1/2$. *For each* $f \in L^1(G//K, \omega)$ *and* $\lambda \in \Sigma(\alpha)^\circ$, *we define the function* $T_\lambda f : (1, \infty) \to \mathbb{C}$ *by*

$$T_\lambda f(\tau)$$

$$= \begin{cases} Q_\lambda(\tau) \int_\tau^\infty f(x) P_\lambda(x)\, dx - P_\lambda(\tau) \int_\tau^\infty f(x) Q_\lambda(x)\, dx, & \mathrm{Re}\,\lambda \geq -\tfrac{1}{2}, \\[2mm] Q_{-\lambda-1}(\tau) \int_\tau^\infty f(x) P_\lambda(x)\, dx - P_\lambda(\tau) \int_\tau^\infty f(x) Q_{-\lambda-1}(x)\, dx, & \mathrm{Re}\,\lambda \leq -\tfrac{1}{2}. \end{cases}$$

Here, we use the coordinatization $\tau = \cosh(2\zeta)$, *as in §1. For the function* $T_\lambda f$, *with* $\lambda \in \Sigma(\alpha)^\circ$, *we have*

(1) $T_\lambda f \in L^1(G//K, \omega)$,

(2) $\widehat{T_\lambda f}(s) = \dfrac{\hat{f}(\lambda) - \hat{f}(s)}{\lambda(\lambda+1) - s(s+1)}, \quad s \in \Sigma(\alpha) \setminus \{\lambda, -\lambda - 1\}$,

(3) $\|T_\lambda f\|_{L^1(\omega)} \leq C_\omega \dfrac{\|f\|_{L^1(\omega)}}{(\mathrm{Re}\,\lambda + \alpha + 1)(\alpha - \mathrm{Re}\,\lambda)}$,

where in (3), C_ω *is a constant depending only on* ω.

PROOF. In the proof of the lemma, we use the notation C_α to denote some absolute constant depending on α; the correct value of C_α varies in different parts of the proof.

First note that for fixed $\tau > 1$, (3.9) implies that $T_\lambda f(\tau)$ depends analytically on λ for $\lambda \in \Sigma(\alpha)^\circ$, and moreover, for such λ, we have $T_\lambda f = T_{-\lambda-1} f$.

We start by proving (3). By Proposition 1.5, it is sufficient to prove the statement with ω replaced by the weight function $\tilde{\omega}_\alpha(x) = P_\alpha(x)$. Choose some number λ_α in the open interval $(-1/2, \min(0, \alpha))$. First, we shall assume $\mathrm{Re}\,\lambda \geq \lambda_\alpha$. By Lemma 1.3, the inequality (3.5), and a change of the order of integration, we obtain

$$\begin{aligned} \|T_\lambda f\|_{L^1(\omega_\alpha)} &= \int_1^\infty |T_\lambda f(\tau)| P_\alpha(\tau)\, d\tau \\ &\leq \int_1^\infty \Bigg[Q_{\mathrm{Re}\,\lambda}(\tau) \int_\tau^\infty |f(x)| P_{\mathrm{Re}\,\lambda}(x)\, dx \\ &\qquad + P_{\mathrm{Re}\,\lambda}(\tau) \int_\tau^\infty |f(x)| Q_{\mathrm{Re}\,\lambda}(x)\, dx \Bigg] P_\alpha(\tau)\, d\tau \\ &= \int_1^\infty |f(x)| \Bigg[P_{\mathrm{Re}\,\lambda}(x) \int_1^x Q_{\mathrm{Re}\,\lambda}(\tau) P_\alpha(\tau)\, d\tau \\ &\qquad + Q_{\mathrm{Re}\,\lambda}(x) \int_1^x P_{\mathrm{Re}\,\lambda}(\tau) P_\alpha(\tau)\, d\tau \Bigg] dx \\ &= \int_1^\infty |f(x)| H_{(\mathrm{Re}\,\lambda, \alpha)}(x)\, dx; \end{aligned}$$

we recall that the function $H_{(\lambda, v)}$ was defined by (3.15) in §3. If $\alpha \leq 0$ we apply Lemma 3.7 in order to obtain

(5.1) $\|T_\lambda f\|_{L^1(\omega_\alpha)} \leq C_\alpha S(\alpha, \mathrm{Re}\,\lambda) \|f\|_{L^1(\omega_\alpha)}$, for λ with $\operatorname{Re}\lambda \geq \lambda_\alpha$. Similarly, Lemma 3.9 proves (5.1) in the case $\alpha \geq 0$.

Next, we consider the case when $-1/2 \leq \operatorname{Re}\lambda \leq \lambda_\alpha$, $\lambda \neq -1/2$. If we apply the symmetry formula (3.9), and then the triangle inequality, we end up with

$$|T_\lambda f(\tau)| \leq \frac{|\tan\pi\lambda|}{\pi}\left(Q_{-\operatorname{Re}\lambda-1}(\tau)\int_\tau^\infty |f(x)|Q_{\operatorname{Re}\lambda}(x)\,dx\right.$$
$$\left. + Q_{\operatorname{Re}\lambda}(\tau)\int_\tau^\infty |f(x)|Q_{-\operatorname{Re}\lambda-1}(x)\,dx\right),$$

applying (3.9) once more, we arrive at

$$|T_\lambda f(\tau)| \leq \frac{|\tan\pi\lambda|}{\pi}\left[2Q_{\operatorname{Re}\lambda}(\tau)\int_\tau^\infty |f(x)|Q_{\operatorname{Re}\lambda}(x)\,dx\right.$$
$$- \pi\cot(\pi\operatorname{Re}\lambda)\left(Q_{\operatorname{Re}\lambda}(\tau)\int_\tau^\infty |f(x)|P_{\operatorname{Re}\lambda}(x)\,dx\right.$$
$$\left.\left. + P_{\operatorname{Re}\lambda}(\tau)\int_\tau^\infty |f(x)|Q_{\operatorname{Re}\lambda}(x)\,dx\right)\right].$$

Here, and in the rest of the proof, the term containing the factor $\cot(\pi\operatorname{Re}\lambda)$ should be treated as 0 in the case $\operatorname{Re}\lambda = -1/2$.

Hence, by changing the order of integration, we have

$$\|T_\lambda f\|_{L^1(\omega_\alpha)} \leq \frac{|\tan\pi\lambda|}{\pi}\int_1^\infty\left[2Q_{\operatorname{Re}\lambda}(\tau)\int_\tau^\infty |f(x)|Q_{\operatorname{Re}\lambda}(x)\,dx\right.$$
$$- \pi\cot(\pi\operatorname{Re}\lambda)\left(Q_{\operatorname{Re}\lambda}(\tau)\int_\tau^\infty |f(x)|P_{\operatorname{Re}\lambda}(x)\,dx\right.$$
$$\left.\left. + P_{\operatorname{Re}\lambda}(\tau)\int_\tau^\infty |f(x)|Q_{\operatorname{Re}\lambda}(x)\,dx\right)\right]P_\alpha(\tau)\,d\tau$$
$$= \frac{|\tan\pi\lambda|}{\pi}\int_1^\infty |f(x)|\left[2Q_{\operatorname{Re}\lambda}(x)\int_1^x Q_{\operatorname{Re}\lambda}(\tau)P_\alpha(\tau)\,d\tau\right.$$
$$- \pi\cot(\pi\operatorname{Re}\lambda)\left(P_{\operatorname{Re}\lambda}(x)\int_1^x Q_{\operatorname{Re}\lambda}(\tau)P_\alpha(\tau)\,d\tau\right.$$
$$\left.\left. + Q_{\operatorname{Re}\lambda}(x)\int_1^x P_{\operatorname{Re}\lambda}(\tau)P_\alpha(\tau)\,d\tau\right)\right]dx,$$
$$= \frac{|\tan\pi\lambda|}{\pi}\int_1^\infty |f(x)|\left[2Q_{\operatorname{Re}\lambda}(x)\int_1^x Q_{\operatorname{Re}\lambda}(\tau)P_\alpha(\tau)\,d\tau\right.$$
$$\left. - \pi\cot(\pi\operatorname{Re}\lambda)H_{(\operatorname{Re}\lambda,\alpha)}(x)\right]dx.$$

The last term inside the brackets, $-\pi\cot(\pi\operatorname{Re}\lambda)H_{(\operatorname{Re}\lambda,\alpha)}$, is bounded by $C_\alpha P_\alpha$ when $\operatorname{Re}\lambda \in (-1/2, \lambda_\alpha]$, this follows from Lemma 3.7 when $\alpha \leq 0$, and from Lemma 3.9 when $\alpha \geq 0$. (Note that the singularity at $\lambda = -1/2$ is cancelled.)

Next, we intend to estimate the first term in the brackets. In the case $\alpha \geq 0$, Lemma 3.1, Lemma 3.10, and Proposition 3.4 imply

$$Q_{\operatorname{Re}\lambda}(x)\int_1^x Q_{\operatorname{Re}\lambda}(\tau)P_\alpha(\tau)\,d\tau \leq P_\alpha(x)Q_{-1/2}(x)\int_1^x Q_{-1/2}(\tau)\,d\tau \leq \pi^2 P_\alpha(x).$$

When $\alpha \in (-1/2, 0)$, we first observe that if $x \leq 2$, then Proposition 3.2, Lemma 3.10, and Proposition 3.4 imply

$$Q_{\mathrm{Re}\,\lambda}(x) \int_1^x Q_{\mathrm{Re}\,\lambda}(\tau) P_\alpha(\tau)\, d\tau \leq \pi^2 \leq \pi^2 2^{-\alpha} P_\alpha(x).$$

In the remaining case $\alpha \in (-1/2, 0)$ and $x \geq 2$, Proposition 3.2, Lemma 3.10 and Proposition 3.4 imply

$$Q_{\mathrm{Re}\,\lambda}(x) \int_1^x Q_{\mathrm{Re}\,\lambda}(\tau) P_\alpha(\tau)\, d\tau \leq \frac{2^{\alpha-1}\pi^{3/2}}{\sqrt{x-1}} \frac{\Gamma(\alpha+1/2)}{\Gamma(\alpha+1)} \int_1^x (\tau-1)^{\alpha-1/2}\, d\tau$$

$$= 2^\alpha \pi^{3/2} \frac{\Gamma(\alpha+1/2)}{(2\alpha+1)\Gamma(\alpha+1)}(x-1)^\alpha \leq \pi^{3/2} \frac{\Gamma(\alpha+1/2)}{(2\alpha+1)\Gamma(\alpha+1)} P_\alpha(x).$$

Here, we used, in a first step, the inequality $x^\alpha \leq (x-1)^\alpha$, and in a second step, that $2^\alpha(x-1)^\alpha \leq x^\alpha \leq P_\alpha(x)$.

Finally, we achieve, for λ with $-1/2 \leq \mathrm{Re}\,\lambda \leq \lambda_\alpha$ and $\lambda \neq -1/2$, the estimate

$$(5.2) \qquad \|T_\lambda f\|_{L^1(\omega_\alpha)} \leq C_\alpha |\tan \pi\lambda| \, \|f\|_{L^1(\omega_\alpha)}.$$

To handle the singularity at the point $\lambda = -1/2$ in (5.2), we first use the identity $T_\lambda f = T_{-\lambda-1} f$, which shows that (5.2) holds for all λ with $-\lambda_\alpha - 1 \leq \mathrm{Re}\,\lambda \leq \lambda_\alpha$ and $\lambda \neq -1/2$. Secondly, we observe, as in [3], that $\lambda \mapsto \|T_\lambda\|_{L^1(\omega_\alpha)}$ is subharmonic on $\Sigma(\alpha)^\circ$. Hence, according to the maximum principle for subharmonic functions applied inside the disc with radius $r_\alpha = |\lambda_\alpha|$ centered at $-1/2$, (5.2) implies

$$\|T_{-1/2} f\|_{L^1(\omega_\alpha)} \leq C_\alpha \|f\|_{L^1(\omega_\alpha)};$$

and by upper semicontinuity, we must then have

$$(5.3) \qquad \|T_\lambda f\|_{L^1(\omega_\alpha)} \leq C_\alpha \|f\|_{L^1(\omega_\alpha)},$$

for all λ is a neighborhood of $-1/2$.

Hence, (3) follows from the estimates (5.1), (5.2), (5.3), together with Proposition 1.5 and the identity $T_\lambda f = T_{-1-\lambda} f$.

Since (1) follows from (3), it only remains to prove (2). According to the identity $T_\lambda f = T_{-\lambda-1} f$, we may assume $\mathrm{Re}\,\lambda \geq -1/2$. From the definition of $T_\lambda f$, we have

$$\widehat{T_\lambda f}(s) = \int_1^\infty Q_\lambda(\tau) P_s(\tau) \int_\tau^\infty f(x) P_\lambda(x)\, dx\, d\tau$$

$$- \int_1^\infty P_\lambda(\tau) P_s(\tau) \int_\tau^\infty f(x) Q_\lambda(x)\, dx\, d\tau.$$

By changing the order of integration, we get

$$\widehat{T_\lambda f}(s) = \int_1^\infty f(x)\left[P_\lambda(x) \int_1^x Q_\lambda(\tau) P_s(\tau)\, d\tau - Q_\lambda(x) \int_1^x P_\lambda(\tau) P_s(\tau)\, d\tau \right] dx.$$

Using (3.13), (3.14), and in a second step (3.17), we find that

$$P_\lambda(x) \int_1^x Q_\lambda(\tau) P_s(\tau)\, d\tau - Q_\lambda(x) \int_1^x P_\lambda(\tau) P_s(\tau)\, d\tau$$

$$= S(\lambda, s)[P_\lambda(x)(1 + G_{(\lambda,s)}(x)) - Q_\lambda(x) F_{\lambda,s}(x)]$$

$$= S(\lambda, s)[P_\lambda(x) - P_s(x)].$$

It follows that

$$\widehat{T_\lambda f}(s) = S(\lambda, s) \int_1^\infty f(x)(P_\lambda(x) - P_s(x))\, dx = \frac{\hat{f}(\lambda) - \hat{f}(s)}{\lambda(\lambda + 1) - s(s + 1)},$$

which completes the proof of the lemma. $\square$

We now summarize the needed properties of the resolvent transform.

THEOREM 5.2. *Suppose* $\omega \in W(\alpha)$, $\alpha > -1/2$. *Let* I *be a closed convolution ideal in* $L^1(G//K, \omega)$. *Fix any* $f \in I$, *and assume that* $g \in L^\infty(G//K, 1/\omega)$ *annihilates* I.

(1) *The resolvent transform* $\mathfrak{R}[g]$ *of* g *is defined and analytic on* $\mathbb{C} \setminus Z_\infty(I)$. *It is given by the formula*

$$\mathfrak{R}[g](\lambda) = \begin{cases} \langle b_\lambda, g \rangle, & \lambda \in \mathbb{C} \setminus \Sigma(\alpha), \\ \dfrac{\langle T_\lambda f, g \rangle}{\hat{f}(\lambda)}, & \lambda \in \Sigma(\alpha)^\circ \setminus Z(f). \end{cases}$$

(2) $\mathfrak{R}[g](\lambda) = \mathfrak{R}[g](-\lambda - 1)$, *for all* $\lambda \in \mathbb{C} \setminus Z_\infty(I)$.
(3) $|\mathfrak{R}[g](\lambda)| \le C_\omega \|g\|_{L^\infty(1/\omega)}/d(\lambda, \partial\Sigma(\alpha))$, *for* $\lambda \in \mathbb{C} \setminus \Sigma(\alpha)$.
(4) $|\langle T_\lambda f, g \rangle| \le C_\omega \|f\|_{L^1(\omega)} \|g\|_{L^\infty(1/\omega)}/d(\lambda, \partial\Sigma(\alpha))$, *for* $\lambda \in \Sigma(\alpha)^\circ$.

Here, C_ω *denotes some positive constant depending only on* ω, *and* $d(\lambda, \partial\Sigma(\alpha))$ *denotes the Euclidean distance between* λ *and the boundary of* $\Sigma(\alpha)$.

PROOF. This follows from Lemma 4.1, Lemma 5.1, and the discussion in §2.

$\square$

REMARK. For the counterpart of (4), in the case $\omega = 1$, Ben Natan, Benyamini, Hedenmalm and Weit showed in [**3**] that $|\langle T_\lambda f, g \rangle| \le C\|f\|_{L^1}\|g\|_{L^\infty} d(\lambda, \partial\Sigma_\omega)^{-2}$. Hence our method is sharper than theirs.

6. Results from the theory of analytic functions

In this section, we study resolvent type functions on the strip $S = \{z : |\mathrm{Im}\, z| < \pi/2\}$. Loosely speaking, by a resolvent type function on S we mean a function G which is analytic in $\mathbb{C}$, except perhaps on a finite subset of $\mathrm{clos}(S)$, such that $G = H/F$ in S, where F and H are analytic in S, with F bounded, and such that H and G satisfies a certain growth conditions towards the boundary of S from the inside and the outside of S respectively (cf. the function $\mathfrak{R}[g]$ of Theorem 5.2). We will focus on the behavior of such functions at singular points, located at infinity and on the boundary of S, respectively.

The next theorem is classical and is commonly referred to as the $\log - \log$ theorem. It has its roots from work of Carleman, Levinson, Sjöberg, and Beurling. The following general variant is due to Y. Domar and can be found in Koosis' book [**16**, p. 376].

THEOREM 6.1. *Let* Q *be the rectangle* $\{z = x + iy \in \mathbb{C} : |x| \le 1, |y| \le 1\}$. *Let* $M : (0, 1] \to [e, \infty)$ *be a Lebesgue measurable function. Suppose* f *is an analytic function on* Q *that satisfies*

$$|f(x + iy)| \le M(|x|), \quad x + iy \in Q.$$

If

$$\int_0^1 \log \log M(x)\, dx < \infty,$$

then there is a constant C, depending only on M and $\delta > 0$, such that

$$|f(x+iy)| \le C, \qquad \text{for } |x| \le 1 - \delta,\ |y| \le 1 - \delta.$$

The $\log - \log$ theorem is a powerful tool to get estimates of resolvent type functions. We will use it to get a priori estimates of such functions. These estimates are then smoothed using the Ahlfors-Heins theorem and Phragmén-Lindelöf type arguments. The precise behavior of resolvent type functions near the singularity at infinity, is then obtained using a generalized Phragmén-Lindelöf argument based on Ahlfors distortion inequality. This technique was used earlier by Håkan Hedenmalm (see [**12**]). This will be done in sub-section 6.1.

In sub-section 6.2, we study the behavior of resolvent type functions at a boundary point singularity. To do this, we first map the strip onto it self in such a way that the singularity on the boundary of S goes to infinity, this makes it possible to apply the technique of sub-section 6.1.

6.1. The singularity at infinity. The following lemma is a slight modification of Lemma C.1 in [**2**]. See also the papers of Gurariĭ [**10**], and Hedenmalm [**12**].

LEMMA 6.2. *Let $S = \{z = x + iy \in \mathbb{C} : |y| < \pi/2\}$, and let Z be a finite subset of $\mathrm{clos}(S)$. Suppose F is a bounded analytic function on S which is continuous on $\mathrm{clos}(S)$. Suppose H is analytic in S and satisfies*

$$(6.1) \qquad |H(z)| \le M(d(z, \partial S)), \quad z \in S,$$

where $M : (0, \pi/2) \to (0, \infty)$ is decreasing, and $d(x + iy, \partial S) = \pi/2 - |y|$ is the Euclidean distance of $z = x + iy$ to the boundary ∂S of S. Suppose, furthermore, that G is analytic in $S \setminus Z$, where

$$G(z) = \frac{H(z)}{F(z)}, \quad z \in S.$$

Fix a number R such that $R > \max_{z \in Z} |\mathrm{Re}\, z|$. Then there are constants C_1, C_2, such that for each ε, $\varepsilon \in (0, 1)$,

$$\log |G(x + iy)| \le C_1 + \log M(\varepsilon) + C_2 \frac{\exp(\frac{\pi}{2}|x|/(\frac{\pi}{2} - \varepsilon))}{\cos \frac{\pi}{2} y/(\frac{\pi}{2} - \varepsilon)},$$

for $x + iy \in S$ with $|x| \ge R$ and $|y| < \pi/2 - \varepsilon$.

PROOF. See [**12**]. Note that in $S_\varepsilon = \{z \in \mathbb{C} : |\mathrm{Im}\, z| < \pi/2 - \varepsilon\}$, $\varepsilon \in (0, \pi/2)$, G is a quotient of bounded analytic functions. In the upper half-plane, $\mathbb{C}_+$, we have explicit formulas for factorization of bounded analytic functions. To use this, we consider the conformal map φ_ε, defined by

$$\varphi_\varepsilon : S \ni z \mapsto i \exp\left(\frac{\frac{\pi}{2} z}{\frac{\pi}{2} - \varepsilon}\right),$$

for $0 < \varepsilon < \pi/2$. It is clear that φ_ε maps S_ε onto $\mathbb{C}_+$.

Define

$$F_\varepsilon = F \circ \varphi_\varepsilon^{-1}, \ H_\varepsilon = H \circ \varphi_\varepsilon^{-1}$$

and
$$G_\varepsilon = G \circ \varphi_\varepsilon^{-1} = \frac{H_\varepsilon}{F_\varepsilon}.$$

Then $F_\varepsilon, H_\varepsilon \in H^\infty(\mathbb{C}_+)$. By the factorization theorem for Hardy spaces (see [15, pp. 160-161]), any $f \in H^\infty(\mathbb{C}_+)$ can be factorized

$$f(\zeta) = e^{i\gamma_f} B_f(\zeta) e^{i\alpha_f \zeta} \exp\left(\frac{i}{\pi} \int_\mathbb{R} \frac{1 + \zeta t}{(\zeta - t)(t^2 + 1)} \left(\log |f(t)| dt - d\mu_f(t)\right)\right), \quad \zeta \in \mathbb{C}_+,$$

where $\gamma_f \in \mathbb{R}$, $\alpha_f \geq 0$, B_f is the Blaschke product related to f, and μ_f is a positive singular measure such that

$$\int_\mathbb{R} \frac{d\mu_f(t)}{t^2 + 1} < \infty.$$

The number α_f measures the singularity of f at ∞ and is given by

(6.2) $$\alpha_f = -\limsup_{\xi \to +\infty} \xi^{-1} \log |f(i\xi)|,$$

the Blaschke product, B_f, has the form

$$B_f(\zeta) = \prod_{k=1}^\infty \lambda_k \frac{\zeta - \zeta_k}{\zeta - \overline{\zeta_k}}, \quad \text{where } |\lambda_k| = 1, \, \zeta_k \in Z(f) \text{ and } \zeta \in \mathbb{C}_+.$$

Applying the factorization theorem to the functions F_ε and H_ε, one deduces that

(6.3) $$\alpha_{F_\varepsilon} = \alpha_{H_\varepsilon} = 0.$$

In fact, choose $\varepsilon' \in (\varepsilon, \pi/2)$. It will suffice to show that $\alpha_{F_{\varepsilon'}} = 0$ provided $\alpha_{F_\varepsilon} < \infty$. By (6.2), we have

$$\alpha_{F_{\varepsilon'}} = -\limsup_{\xi \to +\infty} \xi^{-1} \log |F_{\varepsilon'}(i\xi)| = -\limsup_{\xi \to +\infty} \xi^{-1} \log \left|F \circ \varphi_{\varepsilon'}^{-1}(i\xi)\right|$$

$$= -\limsup_{\xi \to +\infty} \xi^{-1} \log \left|F \circ \varphi_\varepsilon^{-1} \circ \varphi_\varepsilon \circ \varphi_{\varepsilon'}^{-1}(i\xi)\right|$$

$$= -\limsup_{\xi \to +\infty} \xi^{-1} \log \left|F_\varepsilon\left(i\xi^{\frac{\pi/2 - \varepsilon'}{\pi/2 - \varepsilon}}\right)\right| = -\limsup_{\xi \to +\infty} \xi^{-1 - \frac{\varepsilon' - \varepsilon}{\pi/2 - \varepsilon'}} \log |F_\varepsilon(i\xi)| = 0.$$

Moreover, since F_ε and H_ε are analytic in a neighborhood of $\mathrm{clos}(\mathbb{C}_+) \setminus \{0\}$, we have

$$\mu_{F_\varepsilon} = \beta_{F_\varepsilon} \delta_0,$$

and

$$\mu_{H_\varepsilon} = \beta_{H_\varepsilon} \delta_0,$$

where $\beta_{F_\varepsilon}, \beta_{H_\varepsilon} \geq 0$.

A calculation shows that

$$\beta_{F_\varepsilon} = -\pi \limsup_{\xi \to 0+} \xi \log |F_\varepsilon(i\xi)|$$

and

$$\beta_{H_\varepsilon} = -\pi \limsup_{\xi \to 0+} \xi \log |H_\varepsilon(i\xi)|.$$

In the same fashion as we derived (6.3), we obtain

(6.4) $$\beta_{F_\varepsilon} = \beta_{H_\varepsilon} = 0.$$

By (6.3) and (6.4), we get for $\varepsilon > 0$ and $\zeta \in \mathbb{C}_+$

$$(6.5) \qquad F_\varepsilon(\zeta) = e^{i\gamma_{F_\varepsilon}} B_{F_\varepsilon}(\zeta) \exp\left(\frac{i}{\pi} \int_\mathbb{R} \frac{1 + \zeta t}{(\zeta - t)(t^2 + 1)} \log |F_\varepsilon(t)|\, dt\right),$$

and

$$(6.6) \qquad H_\varepsilon(\zeta) = e^{i\gamma_{H_\varepsilon}} B_{H_\varepsilon}(\zeta) \exp\left(\frac{i}{\pi} \int_\mathbb{R} \frac{1 + \zeta t}{(\zeta - t)(t^2 + 1)} \log |H_\varepsilon(t)|\, dt\right).$$

Recall that $G_\varepsilon = H_\varepsilon / F_\varepsilon$, hence by (6.5) and (6.6), we have

$$(6.7) \qquad \log |G_\varepsilon(\zeta)| = \log \left|\frac{B_{H_\varepsilon}(\zeta)}{B_{F_\varepsilon}(\zeta)}\right| + \frac{\operatorname{Im}\zeta}{\pi} \int_\mathbb{R} \frac{\log |H_\varepsilon(t)| - \log |F_\varepsilon(t)|}{|\zeta - t|^2}\, dt,$$

for $\zeta \in \mathbb{C}_+ \setminus \varphi_\varepsilon(Z)$ and $\varepsilon \in (0, \pi/2)$. Recall that R is a fixed number greater than $|\operatorname{Re} z|$ for all $z \in Z$. Since G is analytic outside the finite set Z, the term $\log |B_{H_\varepsilon}/B_{F_\varepsilon}|$ has only finitely many singularities in $\mathbb{C}_+$ which are all contained in the set $\varphi_\varepsilon(Z)$. Hence, for $\varepsilon \in (0, 1]$,

$$(6.8) \qquad \log |B_{H_\varepsilon}(\zeta)/B_{F_\varepsilon}(\zeta)| \leq C_1,$$

provided $|\zeta| \geq |\varphi_1(R)|$, $\zeta \in \mathbb{C}_+$.

By (6.1) and the definition of φ_ε, we have

$$|H_\varepsilon(t)| \leq M(\varepsilon), \quad t \in \mathbb{R}.$$

An easy calculation then shows

$$(6.9) \qquad \frac{\operatorname{Im}\zeta}{\pi} \int_\mathbb{R} \frac{\log |H_\varepsilon(t)|}{|\zeta - t|^2}\, dt \leq \log M(\varepsilon),$$

for all $\varepsilon \in (0, \pi/2)$. To estimate the second term in (6.7), we observe that

$$\frac{t^2 + 1}{|\zeta - t|^2} \leq \frac{1 + |\zeta|^2}{(\operatorname{Im}\zeta)^2}, \quad \text{for all } t \in \mathbb{R},$$

so that

$$(6.10) \qquad \left|\frac{\operatorname{Im}\zeta}{\pi} \int_\mathbb{R} \frac{\log |F_\varepsilon(t)|}{|\zeta - t|^2}\, dt\right| \leq \frac{1 + |\zeta|^2}{\pi \operatorname{Im}\zeta}\left(\int_\mathbb{R} \frac{\log^+ |F_\varepsilon(t)|}{t^2 + 1}\, dt + \int_\mathbb{R} \frac{\log^- |F_\varepsilon(t)|}{t^2 + 1}\, dt\right),$$

where $\log^+ |F_\varepsilon| = \max(0, \log |F_\varepsilon|)$ and $\log^- |F_\varepsilon| = |\log |F_\varepsilon|| - \log^+ |F_\varepsilon|$. Since

$$(6.11) \qquad \int_\mathbb{R} \frac{\log^+ |F_\varepsilon(t)|}{t^2 + 1}\, dt \leq \pi \log^+ \sup_{z \in S} |F(z)|,$$

we only have to estimate the second integral to the right of (6.10). To do this, we intend to use the subharmonicity of $\zeta \mapsto \log |F_\varepsilon(\zeta)|$. Consider the function $F_0 = F \circ \varphi_0$. Pick $\zeta_0 \in \mathbb{C}_+$ such that $F_0(\zeta_0) \neq 0$. By continuity, we can find $\varepsilon_0 \in (0, \pi/2)$, such that $|F_\varepsilon(\zeta_0)| \geq |F_0(\zeta_0)|/2 > 0$ for all $\varepsilon \in [0, \varepsilon_0)$. Due to subharmonicity (or the factorization in (6.5) above), we have

$$\log |F_\varepsilon(\zeta_0)| \leq \frac{\operatorname{Im}\zeta_0}{\pi} \int_\mathbb{R} \frac{\log |F_\varepsilon(t)|}{|\zeta_0 - t|^2}\, dt,$$

for all $\varepsilon \in [0, \varepsilon_0)$. It follows that we can find a positive constant C not depending on ε such that

$$(6.12) \qquad \int_\mathbb{R} \frac{\log |F_\varepsilon(t)|}{t^2 + 1}\, dt \geq -C > -\infty,$$

for all $\varepsilon \in [0, \varepsilon_0)$. If $\varepsilon_0 > 1$ we are done, otherwise we pick for each $\varepsilon' \in [\varepsilon_0, 1]$ an element $\zeta_{\varepsilon'} \in \mathbb{C}_+$ such that $F_{\varepsilon'}(\zeta_{\varepsilon'}) \neq 0$. For each such ε', we can find an open interval $I_{\varepsilon'} \subset (0, \pi/2)$, containing ε', such that (6.12) holds uniformly on $I_{\varepsilon'}$ with C depending only on ε' and $\zeta_{\varepsilon'}$. Since $[\varepsilon_0, 1] \subset \bigcup_{\varepsilon \in [\varepsilon_0, 1]} I_\varepsilon$, the compactness of $[\varepsilon_0, 1]$ shows that (6.12) holds for all $\varepsilon \in [0, 1]$ with some $C > 0$ not depending on ε. Hence, by (6.11) and (6.12), we have

$$(6.13) \qquad \int_{\mathbb{R}} \frac{\log^- |F_\varepsilon(t)|}{t^2 + 1} \, dt \leq C + \pi \log^+ \sup_{z \in S} |F(z)|.$$

Combining (6.7), (6.8) (6.9), (6.10), (6.11) and (6.13), we get

$$\log |G_\varepsilon(\zeta)| \leq C_1 + \log M(\varepsilon) + C_2 \frac{|\zeta|^2 + 1}{\mathrm{Im}\,\zeta},$$

for all $\varepsilon \in (0, 1]$ and $\zeta \in \mathbb{C}_+$ with $|\zeta| \geq |\varphi_1(R)|$. Going back to the strip S, the result follows. $\qquad \square$

LEMMA 6.3. *Suppose the conditions of Lemma 6.2 are satisfied. Suppose furthermore, that*

$$\int_0^\infty \log \log M(t) \, dt < \infty,$$

that G is analytic in $\mathbb{C} \setminus Z$, and obeys the estimate

$$(6.14) \qquad |G(z)| \leq M(d(z, \partial S)), \quad z \in \mathbb{C} \setminus \mathrm{clos}(S).$$

Then there are positive numbers R and C, such that

$$|G(z)| \leq \exp(C \exp(3|\mathrm{Re}\,z|/2)),$$

for all z with $|z| > R$.

PROOF. Apply Lemma 6.2 with $\varepsilon = \pi/6$. Then we find a constant $C_1 > 0$ such that

$$\log |G(x + iy)| \leq C_1 \exp\left(3|x|/2\right),$$

provided $|y| \leq 3\pi/10$ and $|x|$ is sufficiently large.

Next, we intend use the $\log - \log$ theorem to estimate the absolute value of G on the strips

$$S_+ = \{x + iy : \pi/4 \leq y < \pi/2\},$$

and

$$S_- = \{x + iy : -\pi/2 < y \leq -\pi/4\}.$$

Suppose $x + iy \in S_+ \cup S_-$. If we choose $\varepsilon = (\pi/2 - |y|)/2$, then $\varepsilon \leq \pi/8$, and also

$$\cos \frac{\frac{\pi}{2} y}{\frac{\pi}{2} - \varepsilon} \geq \cos(\pi/2 - \varepsilon) \geq (\pi/2 - |y|)/\pi,$$

which, together with Lemma 6.2, implies

$$(6.15) \qquad \log |G(x + iy)| \leq \log M\left(\frac{\pi/2 - |y|}{2}\right) + C \frac{\exp(4|x|/3)}{\pi/2 - |y|},$$

provided $|y| \in (\pi/4, \pi/2)$ and $|x| \geq R_0$, for some sufficiently large constant R_0; here and in the sequel C denotes some uniform constant independent of any variable at state.

Now, let $n \geq R_0$ be an integer, then for $|x| \in [n, n + 3/2]$, (6.15) implies

(6.16)

$$\log \log |G(x + iy)|^{\exp -\frac{4}{3}n} \leq -\frac{4n}{3} + \log \left(\log M \left(\frac{\pi/2 - |y|}{2} \right) + C \frac{\exp(4|x|/3)}{\pi/2 - |y|} \right)$$

$$\leq C + \log \log M \left(\frac{\pi/2 - |y|}{2} \right) + \log \log \frac{1}{\pi/2 - |y|}.$$

By (6.16), (6.14) and the $\log - \log$ theorem, there is a constant C not depending on n such that

$$|G(x + iy)|^{\exp -4n/3} \leq C,$$

for $x + iy$ with $|x| \in [n, n + 1]$ and $|y| \in (3\pi/10, \pi)$. In conclusion, we have

$$|G(x + iy)| \leq \exp(C \exp(4|x|/3)),$$

for $x + iy$ with $|x| \geq R_0$ and $|y| \in (3\pi/10, \pi)$. Applying (6.14) once more, we see that G is bounded in $\{z \in \mathbb{C} : |\operatorname{Im} z| \geq 2\}$. The assertion follows. $\qquad \square$

LEMMA 6.4. *Let $S = \{z = x + iy \in \mathbb{C} : |y| < \pi/2\}$. Suppose F ($F \not\equiv 0$) is a bounded analytic function on S which is continuous on $\operatorname{clos}(S)$. Suppose H is analytic in S and that*

$$\sup_{x \in \mathbb{R}} |H(x + iy)| < \infty, |y| < \pi/2.$$

If for $z \in S$ with $|\operatorname{Re} z|$ large enough, $G(z) = H(z)/F(z)$ is analytic, and for such z satisfies $|G(z)| \leq \exp(A \exp(\alpha |\operatorname{Re} z|))$, for some numbers $A, \alpha \geq 0$, then for all $\varepsilon > 0$ and all $\delta \in (0, \pi/2)$ there exist numbers $C = C(\varepsilon, \delta)$, $R = R(\varepsilon, \delta)$ such that

(6.17) $$|G(x + iy)| \leq C \exp \left((\delta_\infty^+(F) + \varepsilon) e^x \right), \quad \text{for } x > R,$$

and

(6.18) $$|G(x + iy)| \leq C \exp \left((\delta_\infty^-(F) + \varepsilon) e^{-x} \right), \quad \text{for } x < -R,$$

for all y, with $|y| \leq \pi/2 - \delta$. Here

$$\delta_\infty^+(F) = -\limsup_{x \to \infty} e^{-x} \log |F(x)|$$

and

$$\delta_\infty^-(F) = -\limsup_{x \to \infty} e^{-x} \log |F(-x)|.$$

PROOF. We first investigate $G(z)$ for $\operatorname{Re} z \geq 0$. By the Ahlfors-Heins theorem (see [4, Theorem 7.2.6]), we have

$$\limsup_{x \to \infty} e^{-x} \log |F(x + iy)| = -\delta_\infty^+(F) \cos y,$$

for almost all $y \in (-\pi/2, \pi/2)$. Hence, for every $\varepsilon > 0$ and almost all y, we can find a number $R = R(\varepsilon, y)$ such that

(6.19) $$|G(x + iy)| \leq h(y) \exp \left((\delta_\infty^+(F) + \varepsilon) e^x \right), \quad \text{for all } x > R,$$

where $h(y) = \sup_{x \in \mathbb{R}} |H(x + iy)|$.

Fix $\delta \in (0, \pi/2)$. Choose an increasing sequence $y_1, y_2, \ldots, y_n \in (-\pi/2, \pi/2)$, such that $(-\pi/2 + \delta, \pi/2 - \delta) \subset (y_1, y_n)$, $y_{j+1} - y_j < \pi/2\alpha$ $(j = 1, \ldots, n-1)$, and such that (6.19) holds for each y_j $(j = 1, \ldots, n)$. Put

$$G_j(z) = G(z)/K \exp\left((\delta_\infty^+(F) + \varepsilon)\frac{e^{z-(y_j+y_{j+1})i/2}}{\cos(y_2 - y_1)/2}\right)$$

where $K = \max(h(y_1), h(y_2), \ldots, h(y_n))$.

When $y = y_j$ or $y = y_{j+1}$, we have

$$\operatorname{Re}\frac{e^{x+iy-(y_j+y_{j+1})i/2}}{\cos(y_{j+1} - y_j)/2} = e^x;$$

hence $|G_j(x + iy)| \leq 1$ for $x \geq R$ and $y \in \{y_j, y_{j+1}\}$. Since $|G_j(x + iy)| \leq \exp(A'e^{\alpha x})$, the Phragmén-Lindelöf principle applies to G_j in the half-strip $S_j = \{z = x + iy : x > R, y_j \leq y \leq y_{j+1}\}$ $(j = 1, \ldots, n-1)$.

This proves (6.17); the proof of (6.18) is similar. $\square$

COROLLARY 6.5. *Suppose the conditions of Lemma 6.3 are satisfied. Then for all $\varepsilon > 0$, there exist numbers $C = C(\varepsilon)$, $R = R(\varepsilon)$ such that*

$$|G(z)| \leq C \exp\left((\delta_\infty^+(F) + \varepsilon)e^x\right), \quad \text{for } z = x + iy, \ x \geq 0, \ |z| > R,$$

and

$$|G(z)| \leq C \exp\left((\delta_\infty^-(F) + \varepsilon)e^{-x}\right), \quad \text{for } z = x + iy, \ x < 0, \ |z| > R.$$

The notations $\delta_\infty^+(\cdot)$ and $\delta_\infty^-(\cdot)$ were introduced in Lemma 6.4.

PROOF. According to (6.14), G is bounded in $\{z \in \mathbb{C} : |\operatorname{Im} z| \geq 2\}$. By Lemma 6.3, we have $|G(z)| \leq \exp\left(C \exp(3|\operatorname{Re} z|/2)\right)$, provided $|\operatorname{Re} z|$ is large enough. A proper choice of δ in Lemma 6.4 shows that G satisfies the desired growth when $|\operatorname{Im} z| \leq 1$. Hence, after a normalization, the statement follows from the Phragmén-Lindelöf principle. $\square$

The following generalized Pragmén-Lindelöf type theorem is implicitly contained in [12].

PROPOSITION 6.6. *Let $\theta \geq 0$ be a bounded continuous function on $(0, \infty)$ such that*

$$\int_0^\infty \theta(t)\,dt < \infty.$$

Suppose f is analytic in the domain $\Omega = \{z = x + iy : x > 0, |y| < \pi/2 + \theta(x)\}$, and satisfies

$$\limsup_{\substack{z \to \zeta \\ z \in \Omega}} |f(z)| \leq 1$$

for all $\zeta \in \partial\Omega$. Then

$$|f(z)| \leq 1$$

for all $z \in \Omega$, or there exist a positive constant C such that

$$\sup_{\substack{\operatorname{Re}\zeta = \operatorname{Re} z \\ \zeta \in \Omega}} |f(\zeta)| \geq \exp(Ce^{\operatorname{Re} z}),$$

when the real part of $z \in \Omega$, $\operatorname{Re} z$, is sufficiently large.

PROOF. Suppose there exist some $z_0 = x_0 + iy_0 \in \Omega$ such that $|f(z_0)| > 1$; otherwise we are done.

For $\xi > x_0$ we write $\Theta_\xi = \{z \in \Omega : \operatorname{Re} z = \xi\}$. Let

$$B(\xi) = \sup \{ \log |f(z)| : z \in \Theta_\xi \}.$$

Let $\omega(z, \Theta_\xi)$ be the harmonic measure on $\Omega_\xi = \Omega \cap \{z : \operatorname{Re} z < \xi\}$, with

$$\lim_{\substack{z \to \zeta \\ z \in \Omega_\xi}} \omega(z, \Theta_\xi) = \begin{cases} 0, & \zeta \in \partial\Omega_\xi \setminus \operatorname{clos}(\Theta_\xi), \\ 1, & \zeta \in \Theta_\xi. \end{cases}$$

Then, since $\log |f|$ is subharmonic, we have

$$(6.20) \qquad 0 < \log |f(z_0)| \leq \int_{\partial\Omega_\xi} \log |f(\zeta)| d\omega(z_0, \Theta_\xi) \leq B(\xi)\omega(z_0, \Theta_\xi),$$

where the expression under the integral sign is the upper semicontinuous function defined by f on the boundary of Ω.

According to the first distortion inequality of Ahlfors ([1], [11]), we have the estimate

$$\begin{aligned} \omega(z_0, \Theta_\xi) \quad &\leq \quad \frac{4}{\pi} \arctan \exp \left(4\pi - \pi \int_{x_0}^\xi \frac{dt}{\pi + 2\theta(t)} \right) \\ &\leq \quad C_1 \exp \left(-\int_{x_0}^\xi \frac{dt}{1 + \frac{2}{\pi}\theta(t)} \right) \\ &\leq \quad C_2 \exp \left(-\int_{x_0}^\xi 1 - \theta(t)\, dt \right) \leq C_3 e^{-\xi} \end{aligned}$$

provided $\int_{x_0}^\xi (\pi + 2\theta(t))^{-1} dt > 2$, which is the case when $\xi > 2(\pi + 2\sup\theta) + x_0$. Here C_1, C_2, C_3 are numerical quantities not depending on ξ. Together with (6.20), this shows that

$$B(\xi) \geq C_3^{-1} \log |f(z_0)| e^\xi.$$

The statement now follows from the definition of B. $\qquad \square$

COROLLARY 6.7. *Suppose $M : (0, \infty) \to (e, \infty)$ is a continuously differentiable decreasing function with*

$$\lim_{t \to 0+} t \log \log M(t) < \infty,$$

and

$$\int_0^\infty \log \log M(t)\, dt < \infty.$$

Let $\kappa_0 > 0$ and let $S' = \{z : \operatorname{Re} z > 0, |\operatorname{Im} z| < \pi/2 + \kappa_0\}$. Suppose G is analytic in S' and continuous on $\operatorname{clos}(S')$. Suppose, furthermore, that

$$(6.21) \qquad |G(z)| \leq M(d(z, S)), \quad z \in S' \setminus S,$$

and

$$|G(z)| = O(\exp(\varepsilon e^{\operatorname{Re} z})), \quad \text{as } \operatorname{Re} z \to \infty, z \in S',$$

holds for all $\varepsilon > 0$. Then G is bounded in S'.

PROOF. Without loss of generality we may assume $\lim_{t\to 0^+} M(t) = \infty$. Put $g(z) = G(z) \exp(-2e^{z/2})$.

The function $\lambda = \log\log M$, is strictly decreasing in some finite interval $I = (0, a)$, hence it is invertible on I. Let θ be the decreasing function

$$\theta : x \mapsto \min(\kappa_0, \lambda^{-1}(x/2)).$$

Then $\theta : (0, \infty) \to I$ is continuous, and

$$\int_0^\infty \theta(x)\, dx < \infty.$$

Let

$$\Omega = \{z = x + iy : x > 0,\ |y| \le \pi/2 + \theta(x)\}.$$

Now, $M(\theta(x)) \le \exp(e^{x/2}))$, and hence, since $\theta(x) \to 0$ as $x \to +\infty$, we see that g is bounded on $\partial\Omega$; Proposition 6.6 shows that g is bounded in Ω. It follows that

$$|G(z)| \le \sup_\Omega |g| \exp(2e^{x/2}), \quad \text{for all } z = x + iy \in \Omega.$$

Since M is decreasing, (6.21) shows that

$$|G(z)| \le M(|\operatorname{Im} z| - \pi/2) \le M(\theta(x)) \le \exp(e^{x/2}), \quad \text{for } z \in S' \setminus \Omega.$$

Finally, (6.21) shows that G is bounded in $S' \setminus \{z : |\operatorname{Im} z| < \min(\frac{1}{2}(\pi + \kappa_0), \frac{3}{4}\pi)\}$, and since G satisfies the above estimate, the Phragmén-Lindelöf principle shows that G is bounded in S'. $\qquad\square$

THEOREM 6.8. *Suppose the conditions of Lemma 6.3 hold. Suppose, furthermore, that M satisfies the conditions of Corollary 6.7, and that $\delta_\infty^+(F) = \delta_\infty^-(F) = 0$, where $\delta_\infty^+(\cdot)$ and $\delta_\infty^-(\cdot)$ were introduced in Lemma 6.4.*

Then there exist constants $R, C > 0$, such that

$$|G(z)| \le C, \quad |z| > R.$$

PROOF. By Corollary 6.5 and a proper translate of G, Corollary 6.7 shows that G is bounded when $|\operatorname{Im} z| < 2$ and $\operatorname{Re} z$ is large enough. The same argument applied to $z \mapsto G(-z)$ shows G is bounded when $|\operatorname{Im} z| < 2$ and $-\operatorname{Re} z$ is large enough. Since M is decreasing, (6.14) shows that G is bounded in a neighborhood of ∞. $\qquad\square$

6.2. Singularity at boundary points. Consider a point on ∂S. In order to get the needed estimate of the resolvent transform in a neighborhood of such a point, we consider (as in [2]) the rotation of S, as defined below. The method we use to get the desired estimate is similar to that of the preceding paragraph, although the setting requires slightly different arguments.

Consider the conformal maps $\varphi : z \mapsto ie^z$ and $\varpi : \zeta \mapsto \frac{\zeta - i}{\zeta + i}$, which maps the strip S onto the upper half-plane, and the upper half-plane onto the unit disc $\mathbb{D}$, respectively. Let us further write ϱ for the right angled counterclockwise rotation of $\mathbb{D}$, i.e. $\varrho : w \mapsto iw$. The rotation of S is then defined by

$$\beta = \varphi^{-1} \circ \varpi^{-1} \circ \varrho \circ \varpi \circ \varphi.$$

Here, we choose the branch of φ^{-1} such that $|\operatorname{Im} \beta(z)| < \pi$, for all $z \in \mathbb{C}$.

For $z = x + iy$, we have the explicit formula

$$\beta(z) = \log \frac{ie^z + 1}{e^z + i} = \log \frac{2e^x \cos y + i(e^{2x} - 1)}{2e^x \sin y + e^{2x} + 1},$$

when $x > 0$ we thus have

$$\beta(x + iy) = \begin{cases} \dfrac{1}{2} \log \dfrac{e^{2x} - 2e^x \sin y + 1}{e^{2x} + 2e^x \sin y + 1} + i \arctan \dfrac{\sinh x}{\cos y}, & |y| < \dfrac{\pi}{2}, \\[4mm] \dfrac{1}{2} \log \dfrac{e^{2x} - 2e^x \sin y + 1}{e^{2x} + 2e^x \sin y + 1} + i \left(\dfrac{\pi}{2} - \arctan \dfrac{\cos y}{\sinh x} \right), & \dfrac{\pi}{2} < |y| < \pi. \end{cases}$$

The needed properties of β are summarized in the following lemma.

LEMMA 6.9. *The function β has the following properties:*

(1) *Let $\varepsilon \in (0, \pi/2)$ and let $S_\varepsilon = \{z \in \mathbb{C} : |\mathrm{Im}\, z| \le \pi/2 - \varepsilon\}$. $\beta(S_\varepsilon)$ is convex, and contained in the rhombus with vertices at $\pm \frac{\pi}{2} i$ and with the angle $\pi - 2\varepsilon$ at these corners.*

(2) *There is number $R_0 > 0$, such that*

$$e^{-x} \le \left| \beta(x + iy) - \frac{\pi}{2} i \right| \le 3e^{-x}$$

for all $x \ge R_0$ and all $y \in (-\pi, \pi)$.

(3) *There is a number $\kappa_0 > 0$, such that*

$$\frac{|y - \pi/2|}{2 \sinh x} \le \left| \mathrm{Im}\, \beta(x + iy) - \frac{\pi}{2} \right| \le 2 \frac{|y - \pi/2|}{\sinh x}$$

provided $|y - \pi/2| \le \kappa_0$ and $x > 1$.

OUTLINE OF PROOF. Let us apply β to the boundary lines of S_ε. It is clear that the resulting curves meet at the points $\frac{\pi}{2} i$ and $-\frac{\pi}{2} i$, corresponding to $+\infty$ and $-\infty$ respectively. A calculation shows that tangents meet with the angle $\pi - 2\varepsilon$.

We will now show that $\beta(S_\varepsilon)$ is convex. To this end, we show that u is convex with respect to v, where $u(x) = \mathrm{Re}\, \beta(x + i(\frac{\pi}{2} - \varepsilon))$ and $v = \mathrm{Im}\, \beta(x + i(\frac{\pi}{2} - \varepsilon))$. The second derivative of u with respect to v is given by

$$\frac{d^2 u}{dv^2} = \frac{d}{dv} \left(\frac{du}{dx} \Big/ \frac{dv}{dx} \right) = \frac{d}{dx} \left(\frac{du}{dx} \Big/ \frac{dv}{dx} \right) \Big/ \frac{dv}{dx}.$$

After some computations, we arrive at

$$\frac{d^2 u}{dv^2} = \frac{\cot \varepsilon}{\cosh^2 x} \frac{\cosh^2 x - \cos^2 \varepsilon}{\sin \varepsilon \cosh x},$$

which is positive. The convexity of $\beta(S_\varepsilon)$ now follows from the identity $\beta(z) = -\beta(-z)$. This proves (1) apart from the indicated calculations.

Since, $\lim\limits_{x \to \infty} e^x |\beta(x + iy) - \frac{\pi}{2} i| = 2$, uniformly in y, (2) is immediate.

Similarly, $\lim\limits_{y \to \pi/2} \sinh x (\mathrm{Im}\, \beta(x + iy) - \pi/2)/(y - \pi/2) = 1$, uniformly in x, and hence (3) follows. $\qquad \square$

The next lemma corresponds to Lemma 6.2.

LEMMA 6.10. *Let β be as above, and suppose that the conditions of Lemma 6.2 are fulfilled. Assume furthermore that $F(z) \to 0$ as $|z| \to \infty$ in S, and that*

$$\left|\beta(z) - \frac{\pi}{2}i\right| \left|\beta(z) + \frac{\pi}{2}i\right| M(d(\beta(z), \partial S)) \leq M_0(d(z, \partial S)), \quad z \in S,$$

where M_0 is some function that shares the same properties as M does.

Let $G_0(z) = \left(z - \frac{\pi}{2}i\right)\left(z + \frac{\pi}{2}i\right) G(z)$. Fix $R \in \mathbb{R}$, with $R > \max\{|\mathrm{Re}\,\beta(z)| : z \in Z, z \neq \pm\frac{\pi}{2}i\}$. Then there are constants C_1, C_2, such that for each ε, $\varepsilon \in (0, 1)$,

$$\log|G_0(\beta(x + iy))| \leq C_1 + \log M_0(\varepsilon) + C_2 \frac{\exp(\frac{\pi}{2}|x|/(\frac{\pi}{2} - \varepsilon))}{\cos\frac{\pi}{2}y/(\frac{\pi}{2} - \varepsilon)},$$

for $x + iy \in S$ with $|x| \geq R$ and $|y| \leq \pi/2 - \varepsilon$.

PROOF. This follows from Lemma 6.2. □

REMARK. Since, by Lemma 6.9, β maps $z = x + iy$ into the rhombus with vertices in $\frac{\pi}{2}i$ and $-\frac{\pi}{2}i$ with the angle $|y|$ in these corners, we have

$$\left|\beta(z) - \frac{\pi}{2}i\right| \left|\beta(z) + \frac{\pi}{2}i\right| \leq \frac{d(\beta(z), \partial S)^2}{\cos^2|y|} = \frac{d(\beta(z), \partial S)^2}{\sin^2\left(\frac{\pi}{2} - |y|\right)} \leq \frac{\pi^2}{4} \frac{d(\beta(z), \partial S)^2}{d(z, \partial S)^2}.$$

Hence the conditions on M and M_0 are satisfied when $M(t) \leq C_1/t^2$ and $M_0(t) \geq C_2/t^2$, for some constants $C_1, C_2 > 0$.

LEMMA 6.11. *Suppose the assumptions of Lemma 6.10 hold. Suppose, furthermore, that*

$$\int_0^1 \log\log M_0(t)\, dt < \infty,$$

and that there exists positive numbers R_0 and κ_0, $\kappa_0 < 1$ such that

$$(6.22) \qquad e^{-x} M(d(\beta(x + iy), \partial S)) \leq M_0(d(x + iy, \partial S)),$$

provided $x \geq R_0$ and $\pi/2 < |y| \leq \pi/2 + \kappa_0$. Then there are positive numbers R and C, such that

$$(6.23) \qquad |G_0(\beta(z))| \leq \exp(C\exp(3\,\mathrm{Re}\,z/2)),$$

for all z with $|\mathrm{Im}\,z| \leq \pi/2 + \kappa_0$ and $\mathrm{Re}\,z \geq R$. Moreover,

$$(6.24) \qquad |G_0(\beta(z))| \leq 10 M_0(d(z, \partial S))$$

when $|\mathrm{Im}\,z| \in (\pi/2, \pi/2 + \kappa_0]$ and $\mathrm{Re}\,z \geq R$.

PROOF. Apply Lemma 6.10 with $\varepsilon = \pi/6$; then we find a constant $C_1 > 0$ such that

$$(6.25) \qquad \log|G_0(\beta(x + iy))| \leq C_1 \exp(3|x|/2)$$

provided $|y| \leq 3\pi/10$ and $|x|$ is sufficiently large.

As in the proof of Lemma 6.3, Lemma 6.10 implies

$$(6.26) \qquad \log|G_0(\beta(x + iy))| \leq \log M_0\left(\frac{\pi/2 - |y|}{2}\right) + C\frac{\exp(4|x|/3)}{\pi/2 - |y|}$$

provided $|y| \in (\pi/4, \pi/2)$ and $|x| \geq R_0$, for some sufficiently large constant R_0.

By Lemma 6.9 and (6.22), we have

$$(6.27) \quad |G_0(\beta(x+iy))| \le \left|\beta(x+iy) - \frac{\pi}{2}i\right| \left|\beta(x+iy) + \frac{\pi}{2}i\right| M(d(\beta(x+iy), \partial S))$$

$$\le 10e^{-x} M(d(\beta(x+iy), \partial S)) \le 10 M_0(|y| - \pi/2)$$

provided $\pi/2 < |y| \le \pi/2 + \kappa_0$ and $x > R_1$, for some sufficiently large constant $R_1 > 0$.

Now, let $n \ge \max(R_0, R_1)$ be an integer; then for $x \in [n, n + 3/2]$, (6.26) and (6.27) show that

$$\log\log |G_0(\beta(x+iy))|^{\exp{-4n/3}}$$

$$\le \begin{cases} C + \log\log M_0\left(\dfrac{\pi/2 - |y|}{2}\right) + \log\dfrac{1}{\pi/2 - |y|}, & \dfrac{\pi}{4} < |y| < \dfrac{\pi}{2}, \\[2ex] C + \log\log M_0(|y| - \pi/2), & \dfrac{\pi}{2} < |y| \le \dfrac{\pi}{2} + \kappa_0, \end{cases}$$

By the $\log - \log$ theorem, it follows that

$$|G_0(\beta(x+iy))| \le \exp(C\exp(4x/3)),$$

provided x is large enough and $|y| \in (3\pi/10, \pi/2 + \kappa_0/2)$. Applying (6.27) once more, we see that G_0 is bounded when $|y| \in [\pi/2 + \kappa_0/2, \pi/2 + \kappa_0]$ and x is large enough. Hence the estimate (6.25) holds when x is large enough and $|y| \le \pi/2 + \kappa_0$. $\qquad\square$

REMARK. That condition (6.22) is fulfilled when $M(t) \le C/t$ and $M_0(t) \ge C/t$ follows from Lemma 6.9.

COROLLARY 6.12. *Suppose the assumptions of Lemma 6.11 hold. Then, for all $\varepsilon > 0$, there exist numbers $C = C(\varepsilon)$, $R = R(\varepsilon)$ such that*

$$|G_0(\beta(z))| \le \exp\left((\delta_\infty^+(F \circ \beta) + \varepsilon)e^{\operatorname{Re} z}\right),$$

for all z, with $\operatorname{Re} z > R$ and $|\operatorname{Im} z| \le \pi/2 + \kappa_1$, for some $\kappa_1 > 0$ not depending on ε. The notation $\delta_\infty^+(\cdot)$ is that of Lemma 6.4.

PROOF. Since G_0 satisfies the estimate (6.23), we can apply the proof of Lemma 6.4 (with $H(z)$ replaced $(\beta(z) + \frac{\pi}{2}i)(\beta(z) + \frac{\pi}{2}i)H(\beta(z))$ and F replaced by $F \circ \beta$), to get the desired estimate for z with $|\operatorname{Im} z| \le 1$ and $\operatorname{Re} z > R(\varepsilon)$.

Now, (6.24) shows that $G_0 \circ \beta$ is bounded on the lines $|\operatorname{Im} z| = \frac{\pi}{2} + \kappa_1$ when $\operatorname{Re} z$ is large enough, here we choose $\kappa_1 = \min(\kappa_0, 1/3)$, where κ_0 is the number appearing in Lemma 6.11. Since G_0 satisfies the desired estimate on the lines $|\operatorname{Im} z| = 1$, the choice of κ_1 shows that we can, after a normalization of G_0, apply a Phragmén-Lindelöf argument using (6.23) to achieve the desired estimate. $\qquad\square$

THEOREM 6.13. *Suppose the assumptions of Lemma 6.11 hold. Suppose, furthermore, that M_0 satisfies the conditions of Corollary 6.7, and that $\delta_\infty^+(F \circ \beta) = 0$, where $\delta_\infty^+(\cdot)$ is the quantity introduced in Lemma 6.4.*

Then there is a neighborhood U of the point $\frac{\pi}{2}i$, such that G_0 is bounded in U. In particular, G has at most a simple pole at $\frac{\pi}{2}i$.

PROOF. By Corollary 6.12 and (6.24), G_0 satisfies the conditions of Corollary 6.7. Hence, we can find constants C, R, κ_0 so that $|G_0(\beta(z))| \le C$ when

$z \in U'$, and $U' = \{z : \operatorname{Re} z > R, |\operatorname{Im} z| < \pi/2 + \kappa_0\}$. Now, β maps U' onto a neighborhood U of $\frac{\pi}{2}i$, and we are done.   □

7. The proof of the completeness theorem

We will now continue the discussion of §2 and complete the proof of Theorem 2.1. Let us recall the setting: $g \in L^\infty(G//K, 1/\omega)$ annihilates $I(\mathfrak{S})$, and

$$Z_\infty(I(\mathfrak{S})) = \{s_1, s_2, \ldots, s_n, \infty\}.$$

It is assumed that $s_1, s_2, \ldots, s_n$ lie on the boundary of $\Sigma(\alpha)$. Since $\hat{f}(s) = \hat{f}(-1-s)$, the set $\{s_1, s_2, \ldots, s_n\}$ must be symmetric with respect to the map $s \mapsto -1 - s$. Moreover, it is assumed that

(7.1) $\delta_\infty(\mathfrak{S}) = \delta_{s_j}(\mathfrak{S}) = 0, \quad j = 1, 2, \ldots, n.$

Here $\delta_\infty(\mathfrak{S}) = \inf\{\delta_\infty(f) : f \in \mathfrak{S}\}$, where $\delta_\infty(f)$ is given by (2.1); similarly, $\delta_{s_j}(\mathfrak{S}) = \inf\{\delta_{s_j}(f) : f \in \mathfrak{S}\}$, where $\delta_{s_j}(f)$ is given by (2.2).

That the condition (7.1) is necessary in order to have $I(\mathfrak{S}) = L^1(G//K, \omega)$ follows from the Beurling-Rudin theorem (see [12]). The difficult part of Theorem 2.1 is the sufficiency of (7.1).

The map $\phi_\alpha : z \mapsto i\pi(z + 1/2)/(2\alpha + 1)$ maps $\Sigma(\alpha)^\circ$ onto the strip $S = \{z : |\operatorname{Im} z| < \pi/2\}$. The number $\delta_\infty(f)$ then coincides with the numbers $\delta_\infty^+(\hat{f} \circ \phi_\alpha^{-1})$ and $\delta_\infty^-(\hat{f} \circ \phi_\alpha^{-1})$ considered in §6. Hence, by Theorem 5.2 and Theorem 6.8, we conclude that $\mathfrak{R}[g]$ is bounded in a neighborhood of infinity, and by Theorem 5.2, it has to be zero there (consider the Laurent series of $\mathfrak{R}[g]$). Consider a boundary point s_j which, by symmetry, we may assume to lie on the ray $\alpha + i\mathbb{R}$. If τ_j is the translation $z \mapsto z + \operatorname{Re}(\phi_\alpha(s_j))$ mapping $\frac{\pi}{2}i$ to $\phi_\alpha(s_j)$, then the number $\delta_{s_j}(f)$ corresponds to the number $\delta_\infty^+(\hat{f} \circ \phi_\alpha^{-1} \circ \tau_j \circ \beta)$ of §6, and we can apply Theorem 6.13 to see that $\mathfrak{R}[g]$ must have a simple pole at s_j. The proof is complete.   □

Acknowledgements. The author would like to thank his licentiate thesis advisor, Håkan Hedenmalm, for introducing him to the problem of the paper and also for his careful reading of the manuscript.

References

[1] L. Ahlfors, *Untersuchungen zur Theorie der konformen Abbildungen und der ganzen Funktionen*, Acta Soc. Sci. Fenn. N. S. 1. 9 (1930), 3–40.

[2] Y. Ben Natan, *Wiener's tauberian theorem on $G//K$ and harmonic functions on G/K*, preliminary manuscript, 1999.

[3] Y. Ben Natan, Y. Benyamini, H. Hedenmalm and Y. Weit, *Wiener's tauberian theorem for spherical functions on the automorphism group of the unit disk*, Ark. Mat. **34** (1996), 199–224.

[4] R. P. Jr. Boas, *Entire Functions*, Academic Press, 1954.

[5] A. Borichev and H. Hedenmalm, *Completeness of translates in weighted spaces on the half-line*, Acta Math. **174** (1995), 1–84.

[6] Y. Domar, *On the analytic transform of bounded linear functionals on certain Banach algebras*, Studia Math. **53** (1975), no. 3, 203–224.

[7] L. Ehrenpreis and F. I. Mautner, *Some properties of the Fourier transform on semi-simple Lie Groups I*, Ann. of Math. **61** (1955), 406–439.

[8] L. Ehrenpreis and F. I. Mautner, *Some properties of the Fourier-transform on semisimple Lie Groups III*, Trans. Amer. Math. Soc. **90** (1959), 431–484.

[9] A. Erdélyi, W. Magnus, F. Oberhettinger and F. G. Tricomi, *Higher Transcendental Functions. Vol. I*, Robert E. Krieger Publishing Co., 1981.

[10] V. P. Gurariĭ, *Harmonic analysis in spaces with weight*, Trudy Moskov. Mat. Obšč. **35** (1976), 21–76. (English translation in Trans. Moscow Math. Soc. **35** (1979), 21–75.)

[11] K. Haliste, *Estimates of harmonic measures*, Ark. Mat. **6** (1965), 1–31.

[12] H. Hedenmalm, *On the primary ideal structure at infinity for analytic Beurling algebras*, Ark. Mat. **23** (1985), no. 1, 129–158.

[13] H. Hedenmalm, *Translates of functions of two variables*, Duke Math. J. **58** (1989), no. 1, 251–297.

[14] S. Helgason, *Groups and Geometric Analysis*, Pure and Appl. Math., vol. 113, Academic Press, Orlando, Fla., 1984.

[15] P. Koosis, *Introduction to H_p Spaces*, London Math. Soc. Lecture Notes, vol. 40, Cambridge Univ. Press, Cambridge-New York, 1980.

[16] P. Koosis, *The Logarithmic Integral*. I, Cambridge Stud. in Adv. Math., vol. 12, Cambridge Univ. Press, Cambridge-New York, 1988.

[17] S. Lang, $SL_2(R)$, GTM, vol. 105, Springer-Verlag, 1985.

[18] N. N. Lebedev, *Special Functions and their Applications*, Dover Publ., New York, 1972.

[19] H. Leptin, *Ideal theory in group algebras of locally compact groups*, Invent. Math. **31** (1975/76), no. 3, 259–278.

[20] L. Loomis, *An Introduction to Abstract Harmonic Analysis*, D. Van Nostrand Company, Princeton, N.J., 1953.

[21] F. W. J. Olver, *Asymptotics and Special Functions*, Academic Press, New York and London, 1974.

[22] H. Reiter, *Classical Harmonic Analysis and Locally Compact Groups*, Clarendon Press, Oxford, 1968.

[23] M. Sugiura, *Unitary Representations and Harmonic Analysis, an Introduction*, North-Holland/Kodansha, 1990.

[24] E. T. Whittaker and G. N. Watson, *A Course of Modern Analysis*, Cambridge Math. Lib., Cambridge Univ. Press, Cambridge, 1996.

CENTRE FOR MATHEMATICAL SCIENCES, MATHEMATICS (FACULTY OF SCIENCE), UNIVERSITY OF LUND, BOX 118, S-221 00 LUND, SWEDEN
E-mail address: Anders.Dahlner@math.lu.se

Contemporary Mathematics
Volume **404**, 2006

Domination on Sets and in Norm

Walter K. Hayman

Dedicated to Boris Korenblum on his 80th birthday

Much of the work described in this survey was suggested by Boris Korenblum or arose out of joint work with him. Collaboration with Boris has given me great happiness and it is a pleasure to pay tribute to him in this survey.

1. Introduction

Suppose that B is a space of analytic functions f in the unit disk

$$\Delta = \{z \in \mathbb{C} : |z| < 1\}$$

with norm $\|f\|_B$, where $\mathbb{C}$ denotes the set of all complex numbers. Let E be a subset of Δ and let f, g be two functions in B such that

$$(1) \qquad |f(z)| \leq |g(z)|, \qquad z \in E.$$

In [**3**] Danikas and I ask two questions.

    1. What conditions on E ensure that (1) implies

$$(2) \qquad \|f\|_B \leq \|g\|_B?$$

      In this case we call E a "dominating set".

    2. If E is not dominating, can (1) still imply some restrictions on $\|f\|_B$?

2. The case $B = H^p$

Following Bonsall [**1**], we call E nontangentially dense (n.t.d.) if and only if almost every point ζ of the unit circle $\partial\Delta$ is a nontangential limit point of E. In this case there exists, for almost every ζ on $\partial\Delta$, a sequence $\{z_n\}$ in E such that

$$z_n \to \zeta \quad \text{and} \quad |z_n - \zeta| = O(1 - |z_n|)$$

as $n \to \infty$.

The following result is proved by Brown, Shields, and Zeller [**2**] if $p = \infty$, and by Danikas and Hayman [**3**] for general p.

THEOREM 1. *The set E is dominating for H^p, where $0 < p \leq \infty$, if and only if E is nontangentially dense.*

2000 *Mathematics Subject Classification.* Primary 30D55.

It is almost immediate that n.t.d. implies domination. In fact, if $f \in H^p$, it follows from a classical theorem of Fatou that, for almost all θ, the nontangential limit $f(e^{i\theta})$ exists. Further,

$$\|f\|_p = \sup_\theta |f(e^{i\theta})|$$

for $p = \infty$, and

$$\|f\|_p = \left[\frac{1}{2\pi} \int_0^{2\pi} |f(e^{i\theta})|^p \, d\theta \right]^{\frac{1}{p}}$$

for $0 < p < \infty$. If E is n.t.d. and $|f(z)| \leq |g(z)|$ on E, we deduce by taking limits that

$$|f(e^{i\theta})| \leq |g(e^{i\theta})|$$

for almost all θ, and hence $\|f\| \leq \|g\|$ in H^p.

The reverse implication takes a little more effort. We use the following result of Bonsall [1].

THEOREM 2. *The following two conditions are equivalent.*

(i) *E is n.t.d.*
(ii) *For every function $h(z)$ bounded and harmonic in Δ*

$$\sup_{z \in E} h(z) = \sup_{z \in \Delta} h(z).$$

Suppose now that E is not n.t.d. Then there exists a bounded harmonic function $h(z)$ such that

$$a = \sup_{z \in E} h(z) < b = \sup_{z \in \Delta} h(z).$$

Making a linear transformation if necessary, we assume that $a = 0$, $b = 1$ and so

$$h(z) \leq 0 \ (z \in E), \quad \sup_{z \in \Delta} h(z) = 1, \quad \inf_{z \in \Delta} h(z) = -m,$$

where $m > 0$.

Let $\hat{h}(z)$ be a harmonic conjugate of $h(z)$. We write

$$H(z) = h(z) + i \, \hat{h}(z).$$

Then

$$\left| e^{H(z)} \right| = e^{h(z)} \leq e, \qquad z \in \Delta,$$

so that $h(z)$ has angular boundary values almost everywhere on $\partial\Delta$. Also, since $h(\rho e^{i\phi})$ converges boundedly to $h(e^{i\phi})$, $h(z)$ is the Poisson integral of its boundary values,

$$(3) \qquad\qquad h(z) = \frac{1}{2\pi} \int_0^{2\pi} \frac{(1 - |z|^2) h(e^{i\theta}) \, d\theta}{|e^{i\theta} - z|^2}, \qquad z \in \Delta.$$

We now suppose that $g \in H^p$ and define

$$g_\alpha(z) = g(z) e^{\alpha H(z)}, \qquad z \in \Delta,$$

where $\alpha > 0$, and proceed to show that, if α is suitably chosen, $\|g_\alpha\|$ becomes arbitrarily large. Assuming this, let M be a nonnegative number and choose α such that $\|g_\alpha\| > M$. Multiplying by the constant $c = M/\|g_\alpha\|$, we see that $f = c g_\alpha$ satisfies

$$|f(z)| = |c g_\alpha(z)| \leq |g(z)|, \qquad z \in E,$$

but $\|f\| = M$. Thus in this case E is not dominating and the answer to question 2 is No, since M is arbitrary, subject to $M \geq 0$.

It remains to show that $\|g_\alpha\|$ becomes large if α is large. Suppose first that $p = \infty$. Then, since $\sup h > 0$, it follows from (3), that $h(e^{i\theta}) > 0$ on a set of positive measure. Also, by a theorem of F. and M. Riesz [9], $|g(e^{i\theta})| > 0$ almost everywhere. We now choose θ such that

$$|g(e^{i\theta})| > 0, \qquad h(e^{i\theta}) > 0,$$

and the nontangential limits exist at $e^{i\theta}$. Then

$$g_\alpha(e^{i\theta}) = g(e^{i\theta}) \exp\left[\alpha h(e^{i\theta})\right] \to \infty$$

as $\alpha \to \infty$, so that

$$\|g_\alpha\| = \sup_\theta |g_\alpha(e^{i\theta})| \to \infty$$

as $\alpha \to \infty$.

If $p < \infty$, we choose a set E of positive measure m on which

$$h(e^{i\theta}) > \eta > 0, \qquad |g(e^{i\theta})| > \epsilon > 0.$$

We can do this since $|g| > 0$ almost everywhere and $h > 0$ on a set of positive measure. Now

$$\|g_\alpha\| \geq \left[\frac{1}{2\pi} \int_E |g_\alpha(e^{i\theta})|^p \, d\theta\right]^{\frac{1}{p}} \geq \left[\frac{m}{2\pi}(\epsilon e^{\alpha\eta})^p\right]^{\frac{1}{p}} \to \infty$$

as $\alpha \to \infty$, and the conclusion follows again.

3. The Disk Algebra

We recall that the disk algebra A consists of all functions f analytic in Δ and continuous in $\overline{\Delta}$. Also,

$$\|f\|_A = \sup_{z \in \Delta} |f(z)| = \sup_{\zeta \in \partial\Delta} |f(\zeta)|$$

by the maximum principle. The following result is proved in [3].

THEOREM 3. *The set E is dominating in A if and only if the closure $\overline{E}$ of E contains $\partial\Delta$. If E is not dominating and $g \neq 0$, then for every positive c, there exists f in A such that $|f(z)| \leq |g(z)|$ in E and $\|f\| = c$.*

Thus the answer to question 2 is again No.

Evidently if $\partial\Delta \subset \overline{E}$ and $|f(z)| \leq |g(z)|$ on E, then $|f(z)| \leq |g(z)|$ on $\partial\Delta$, so that $\|f\| \leq \|g\|$.

Suppose contrary to this that $z_0 \in \partial\Delta$ and that z_0 is not a limit point of E. Thus

$$\sup_{z \in E} |z + z_0| = m < 2.$$

In fact, the supremum is attained for some z in $\overline{E}$ and so in $\overline{\Delta}$, but $|z + z_0| < 2$ in Δ except for $z = z_0$. Consider now

$$f(z) = g_N(z) = g(z)\left[(z + z_0)/m\right]^N,$$

where N is a positive integer. Then $|f(z)| \leq |g(z)|$ on E. However, we can find z in Δ such that $|z + z_0| > m$ and $g(z) \neq 0$. Then $g_N(z) \to \infty$ as $N \to \infty$ and Theorem 3 is proved, for we may choose

$$f(z) = cg(z)/\|g_N\|,$$

where $\|g_N\| > c$.

4. Bergman Spaces

Finally we consider the space B_p of functions f analytic in Δ and such that

$$\|f\|_{B_p} = \left[\frac{1}{\pi} \int_\Delta |f(z)|^p \, dA(z) \right]^{\frac{1}{p}} < \infty,$$

where dA is area measure on Δ. In this case it is much harder to characterise dominating sets. Consider the following example due to Korenblum [6]. We define

$$f(z) = 1, \quad g(z) = az, \quad E_a = \{z : 1/a < |z| < 1\},$$

where $1 < a < \sqrt{2}$. Then

$$|f(z)| \le |g(z)|, \qquad z \in E_a,$$

but in B_2

$$\|g\|^2 = a^2/2 < 1 = \|f\|^2.$$

Thus E_a is not dominating in this range. An example due to Rainer Martin (see p. 486 of [6]) shows that E_a is not dominating if a is slightly larger than $\sqrt{2}$. However, we shall show at the end of this section that the answer to our second question is Yes for E_a whenever $a > 1$.

Korenblum conjectured that if a is sufficiently large then E_a is dominating. This was proved by Korenblum, O'Neil, Richards, and Zhu in [7] if one of f and g is a monomial and $a = \sqrt{3}$. The conclusion was also proved by Korenblum and Richards in [8] under a separation condition. I proved Korenblum's conjecture in [4] with $a = 25$ when $p = 2$, and Hinkkanen [5] proved it for $1 \le p < \infty$ and $c = 0.15724$, where $c = 1/a$. I would like at this point to thank Boris for years of collaboration and stimulating suggestions on this and other matters.

The following is a sketch of the argument in [4].

LEMMA 4. *Suppose that $\omega(z)$ is analytic in the annulus $e^{-2} < |z| < 1$ and satisfies $|\omega(z)| < 1$ there. Suppose further that*

$$\sup\{|\omega(z)| : |z| = 1/e\} = 1 - 2\delta,$$

where $0 < \delta < 1/2$. Then for $e^{-3/2} < |z| < e^{-1/2}$ we have

  (i) $|\omega(z)| < 1 - \delta$.
  (ii) $|\omega(z) - e^{i\lambda}| < A_0\delta$, *where λ is real and A_0 is a positive absolute constant.*

Unfortunately the constant A_0, on which c depends, appears to be rather large.

We note that (i) is a consequence of Hadamard's convexity theorem for the maximum modulus, since

$$(1 - 2\delta)^{1/2} < 1 - \delta.$$

The proof of (ii) is a little more complicated and we refer the reader to [4]. We proceed to deduce Korenblum's conjecture from Lemma 4 with

$$c = 1/a = \left[3(e^2 - 1)/(4e^3 A_0) \right]^{1/2}.$$

Suppose then that

$$|f(z)| \le |g(z)| \quad \text{for} \quad c < |z| < 1,$$

where $c \le e^{-2}$. We apply Lemma 4 to the function
$$\omega(z) = f(z)/g(z).$$
If
$$\sup\{|\omega(z)| : |z| = 1/e\} = 1,$$
then ω is a unimodular constant, say $e^{i\lambda}$, and $f = ge^{i\lambda}$, and there is nothing to prove. So we may assume that $\delta > 0$ in Lemma 4.

We note that
$$\int_\Delta (|f(z)|^2 - |g(z)|^2)\, dA(z) \le I_1 + I_2,$$
where
$$I_1 = \int_{e^{-3/2}<|z|<e^{-1/2}} (|f(z)|^2 - |g(z)|^2)\, dA(z)$$
and
$$I_2 = \int_{|z|<c} (|f(z)|^2 - |g(z)|^2)\, dA(z).$$
We define $r_0 = e^{-3/2}$ and deduce from Lemma 4 and $0 < \delta \le 1/2$ that

$$
\begin{aligned}
-I_1 &\ge \int_{e^{-3/2}}^{e^{-1/2}} r\, dr \int_0^{2\pi} (|g(re^{i\theta})|^2 - |f(re^{i\theta})|^2)\, d\theta \\
&\ge (1 - (1-\delta)^2) \int_{e^{-3/2}}^{e^{-1/2}} r\, dr \int_0^{2\pi} |g(re^{i\theta})|^2\, d\theta \\
&\ge \frac{3}{2}\delta \cdot \frac{1}{2}\left(\frac{1}{e} - \frac{1}{e^3}\right) \int_0^{2\pi} |g(r_0 e^{i\theta})|^2\, d\theta.
\end{aligned}
$$

(4)

We need to obtain an upper bound for I_2. In polar coordinates we have
$$I_2 = \int_0^c r\, dr \int_0^{2\pi} (|f(re^{i\theta})|^2 - |g(re^{i\theta})|^2)\, d\theta.$$
We obtain for $0 \le r \le c$, integrating from 0 to 2π,

$$
\begin{aligned}
\int (|f(re^{i\theta})|^2 - |g(re^{i\theta})|^2)\, d\theta &\le \int |g(re^{i\theta})^2 e^{2i\lambda} - f(re^{i\theta})^2|\, d\theta \\
&= \int |g(re^{i\theta})e^{i\lambda} - f(re^{i\theta})||g(re^{i\theta})e^{i\lambda} + f(re^{i\theta})|\, d\theta \\
&\le \left[\int |g(re^{i\theta})e^{i\lambda} - f(re^{i\theta})|^2\, d\theta \int |g(re^{i\theta})e^{i\lambda} + f(re^{i\theta})|^2\, d\theta\right]^{1/2} \\
&\le \left[\int |g(r_0 e^{i\theta})e^{i\lambda} - f(r_0 e^{i\theta})|^2\, d\theta \int |g(r_0 e^{i\theta})e^{i\lambda} + f(r_0 e^{i\theta})|^2\, d\theta\right]^{1/2}.
\end{aligned}
$$

The last inequality holds since
$$\int_0^{2\pi} |\phi(re^{i\theta})|^2\, d\theta$$
is an increasing function of r whenever ϕ is analytic in Δ.

We now apply (ii) and deduce that, when $|z| = r_0$,
$$|g(z)e^{i\lambda} - f(z)| = |g(z)(e^{i\lambda} - \omega(z))| < A_0\delta|g(z)|,$$
and
$$|g(z)e^{i\lambda} + f(z)| = |g(z)(e^{i\lambda} + \omega(z))| \le 2|g(z)|.$$

Thus we obtain finally

$$\int_0^{2\pi} (|f(re^{i\theta})|^2 - |g(re^{i\theta})|^2)\, d\theta \le 2A_0\delta \int_0^{2\pi} |g(r_0e^{i\theta})|^2\, d\theta.$$

Hence

(5) $$I_2 \le A_0\delta c^2 \int_0^{2\pi} |g(r_0e^{i\theta})|^2\, d\theta.$$

Comparing (4) and (5), we see that

$$I_1 + I_2 \le 0,$$

if

$$A_0\delta c^2 = 3\delta(e^2 - 1)/(4e^3),$$

or

$$c = \left[3(e^2 - 1)/(4e^3 A_0)\right]^{1/2}.$$

Thus Korenblum's conjecture holds with this value of c.

We now consider our second question. In Bergman spaces, unlike the cases considered previously, we can give a positive answer for the sets E_α although, as we saw earlier, these sets are not dominating when $a \le \sqrt{2}$.

We write $c = 1/a$ and suppose that

$$|f(z)| \le |g(z)| \quad \text{for} \quad c < |z| < 1.$$

This implies that

$$I_p(f, \rho) = \left[\frac{1}{2\pi} \int_0^{2\pi} |f(\rho e^{i\theta})|^p\, d\theta\right]^{1/p} \le I_p(g, \rho)$$

for $c \le \rho < 1$. Also, if $r < c < \rho$, we have

$$I_p(f, r) \le I_p(f, c) \le I_p(g, c) \le I_p(g, \rho).$$

Thus

$$\frac{1}{2\pi} \int_0^c \int_0^{2\pi} |f(re^{i\theta})|^p r\, dr\, d\theta = \int_0^c I_p(f, r)^p r\, dr$$

$$\le I_p(g, c)^p \int_0^c r\, dr = (c^2/2) I_p(g, c)^p;$$

and

$$\|g\|_{B_p}^p \ge 2 \int_c^1 I_p(g, \rho)^p \rho\, d\rho \ge (1 - c^2) I_p(g, c)^p.$$

We deduce that

$$\|f\|_{B_p}^p = 2 \left[\int_0^c + \int_c^1\right] I_p(f, r)^p r\, dr$$

$$\le c^2 I_p(g, c)^p + \|g\|_{B_p}^p \le \left(1 + \frac{c^2}{1 - c^2}\right) \|g\|_{B_p}^p,$$

that is,

$$\|f\|_{B_p} \le (1 - c^2)^{-1/p} \|g\|_{B_p}.$$

But E_a is not dominating if a is close to 1.

For most other sets we do not know, whether they are dominating in Bergman spaces. For instance, can a set be dominating if the complement has limit points on $\partial\Delta$? In particular, can a sequence $\{z_n\}$ tending to the boundary ever be a dominating set? The latter question was raised by N. Dudley-Ward.

Acknowledgement. I am greatly indebted to the editor, Professor Alexander Borichev, for a careful reading of the typescript, which resulted in numerous corrections. He also supplied the following positive answer to my first question above.

Let us choose a family of disjoint discs D_n in the unit disc D accumulating to the unit circle, and denote by D_n^* concentric discs of radii, say, $1/25$ that of D_n. Put $E = D \setminus \cup D_n^*$. Then the closure of the complement of E contains the whole unit circle. On the other hand, the set E is dominating: if $|f| \leq |g|$ on E, then

$$\int_D |f(z)|^2 dA(z) = \int_{D \setminus \cup D_n} |f(z)|^2 dA(z) + \sum_n \int_{D_n} |f(z)|^2 dA(z).$$

Applying the above results to $f|D_n$, $g|D_n$, for every n, we obtain that

$$\int_{D_n} |f(z)|^2 dA(z) \leq \int_{D_n} |g(z)|^2 dA(z).$$

Since $|f| \leq |g|$ on $D \setminus \cup D_n$, we conclude that

$$\int_D |f(z)|^2 dA(z) \leq \int_D |g(z)|^2 dA(z).$$

References

[1] F. F. Bonsall, *Domination of the supremum of a bounded harmonic function by its supremum over a countable set*, Proc. Edinburgh Math. Soc. **30** (1987), 471–477.

[2] L. Brown, A. Shields and K. Zeller, *On absolutely convergent exponential sums*, Trans. Amer. Math. Soc. **96** (1960), 162–183.

[3] N. Danikas and W. K. Hayman, *Domination on sets and in H^p*, Results Math. **34** (1998), 85–90.

[4] W. K. Hayman, *On a conjecture of Korenblum*, Analysis **19** (1999), 195–205.

[5] A. Hinkkanen, *On a maximum principle in Bergman space*, J. Analyse Math. **79** (1999), 335–344.

[6] B. Korenblum, *A maximum principle for the Bergman space*, Publ. Math. **35** (1991), 479–486.

[7] B. Korenblum, R. O'Neil, K. Richards and K. Zhu, *Totally monotone functions with applications in the Bergman space*, Trans. Amer. Math. Soc. **337** (1993), 795–806.

[8] B. Korenblum and K. Richards, *Majorization and domination in the Bergman space*, Proc. Amer. Math. Soc. **117** (1993), 153–158.

[9] F. Riesz and M. Riesz, *Über die Randwerte einer analytischen Funktion*, C.R. 4e Congr. Math. Scand. Stockholm (1916), 27–44.

DEPARTMENT OF MATHEMATICS, IMPERIAL COLLEGE LONDON, SOUTH KENSINGTON CAMPUS, LONDON SW7 2AZ, UK

Contemporary Mathematics
Volume **404**, 2006

Extensions of the Asymptotic Maximum Principle

Charles Horowitz and Bernard Pinchuk

ABSTRACT. A function f analytic in the unit disc U is called radially asymptotically bounded if

$$\varlimsup_{r \to 1} |f(re^{i\theta})| \leq M < \infty, \quad 0 \leq \theta < 2\pi.$$

In [**3**], B. Korenblum showed that this condition implies that $|f(z)| \leq M$ in U if f satisfies an appropriate à priori growth restriction. In this paper, we replace radial asymptotic boundedness by angular or nontangential asymptotic boundedness, and determine the sharp growth rate which ensures the validity of the maximum principle in each case.

1. Introduction

B. Korenblum [**3**] has recently proved an asymptotic maximum principle for analytic functions in a disc. He considered functions f analytic in the unit disc U such that for every $\theta \in [0, 2\pi)$

$$(1.1) \qquad \varlimsup_{r \to 1} |f(re^{i\theta})| \leq M < \infty.$$

He proved that if it is known à priori that

$$(1.2) \qquad \log^+ |f(z)| = o\left(\frac{1}{(1-|z|)^2}\right), \quad |z| \to 1,$$

then (1.1) implies that $|f(z)| \leq M$ throughout the disc. On the other hand, the functions

$$(1.3) \qquad f(z) = \exp\left[ci\left(\frac{1+z}{1-z}\right)^2\right], \quad c > 0$$

indicate that the asymptotic maximum principle fails if we replace little o by big O in (1.2).

In this note, we extend Korenblum's result in various ways. In Section 2, we consider angular boundedness and, in Section 3, nontangential boundedness in place of the radial condition (1.1).

We thank Prof. Korenblum for many helpful suggestions regarding this paper.

2000 *Mathematics Subject Classification*. Primary 30C80, 30D40.

This research is partially supported by the Gelbart Research Institute at Bar-Ilan University, Israel.

2. Angular boundedness conditions

DEFINITION 2.1. For each point ξ on the unit circle $\mathbb{T}$ and for $0 < \alpha < \pi/2$, let

$$(2.1) \qquad \Gamma_\alpha(\xi) = \{z \in U : -\alpha \le \arg(1 - \bar{\xi}z) \le \alpha\}$$

be the standard Stolz angle or cone of opening 2α at ξ.

THEOREM 2.1. *Suppose f is analytic in U and for a certain $\alpha \in (0, \pi/2)$ and for every $\xi \in \mathbb{T}$*

$$(2.2) \qquad \varlimsup_{\substack{z \to \xi \\ z \in \Gamma_\alpha(\xi)}} |f(z)| \le M < \infty.$$

Then if

$$(2.3) \qquad \log^+ |f(z)| = o\left(\frac{1}{(1 - |z|)^{2\pi/(\pi - 2\alpha)}}\right); \quad |z| \to 1,$$

it follows that $|f(z)| \le M$ in the entire disc. However, there exist unbounded functions f analytic in U such that (2.2) is satisfied and

$$(2.4) \qquad \log^+ |f(z)| = O\left(\frac{1}{(1 - |z|)^{2\pi/(\pi - 2\alpha)}}\right).$$

PROOF. Assuming (2.2) and (2.3), we choose any number $A > M$. By a Baire category argument as in [**3**], we find that the open set

$$(2.5) \qquad \begin{aligned} G = \{\xi \in \mathbb{T} : \ & \xi \text{ has a neighborhood } N_\xi \text{ such that in} \\ & N_\xi \cap U \ |f(z)| < A\} \end{aligned}$$

is dense in $\mathbb{T}$. It is easily seen that our theorem will be proved if we can establish the following claim:

$$(2.6) \qquad F \equiv \mathbb{T} \setminus G \text{ is empty.}$$

We proceed to prove (2.6) by contradiction. Assuming that F is nonempty, we choose $\tilde{A}$ satisfying $M < \tilde{A} < A$. By another Baire category argument as in [**3**], we find that there must be some open arc $I \subset \mathbb{T}$ and a number r, $0 < r < 1$ such that $F \cap I \ne \emptyset$ and for every $\xi \in F \cap I$, $|f(z)| \le \tilde{A}$ in the domain

$$\Gamma_\alpha(\xi) \cap \{r \le |z| < 1\}.$$

It now follows that for some $B > \tilde{A}$

$$(2.7) \qquad |f(z)| \le B \text{ for all } z \in \Gamma_\alpha(\xi), \ \xi \in K \equiv (\bar{I} \setminus I) \cup (I \cap F).$$

$\square$

CLAIM 2.1. *$|f(z)| \le B$ in the entire sector S of U abutting on I.*

PROOF OF CLAIM. Note that $\bar{I} \setminus K$ is an open subset of $\mathbb{T}$ which can be expressed as a union of disjoint open arcs $\bar{I} \setminus K = \bigcup\limits_{n=1}^{\infty} (e^{ia_n}, e^{ib_n})$. We define S_n to be the closed sector of U abutting on (e^{ia_n}, e^{ib_n}). To establish Claim 2.1, it certainly suffices to prove that $|f(z)| \le B$ in each S_n. Now in $\Gamma_\alpha(e^{ia_n})$ and $\Gamma_\alpha(e^{ib_n})$, we already have this estimate. Furthermore, the arc (e^{ia_n}, e^{ib_n}) belongs to the set G defined in (2.5). Now for an arbitrary fixed n, we conclude as in [**3**] that there exists a smooth curve $\rho = \varphi_n(\theta)$, $a_n \le \theta \le b_n$, joining e^{ia_n} and e^{ib_n} with $|\varphi'_n(\theta)| \le 1/2$ such that $|f(re^{i\theta})| \le A$ for $a_n \le \theta \le b_n$ and $\varphi_n(\theta) \le r < 1$. To conclude that $|f(z)| \le B$ for all $z \in S_n$, we need the following Phragmén-Lindelöf type Lemma:

LEMMA 2.1. *Let $J = \{z \in \mathbb{T} : a \leq \arg z \leq b\}$ be an arc, and S the closed sector of U abutting on J. Let $\tilde{S} \subset S$ be a region bounded by J and the two segments*

$$R^+ = \{z \in U : \arg(1 - e^{-ib}z) = \alpha\}$$

and

$$R^- = \{z \in U : \arg(1 - e^{-ia}z) = -\alpha\}.$$

Let further $\Gamma \subset \tilde{S}$ be a region bounded by J and a smooth curve γ whose polar equation is $r = \varphi(\theta)$, $a \leq \theta \leq b$, $\varphi(a) = \varphi(b) = 1$. Suppose that v is a subharmonic function on S continuous on $S \setminus J$ such that

$$v(z) \leq K \quad on \quad \Gamma \cup (S \setminus \tilde{S})$$

$$\varlimsup_{\substack{|z| \to 1 \\ z \in S}} (1 - |z|)^{2\pi/(\pi - 2\alpha)} v(z) = 0,$$

then

$$\sup_{z \in S} v(z) \leq K.$$

Before proving the Lemma, we explain how it enables us to complete the proof of Theorem 2.1. Applying Lemma 2.1 with $v(z) = \log|f(z)|$ to each region S_n, together with the known boundedness of f in the cones $\Gamma_\alpha(e^{ia_n})$ and $\Gamma_\alpha(e^{ib_n})$, we conclude that $|f(z)| \leq B$ in the entire sector S. In particular, f has nontangential limits $f^*(z)$ almost everywhere on the frontier ∂S, and can be represented in S as a Poisson-type integral of its boundary values. Now (1.1) implies that $|f^*(z)| \leq M$ almost everywhere on the interval I. Also for some $r_0 \in (0, 1)$, we have $|f(z)| < \tilde{A} < A$ for all z on the boundary rays of S such that $|z| > r_0$. It follows immediately $|f(z)| < A$ in a neighborhood of I. In particular, $F \cap I = \emptyset$ contrary to our assumption. This contradiction proves Theorem 2.1. □

It remains only to establish Lemma 2.1. To that end, we follow the proof of the lemma in Section 3 of [**3**] with appropriate modifications.

It will be convenient to apply a logarithmic map that takes the sector S to the half-strip $\{\operatorname{Im} z \geq 0, -1 \leq \operatorname{Re} z \leq 1\}$. An elementary computation shows that the image of the region $\tilde{S}$ is included in the open triangle Ω with vertices at $(\pm 1, 0)$ and $(0, \cot\alpha)$. Lemma 2.1 will then follow from the following:

LEMMA 2.2. *Let Ω denote an open triangle with vertices at $(\pm 1, 0)$ and $(0, \operatorname{ctg}\alpha)$. Let $\ell^\pm$ denote the boundary segments of Ω from $(\pm 1, 0)$ to $(0, \operatorname{ctg}\alpha)$, excluding their base points, and assume that $V : \Omega \to [-\infty, \infty)$ is subharmonic in Ω and extends to be continuous on $\ell^- \cup \ell^+ \cup \Omega$. Furthermore, assume*

(2.8)   *there is a region Γ between the segment $[-1, 1]$ and a smooth curve $y = \varphi(x)$, $x \in [-1, 1]$ with $\varphi(-1) = \varphi(1) = 0$ and $\varphi(x) > 0$ for $x \in (-1, 1)$,*

such that

(2.9) $$\sup\{V(z) : z \in \ell^- \cup \ell^+ \cup \Gamma\} = M < \infty,$$

(2.10) $$\varlimsup_{y \to 0} y^{2\pi/(\pi - 2\alpha)} V(x + iy) = 0, \quad x + iy \quad in \quad \Omega.$$

Then $\sup\limits_{z \in \Omega} V(z) = M$.

PROOF. We first modify V by subtracting a harmonic function which is positive in Ω. Set $p = \frac{2\pi}{\pi - 2\alpha}$ and define

(2.11)
$$\psi(x, y) = -\operatorname{Im}(z + 1)^{-p} \quad (z = x + iy).$$

Thus ψ is harmonic and positive in Ω and vanishes on $[-1, 1]$ and ℓ^-. Similarly,

$$\psi(-x, y) = -\operatorname{Im}(1 - \bar{z})^{-p}$$

is harmonic and positive on Ω and vanishes on ℓ^+ and $[-1, 1]$. For $a > 0$, define

$$V_a(z) = V(z) - a(\psi(x, y) + \psi(-x, y)).$$

Fix $\beta \in (\alpha, \pi/2)$ and define a subregion

$$\Omega_1 = \{z = x + iy \in \Omega : y \geq \min((1 - x)\operatorname{ctg}\beta, (1 + x)\operatorname{ctg}\beta)\}.$$

Now from (2.10) and (2.11), we easily conclude that V_a is bounded on the boundary segments of Ω_1. Since the base angles in Ω_1 are less than $\pi/2 - \alpha$, the condition (2.10) allows us to apply the classical Phragmén-Lindelöf theorem to conclude that $V_a(z)$ is bounded above in Ω_1. It is also continuous on $\overline{\Omega}_1$ except perhaps at ± 1. To make it continuous on $\overline{\Omega}_1$, we adjust as follows: for $\varepsilon_1 > 0$, we define

$$V_{a,\varepsilon_1}(z) = V_a(z) + \log\left|\frac{z + 1}{z + 1 + \varepsilon_1}\right| + \log\left|\frac{1 - z}{1 - z + \varepsilon_1}\right|$$

which is continuous on $\overline{\Omega}_1$ with value $-\infty$ at ± 1.

The next step requires that the curve $\gamma : y = \varphi(x)$ described in (2.8) should satisfy $|\varphi'(x)| \leq \frac{1}{2}\operatorname{ctg}\beta$ (which is possible since φ can be replaced by any smaller function). With this assumption made, we see that $\Gamma \cap \Omega_1 = \emptyset$ and also that every straight line of slope m, $|m| \geq \operatorname{ctg}\beta$ intersects the boundary of the region Ω_2 lying between Ω_1 and Γ in at most two points.

Now let $0 < \varepsilon_2 < 1$ and define

$$V_{a,\varepsilon_1,\varepsilon_2}(z) = V_{a,\varepsilon_1}(z) - a(\psi(x - \varepsilon_2, y) + \psi(-\varepsilon_2 - x, y))$$

which is subharmonic in a triangular region with base $[-1 + \varepsilon_2, 1 - \varepsilon_2]$ and base angles $\pi/2 - \alpha$.

We now apply the classical maximum principle to $V_{a,\varepsilon_1,\varepsilon_2}$ in the region Ω_{ε_2} whose boundary $\partial\Omega_{\varepsilon_2}$ is outlined below, where an additional angle β_1, satisfying $\alpha < \beta_1 < \beta$ has been employed.

We describe the various parts of $\partial\Omega_{\varepsilon_2}$:

$B_1 = $ the segment of γ between the lines $\lambda_{\varepsilon_2}^+ : y = (1 - \varepsilon_2 - x)\operatorname{ctg}\beta_1$

and $\lambda_{\varepsilon_2}^- : y = (x + 1 - \varepsilon_2)\operatorname{ctg}\beta_1$.

$B_2 = $ the segments of $\lambda_{\varepsilon_2}^{\pm}$ which lie in Ω_2

$B_3 = $ parts of $\ell_{\varepsilon_2}^+ : y = (1 - \varepsilon_2 - x)\operatorname{ctg}\alpha$

and $\ell_{\varepsilon_2}^- : y = (x + 1 - \varepsilon_2)\operatorname{ctg}\alpha$ which lie in Ω_1.

$B_4 = $ the two segments of $\partial\Omega_1$ lying between $\ell_{\varepsilon_2}^-$

and $\lambda_{\varepsilon_2}^-$ and between $\ell_{\varepsilon_2}^+$ and $\lambda_{\varepsilon_2}^+$, respectively.

We thus obtain that for z in Ω_{ε_2} $V_{a,\varepsilon_1,\varepsilon_2}(z)$ is bounded by its maximum on the above boundary segments. Now let $\varepsilon_2 \to 0$ holding ε_1 and a fixed. Thus B_2 is drawn to the x-axis so that by (2.10)

$$V(z) - a(\psi(x - \varepsilon_2, y) + (\psi(\varepsilon_2 - x, y))$$

tends to $-\infty$ uniformly on B_2. It follows immediately that also $V_{a,\varepsilon_1,\varepsilon_2}$ tends to $-\infty$ uniformly on this part of the boundary. Furthermore, as $\varepsilon_2 \to 0$ the segments of B_4 are drawn to the points ± 1 along $\partial \Omega_1$. Here $V_{a,\varepsilon_1}(z) \to -\infty$ and so also $V_{a,\varepsilon_1,\varepsilon_2} \to -\infty$. Furthermore, the maximum of $V_{a,\varepsilon_1,\varepsilon_2}$ on B_3 tends to its maximum on $\ell^+ \cup \ell^-$ by continuity on $\overline{\Omega}_1$ and this maximum is bounded by $\max\{V(z) : z \in \ell^+ \cup \ell^-\}$. Similarly, the values of $V_{a,\varepsilon_1,\varepsilon_2}$ on B_1 certainly remain under $\max\limits_{z \in \gamma} V(z)$. Finally, we observe that as $\varepsilon_2 \to 0$ the region Ω_{ε_2} expands to include all of $\Omega \setminus \Gamma$, and in this region $\lim\limits_{\varepsilon_2 \to 0} V_{a,\varepsilon_1,\varepsilon_2}(z) = V_{2a,\varepsilon_1}(z)$. It follows that for all $z \in \Omega \setminus \Gamma$

$$V_{2a,\varepsilon_1}(z) \leq \max_{\xi \in \ell^+ \cup \ell^- \cup \gamma} V(\xi).$$

Since the right side is independent of a and ε_1, we can let them both tend to zero to conclude

$$\sup_{z \in \Omega \setminus \Gamma} V(z) \leq \sup_{\xi \in \ell^+ \cup \ell^- \cup \gamma} V(\xi)$$

from which we immediately obtain

$$\sup_{z \in \Omega} V(z) \leq M,$$

completing the proof of Lemma 2.2 and the positive assertion of Theorem 2.1.

In order to show that Theorem 2.1 is sharp, we need to construct an appropriate example, analogous to that in [3] where the function

$$f(z) = \exp\left[i\left(\frac{1+z}{1-z}\right)^2\right]$$

was used. The obvious generalization

$$g(z) = \exp\left[i\left(\frac{1+z}{1-z}\right)^{2\pi/(\pi - 2\alpha)}\right]$$

will not work since this g is not bounded in the Stolz angle $\Gamma_\alpha(1)$.

However, a suitable example can be constructed as follows: for $0 < q < 2$, we consider the Mittag-Leffler functions

$$E_q(z) = \sum_{n=0}^{\infty} \frac{z^n}{\Gamma(1 + qn)}.$$

The essential properties of E_q are as follows (see [2], pp. 272-275):

(2.12) E_q is an entire function of order $\dfrac{1}{q}$.

(2.13) In the sector $\dfrac{1}{2}q\pi \leq |\arg z| \leq \pi$, E_q is uniformly bounded.

(2.14) In the sector $|\arg z| \leq \dfrac{1}{2}q\pi$, $|E_q(z) - \dfrac{1}{q}e^{z^{1/q}}| < \dfrac{M}{|z|}$.

Returning to our given α in Theorem 2.1, we choose q so that $(2-q)\pi = \frac{3\pi}{2} + \alpha$; i.e., $q = \frac{\pi/2 - \alpha}{\pi}$. Then by inversion and rotation, we obtain a function

$$(2.15) \qquad h(z) = E_q\left(e^{\frac{i\pi q}{2}}\frac{1}{z}\right)$$

which is bounded in the sector

$$q\pi = \frac{\pi}{2} - \alpha \leq \arg z \leq 2\pi.$$

We claim that $h(z)$ gives the desired example in the upper half plane (UHP). To see this, note that h is continuous and uniformly bounded, say by M, on the real axis, except at 0. Thus if $\xi \in \mathbb{R}$, $\xi \neq 0$

$$\varlimsup_{\substack{z \to \xi \\ z \in \Gamma_\alpha(\xi) \subset \text{UHP}}} |h(z)| \leq M.$$

Moreover, h is bounded in $\Gamma_\alpha(0)$ by the essential part of the construction. In summary, h is uniformly bounded asymptotically in every Stolz angle $\Gamma_\alpha(\xi)$, $\xi \in \mathbb{R}$. Furthermore, we claim that in the UHP

$$\log^+ |h(z)| = O(\text{Im } z)^{-2\pi/(\pi - 2\alpha)} = O(\text{Im } z)^{-1/q}$$

(cf. (2.10)). This can be seen as follows: from (2.14) and (2.15) we deduce that when $0 < \arg z < q\pi = \frac{\pi}{2} - \alpha$ (i.e., in the region where h is unbounded)

$$|h(z)| = \left| E_q\left(e^{\frac{i\pi q}{2}}\frac{1}{z}\right) \right| = \left| \frac{1}{q}\exp\left(e^{\frac{i\pi q}{2}}\frac{1}{z}\right)^{1/q} \right| + O(1).$$

Therefore for z near 0 in this region,

$$(2.16) \qquad \log^+ |h(z)| = \text{Re}\left(i\,\frac{1}{z^{1/q}}\right) + O(1) \leq c\left|\frac{1}{z^{1/q}}\right| \leq c\,\frac{1}{(\text{Im } z)^{1/q}}.$$

Wherever $(\text{Im } z)^{1/q}$ is of the same order of magnitude as $|z|^{1/q}$, we obtain a lower bound of the same type for $\log |h(z)|$. In particular, $h(z)$ is unbounded in the UHP and indeed provides an appropriate example in that region.

In order to transfer this example to U, we choose $C > 0$ and a mapping $\varphi : U \to \text{UHP}$ defined by $\varphi(z) = i\,\frac{1-z}{1+z} - C$. Then we let

$$f(z) = h(\varphi(z)).$$

This f becomes unbounded only near the point ξ_0 in ∂U where $\varphi(\xi_0) = 0$. One sees that for C sufficiently large (e.g., $C > \tan\alpha$), the two intersection points of $\partial\Gamma_\alpha(\xi_0)$ with ∂U are both mapped by φ to the negative real axis. It then follows that for such C, $\omega = \varphi(z)$ maps $\Gamma_\alpha(\xi_0)$ into the region

$$\pi/2 - \alpha \leq \arg \omega \leq 2\pi$$

where h is bounded. From this discussion, together with the known properties of h, we conclude that f satisfies condition (2.2) although it is not bounded in U.

The precise growth rate of f can be derived from the estimate (2.16):

$$\log^+ |f(z)| = \log^+ |h(\varphi(z)| \leq c\,\frac{1}{[\text{Im } \varphi(z)]^q} \leq c\,\frac{1}{(1-|z|)^q} = c\,\frac{1}{(1-|z|)^{2\pi/(\pi - 2\alpha)}}$$

as required in (2.4). Thus f provides an appropriate example as described in the statement of Theorem 2.1. $\qquad\qquad\square$

We note that Theorem 2.1 involves Stolz angles, which are not preserved under conformal mappings. However with slight modification, we can prove a related theorem involving a conformally invariant condition.

THEOREM 2.2. *Suppose that f is analytic in U and for a certain $\alpha \in (0, \pi/2)$, the following condition holds:*

(2.17)
    For every sequence $\{z_n\}$ in U which tends to a boundary point ξ
    in such a way that $\varlimsup\limits_{n\to\infty} |\arg(1 - \bar{\xi}z_n)| \leq \alpha$, $\varlimsup\limits_{n\to\infty} |f(z_n)| \leq M < \infty.$

If moreover for every $\varepsilon > 0$

(2.18)
$$\log^+ |f(z)| = O\left(\frac{1}{(1 - |z|)^{2\pi/(\pi - 2\alpha) + \varepsilon}}\right) \text{ as } |z| \to 1,$$

then $|f(z)| \leq M$ throughout U.

PROOF. First we note that for any $M' > M$, the condition of (2.17) implies that for each $\xi \in \partial U$ there is a number $\theta_\xi > \alpha$ such that

$$\varlimsup_{\substack{z \to \xi \\ z \in \Gamma_{\theta_\xi}(\xi)}} |f(z)| \leq M'.$$

To see this, assume the contrary. Then there must be a sequence of Stolz angles at ξ which shrink down to $\Gamma_\alpha(\xi)$ in which f is not asymptotically bounded by M'. It follows immediately that there is a sequence $\{z_n\}$ in U tending to ξ at an angle of α with the radius to ξ such that $|f(z_n)| > M'$ for all n. This contradicts (2.17), and establishes our claim.

We can now proceed as in the proof of Theorem 2.1. Given $A > M$, we obtain an open dense subset $G \subset \partial U$ such that each $\xi \in G$ has a full neighborhood N_ξ such that $|f(z)| \leq A$ in $N_\xi \cap U$. Defining $F = \partial U \setminus G$, it suffices to prove that F is empty. If not then we can reason as in the proof of Theorem 2.1 to find an interval $I \subset \partial U$ which intersects F and a number $B > A$ such that

$$|f(z)| \leq B \text{ for all } z \in \Gamma_{\theta_\xi}(\xi), \ \xi \in (\bar{I} \setminus I) \cup (I \cap F).$$

From here the proof is reduced to the verification of a Phragmén-Lindelöf type Lemma similar to Lemma 2.1 except that this time the base angles of the relevant sector are $\theta_{e^{ia}}$ and $\theta_{e^{ib}}$, both greater than α, in place of α in Lemma 2.2. Now it is easy to see that in this context condition (2.18) is sufficient to prove the boundedness of f in the given sector. The conclusion of the proof is by an elementary potential theory argument as in Theorem 2.1. □

3. Nontangential Boundedness Conditions

In this section, we consider functions f analytic in U such that

for every $\xi \in \partial U$ and for every $\alpha \in (0, \pi/2)$

(3.1)
$$\varlimsup_{\substack{z \to \xi \\ z \in \Gamma_\alpha(\xi)}} |f(z)| \leq M < \infty.$$

Now if for some $p > 0$

(3.2)
$$\log^+ |f(z)| = O\left(\frac{1}{(1 - |z|)^p}\right) \quad \text{as } |z| \to 1,$$

then Theorem 2.1 implies that $|f(z)| \leq M$ throughout U. We wish to investigate the sharpness of this result.

As a first example of a function satisfying (3.1) (in a half plane) yet which is unbounded, consider the function

$$(3.3) \qquad f(z) = \int_0^\infty \frac{e^{(z-2\pi i)t}}{t^t}\, dt.$$

In [4], Newman showed that f is an entire function which is bounded by 1 outside of the half strip $S = \{z : \mathrm{Re}\, z > 0,\ \pi \leq \mathrm{Im}\, z \leq 3\pi\}$. Now define $h(z) = f\left(-\frac{1}{z}\right)$. Then $|h(z)| \leq 1$ unless $-\frac{1}{z} \in S$, i.e., unless z is in the crescent shaped region between two half circles tangent to the real axis at 0. Thus for every real $x_0 \neq 0$ $\lim\limits_{z \to x_0} h(z)$ exists and is bounded by 1. Furthermore, if $\{z_n\}$ is any sequence in the UHP which tends to zero in some Stolz angle $\Gamma_\alpha(0)$,

$$\varlimsup_{n \to \infty} |h(z_n)| \leq 1.$$

Yet h is clearly unbounded in the UHP.

We can estimate the rate of growth of f defined in (3.3) as follows. Clearly, f grows most rapidly on the half line $z = x + 2\pi i$, $x > 0$. There

$$f(x + 2\pi i) = \int_0^\infty \frac{e^{xt}}{t^t}\, dt.$$

By a simple exercise, for all $x, t > 0$

$$\frac{e^{xt}}{t^t} \leq e^{e^x} e^{-t}$$

from which we conclude that when $x > 0$

$$(3.4) \qquad 0 < f(x + 2\pi i) \leq e^{e^x}.$$

By a more delicate analysis, $f(x + 2\pi i)$ is actually of order of magnitude $e^{\frac{x-1}{2}} e^{e^{x-1}}$ as $x \to +\infty$, but (3.4) will suffice for our purposes.

We can construct similar examples of slower growth rate by composing f with other functions. For any φ analytic in the UHP, $f \circ \varphi$ provides a function analytic in the UHP and unbounded only in $\varphi^{-1}(S)$. For $\varepsilon > 0$, we consider

$$(3.5) \qquad \varphi(z) = \varphi_\varepsilon(z) = (\log z)^{1+\varepsilon}$$

where we prescribe $0 \leq \arg z \leq \pi$ in the UHP to define both $\log z$ and $(\log z)^{1+\varepsilon}$.

We wish to know for which $z \in$ UHP, $\varphi(z) \in S$. So we assume that $z \in$ UHP is such that

$$\varphi(z) = \omega = u + iv \in S.$$

By (3.5), this means that

$$z = e^{\omega^p} \quad \text{where} \quad p = \frac{1}{1+\varepsilon} < 1.$$

Now in S, $v = \mathrm{Im}\, \omega$ is bounded and so $\omega \to \infty$ in S only when $u \to \infty$. As u tends to infinity

$$\arg z = \mathrm{Im}\, \omega^p = |\omega|^p \sin(p \arg \omega) \approx u^p \cdot \frac{pv}{u}$$

which tends to zero since $p < 1$. It follows that $f(\varphi(z))$ is unbounded as $z \to \infty$ only on paths "tangential at ∞" to the positive x-axis. Finally, let

$$g(z) = f\left(\varphi\left(-\frac{1}{z}\right)\right).$$

By the above construction, for all $x_0 \neq 0$ in $\mathbb{R}$,

$$(3.6) \qquad \lim_{\substack{z \to x_0 \\ z \in \text{UHP}}} g(z) \text{ exists and is bounded by 1};$$

$$(3.7) \qquad \varlimsup_{\substack{z \to 0 \\ \text{nontangentially} \\ z \in \text{UHP}}} |g(z)| \leq 1.$$

Still g is unbounded in the UHP as $z \to 0$ on the image of the paths described above. Specifically, if $\omega = u + 2\pi i \in S$, $u \gg 0$, let

$$\log^{1+\varepsilon}\left(-\frac{1}{z}\right) = \omega$$

so that

$$z = -e^{-\omega^p} \qquad \left(p = \frac{1}{1+\varepsilon}\right).$$

Reasoning as before, we find that

$$y = \operatorname{Im} z \approx e^{-u^p} \cdot pvu^{p-1}.$$

In view of (3.4), it follows that as $u \to +\infty$ and therefore $z \to 0$ in this manner

$$(3.8) \qquad \log \log |g(z)| \leq u \approx \log^{1+\varepsilon}\left(\frac{1}{y}\right).$$

One easily sees that (3.8) characterizes the maximal growth rate of g.

It remains only to transfer this example to the unit disc. To that end, we define a function $G(z)$ on U by

$$G(z) = g\left(i\,\frac{1-z}{1+z}\right).$$

Regarding G, the condition (3.1) follows immediately from (3.6) and (3.7) since these conditions are conformally invariant. Furthermore, for $z \in U$ near the singular point 1

$$\log \log |G(z)| = \log \log g\left(i\,\frac{1-z}{1+z}\right) \leq c \log^{1+\varepsilon}\left(\frac{1}{\operatorname{Im} i\left(\frac{1-z}{1+z}\right)}\right)$$

$$= c \log^{1+\varepsilon}\left(\frac{|1+z|^2}{1-|z|^2}\right) \leq c \log^{1+\varepsilon}\left(\frac{1}{1-|z|}\right).$$

We can summarize the results of this section in the following theorem:

THEOREM 3.1. *Let f be analytic in U and satisfy* (3.1). *Then if*

$$\log^+ \log^+ |f(z)| = O\left(\log \frac{1}{1-|z|}\right) \qquad as \quad |z| \to 1,$$

$|f(z)| \leq M$ *in U. However for every $\varepsilon > 0$, there exist unbounded analytic functions in U which satisfy* (3.1) *with growth rate*

$$\log^+ \log^+ |f(z)| = O\left[\log^{1+\varepsilon}\left(\frac{1}{1-|z|}\right)\right] \qquad as \quad |z| \to 1.$$

Authors' note. After this paper was completed, we were informed of the work of Berman and Cohn in [1] which contains many of our results, in a more general format. However, their methods are quite different from ours.

References

[1] R. Berman and W. Cohn, *Pragmen-Lindelof theorems for subharmonic functions on the unit disk*, Math. Scand. **62** (1988), 269–293.

[2] L. Bieberbach, *Lehrbuch der Funktionentheorie*, Bd II, 2nd ed., Chelsea, New York, 1945.

[3] B. Korenblum, *Asymptotic Maximum Principle*, Ann. Acad. Sci. Fenn. Math. **27** (2002), 249–255.

[4] D. J. Newman, *An entire function bounded in every direction*, Amer. Math. Monthly **83** (1976), no. 3, 192–193.

DEPARTMENT OF MATHEMATICS, BAR-ILAN UNIVERSITY, 52900 RAMAT GAN, ISRAEL
E-mail address: horowitz@macs.biu.ac.il

DEPARTMENT OF MATHEMATICS, BAR–ILAN UNIVERSITY, RAMAT–GAN, ISRAEL AND DEPARTMENT OF COMPUTER SCIENCE AND MATHEMATICS, NETANYA ACADEMIC COLLEGE, NETANYA, ISRAEL
E-mail address: pinchuk@macs.biu.ac.il

Contemporary Mathematics
Volume **404**, 2006

Singularity Resolution of Weighted Bergman Kernels

Stefan Jakobsson

ABSTRACT. In this paper, we study the singularity of the weighted Bergman kernel K_ω for a domain $\Omega \subset \mathbb{C}$ with C^∞-smooth boundary and a weight ω which is a strictly positive and C^∞-smooth function on the closure of Ω. Let $\mathbb{C}_+$ denote the upper half plane. Suppose that γ is a sub-arc of $\partial\Omega$ and let ψ be a conformal map from $\mathbb{C}_+ \cap N$ to $\Omega \cap N_\gamma$, where N is a neighborhood of an interval $I \subset \mathbb{R}$ and N_γ is a neighborhood of γ, so that ψ maps I onto γ. Then there exist functions $\Psi_1, \Psi_2 \in C^\infty((\overline{\mathbb{C}_+} \cap N)^2)$ such that the Bergman kernel satisfies

$$K_\omega(\psi(z), \psi(\zeta)) = \frac{\Psi_1(z,\zeta)}{(z-\bar\zeta)^2} + \Psi_2(z,\zeta) \log\left(\frac{i}{z-\bar\zeta}\right), \qquad z, \zeta \in \overline{\mathbb{C}_+} \cap N.$$

All derivatives of Ψ_2 on the diagonal set in $I \times I$ are determined by local properties ω at this point, and can be calculated explicitly. The term $\Psi_1(z,\zeta)/(z-\bar\zeta)^2$ is of global nature except for its singular part which involves ω and its first derivative at the corresponding boundary point. We also show that the singularity of the harmonic Bergman kernel is given by

$$Q_\omega(z,\zeta) \equiv 2\operatorname{Re} K_\omega(z,\zeta) \mod C^\infty(\overline\Omega \times \overline\Omega)$$

under the same conditions as above. The proof uses a weighted version of the Friedrichs operator.

1. Introduction

Let Ω be a domain in the complex plane $\mathbb{C}$, and let ω be a continuous and strictly positive function on Ω. We will call ω a *weight* on Ω. The weighted Bergman space $A^2(\Omega, \omega)$ consists of all analytic functions f on Ω such that

$$(1.1) \qquad \|f\|_\omega^2 = \int_\Omega |f(z)|^2 \omega(z)\, dA(z) < +\infty,$$

where

$$dA(z) = dxdy, \qquad z = x + iy.$$

Under the above conditions on the weight, $A^2(\Omega, \omega)$ is a Hilbert space of analytic functions on Ω, with the inner product

$$\langle f, g \rangle_\omega = \int_\Omega f(z)\overline{g(z)}\omega(z)\, dA(z), \qquad f, g \in A^2(\Omega, \omega).$$

2000 *Mathematics Subject Classification.* Primary 30A31, 35S15; Secondary 46E20.

If $\{e_n\}_{n=0}^{\infty}$ is an orthonormal basis in $A^2(\Omega, \omega)$, we define the weighted Bergman kernel as

$$K_\omega(z, \zeta) = \sum_{n=0}^{\infty} e_n(z)\overline{e_n(\zeta)}, \qquad z, \zeta \in \Omega.$$

It can be shown that the sum converges uniformly on compact subsets of $\Omega \times \Omega$ [3, p. 8], and that

$$f(z) = \langle f, K_\omega(\cdot, z)\rangle_\omega = \int_\Omega K_\omega(z, \zeta)f(\zeta)\omega(\zeta)\, dA(\zeta), \qquad z \in \Omega,$$

for all $f \in A^2(\Omega, \omega)$. For this reason, K_ω is also called the reproducing kernel for $A^2(\Omega, \omega)$. The reproducing property determines the Bergman kernel completely, and therefore K_ω does not depend on the particular orthonormal basis chosen. Clearly, the Bergman kernel depends on both the domain and the weight but to avoid cumbersome notation we only indicate the dependence on the weight.

For the case $\Omega = \mathbb{D}$, the unit disk, and ω a continuous weight on $\overline{\mathbb{D}}$, then the Bergman kernel satisfies

$$(1.2) \qquad K_\omega(z, z) = \frac{1}{\pi}\frac{1}{\omega(z)}\frac{1}{(1 - |z|^2)^2} + o\left(\frac{1}{(1 - |z|^2)^2}\right), \qquad \text{as } |z| \to 1.$$

This was, for example, shown in [12], but was probably known before. For instance, in 1965 L. Hörmander proved the corresponding result for the Bergman kernel on a strongly pseudoconvex domain in $\mathbb{C}^n$. The main purpose of this paper is to offer a full analysis of the singularity of the Bergman kernel at the boundary. It is well-known that if both $\partial\Omega$ and the weight are C^∞-smooth, then $K_\omega \in C^\infty(\overline{\Omega} \times \overline{\Omega} \setminus \triangle(\partial\Omega))$, where $\triangle(\partial\Omega) = \{(z, z) : z \in \partial\Omega\}$ is the boundary diagonal. It is therefore enough to study the singularity of the Bergman kernel when both variables are close to the boundary and to each other. We shall always assume that the boundary of Ω is of class C^∞ and that the weight $\omega \in C^\infty(\overline{\Omega})$ is strictly positive. To describe the singularity of the analytic Bergman kernel near a sub-arc γ of the boundary $\partial\Omega$, it is advantageous to make a change of variables. Let ψ be a conformal map from $\mathbb{C}_+ \cap N$ to $\Omega \cap N_\gamma$, where N is a neighborhood of an interval $I \subset \mathbb{R}$ and N_γ is a neighborhood of γ, so that ψ maps I onto γ. Let

$$\omega_\psi(z) = \omega(\psi(z))|\psi'(z)|^2, \qquad z \in \overline{\mathbb{C}_+} \cap N.$$

THEOREM 1.1. *There exist two functions $\Psi_1, \Psi_2 \in C^\infty((\overline{\mathbb{C}_+} \cap N)^2)$, such that the weighted Bergman kernel satisfies*

$$(1.3) \qquad K_\omega(\psi(z), \psi(\zeta)) = \frac{\Psi_1(z, \zeta)}{(z - \bar{\zeta})^2} + \Psi_2(z, \zeta)\log\left(\frac{i}{z - \bar{\zeta}}\right), \qquad z, \zeta \in \overline{\mathbb{C}_+} \cap N.$$

Moreover, $\bar{\partial}_z\Psi_j(z, \zeta)$ and $\partial_\zeta\Psi_j(z, \zeta)$ vanish to infinite order at $z = \zeta \in I$, $j = 1, 2$. We have

$$\frac{\Psi_1(z, \zeta)}{(z - \bar{\zeta})^2} \equiv \frac{1}{\pi}\frac{1}{\omega_\psi(\zeta)}\frac{-1}{(z - \bar{\zeta})^2} + \frac{1}{\pi}\frac{\partial\omega_\psi(\zeta)}{\omega_\psi(\zeta)^2}\frac{1}{z - \bar{\zeta}} \quad \mathrm{mod}\ C^\infty((\overline{\mathbb{C}_+} \cap N) \times I),$$

and there exist functions $\{\gamma_k\}_{k=0}^{\infty}$ on I such that for every integer $n \geq 1$, we have

$$\Psi_2(z, \zeta)\log\left(\frac{i}{z - \bar{\zeta}}\right) \equiv \frac{1}{\pi}\sum_{k=0}^{n}\frac{i^k(z - \bar{\zeta})^k\gamma_k(\zeta)}{k!}\log\left(\frac{i}{z - \bar{\zeta}}\right)$$

$$\mathrm{mod}\ C^{n-1}((\overline{\mathbb{C}_+} \cap N) \times I).$$

For each integer k, the function $\gamma_k(\zeta)$, $\zeta \in I$, is completely determined by the derivatives of the weight ω_ψ at ζ of order at most $k+2$. In particular, the following formula holds:

$$\gamma_0(\zeta) = \frac{\Delta \log \omega_\psi(\zeta)}{\omega_\psi(\zeta)}, \qquad \zeta \in I.$$

In Section 4, we give formulas for calculating $\gamma_k(\zeta)$ for arbitrary positive integers k.

Our result is related to C. Fefferman's singularity resolution for the Bergman kernel in several complex variables for a smooth strongly pseudoconvex domain [9]. Any such domain can be represented as $D = \{z : \psi(z) > 0\} \subset \mathbb{C}^n$, where ψ is a C^∞-smooth plurisubharmonic function which has $\psi' \neq 0$ on the boundary. Fefferman obtained the following expansion of the Bergman kernel:

$$K_D(z, z) = \frac{\phi(z)}{\psi(z)^{n+1}} + \widetilde{\phi}(z) \log \psi(z), \qquad z \in D,$$

where $\phi, \widetilde{\phi} \in C^\infty(\overline{D})$. The function ϕ can be extended to a function $\phi_2 \in C^\infty(\overline{D} \times \overline{D})$ such that $\phi_2(z, z) = \phi(z)$, and analogously for $\widetilde{\phi}$ and ψ, and such that

$$(1.4) \qquad K_D(z, w) = \frac{\phi_2(z, w)}{\psi_2(z, w)^{n+1}} + \widetilde{\phi}_2(z, w) \log \psi_2(z, w)$$

holds for all $z, w \in D$ provided that $|z - w| < \varepsilon$ and $\mathrm{dist}(z, \partial D) < \varepsilon$, with $\varepsilon > 0$ sufficiently small. It is clear that Theorem 1.1 and Fefferman's result have many common features. In fact, we can write the weighted Bergman kernel exactly as in formula (1.4). Moreover, the singular part of the Bergman kernel for D is completely determined by local properties of the boundary, in analogy with Theorem 1.1.

In addition to Fefferman's original proof, there is the alternative proof by Boutet de Monvel–Sjöstrand [7] (see also [2]). In a simplified form, Fefferman's idea is to approximate the domain D at a boundary point to the third order by a domain for which it is possible to give an explicit formula for the Bergman kernel, and then expand K_D in a Neumann series in terms of the known Bergman kernel. The proof of Boutet de Monvel–Sjöstrand uses instead Fourier integral operators and techniques from the $\bar{\partial}$-Neumann problem. In a later work [6], Boutet de Monvel presents an algorithm to compute the singularity of the Bergman kernel. We should also mention D. Catlin's paper [8] where Fefferman's result is generalized to Bergman kernels for holomorphic vector bundles over bounded strictly pseudoconvex manifolds with smooth boundary. In some sense, this result covers Theorem 1.1, but our method is more direct in this special case, and it is also easier to find the explicit expansion.

In the proof of Theorem 1.1, we apply pseudodifferential operators to a boundary value problem. In Section 6, we discuss a different approach to this result which is closer to Fefferman's idea; that is, we expand the weighted Bergman kernel as a Neumann series in terms of known operators.

1.1. Harmonic Bergman kernels. Let $HL^2(\Omega, \omega)$ be the space of all harmonic functions on Ω which meet the integrability condition (1.1). This space also has a reproducing kernel, the harmonic Bergman kernel Q_ω. For the harmonic Bergman kernel, we have the following result.

THEOREM 1.2. *Suppose that the domain Ω has C^∞-smooth boundary, and that $\omega \in C^\infty(\overline{\Omega})$ is strictly positive. Then the harmonic Bergman kernel satisfies*

(1.5) $$Q_\omega(z, \zeta) \equiv 2 \operatorname{Re} K_\omega(z, \zeta) \mod C^\infty(\overline{\Omega} \times \overline{\Omega}).$$

The proof relies on the fact that the space of harmonic functions with well-defined harmonic conjugates on Ω is spanned by the subspaces of analytic and anti-analytic functions, respectively. If Ω is simply connected, then the orthogonal projection onto $HL^2(\Omega, \omega)$ can be decomposed into a sum involving the orthogonal projections onto the subspaces of analytic and anti-analytic functions. All terms in this sum except two (which are the analytic and anti-analytic Bergman projections) contain the composition of the analytic and anti-analytic Bergman projection. We show that such compositions are smoothing operators. With these facts at hand, one can obtain Theorem 1.2. In multiply connected domains, we have harmonic functions without harmonic conjugates, but this does not change the singular part of the harmonic Bergman kernel. The composition of the analytic and anti-analytic Bergman projections is closely related to the Friedrichs operator. The theorem settles a question posed in [**13**, p. 272].

1.2. Applications. One motivation for this investigation is the application of Bergman kernels to the study of Green functions for weighted biharmonic operators $\Delta\omega^{-1}\Delta$ in the unit disk $\mathbb{D}$. In [**12**], the property (1.2) was used to prove that the Green function is positive provided that $\log \omega$ is subharmonic and

$$h(0) = \frac{1}{\pi} \int_{\mathbb{D}} h(z)\omega(z) \, dA(z)$$

for all bounded harmonic functions h on $\mathbb{D}$. By means of the singularity resolution of the Bergman kernel obtained here, the result in [**12**] was improved in [**16**] to give lower bounds on the Green function. Theorem 1.1 might also be interesting for other problems connected to weighted biharmonic Green functions; in particular, we mention Conjecture 11.1 in [**12**], there called the strong maximum principle.

1.3. An outline of the proof of Theorem 1.1. Let G_ω be the Green function G_ω for the differential operator $L_\omega = \bar{\partial}\omega^{-1}\partial$ with Dirichlet boundary condition in a domain $\Omega \subset \mathbb{C}$. Here, $\partial = \frac{1}{2}(\partial_x - i\partial_y)$, and $\bar{\partial} = \frac{1}{2}(\partial_x + i\partial_y)$, $z = x + iy$, are the Wirtinger differential operators. The Green function G_ω solves, for fixed $\zeta \in \Omega$, the boundary value problem

(1.6) $$\begin{cases} \bar{\partial}_z\omega^{-1}(z)\partial_z G_\omega(z, \zeta) = \delta_\zeta(z), & z \in \Omega, \\ G_\omega(z, \zeta) = 0, & z \in \partial\Omega, \end{cases}$$

where δ_ζ is the Dirac measure at ζ. This type of Green function was first considered by P. R. Garabedian in the paper [**10**]. He showed that it is closely related to the weighted Bergman kernel K_ω, as can be seen from the equality

$$K_\omega(z, \zeta) = \frac{1}{\omega(z)\omega(\zeta)} \frac{\partial^2 G_\omega(z, \zeta)}{\partial z \partial \bar{\zeta}}, \qquad z, \zeta \in \Omega, \quad z \neq \zeta.$$

On the other hand, we can also recover the Green function from the Bergman kernel through the formula

$$
G_\omega(z,\zeta) = -\frac{1}{\pi^2} \int_\Omega \frac{\omega(\eta)}{(\bar{z} - \bar{\eta})(\zeta - \eta)} \, dA(\eta)
$$
$$
+ \frac{1}{\pi^2} \iint_{\Omega \times \Omega} K_\omega(\eta, \xi) \frac{\omega(\eta)\omega(\xi)}{(\bar{z} - \bar{\eta})(\zeta - \xi)} \, dA(\eta) \, dA(\xi).
$$

The first term in this formula is the fundamental solution for the operator L_ω, and the second compensates this term to yield the correct boundary values. We will not use this relation directly to prove our main result, but rather use the Poisson-type kernel

$$
P_\omega(z,\zeta) = -\frac{1}{4\omega(\zeta)} \frac{\partial G_\omega(z,\zeta)}{\partial n(\zeta)}, \quad (z,\zeta) \in \Omega \times \partial\Omega,
$$

where $n(\zeta)$ is the inward normal at $\zeta \in \partial\Omega$. The relation to the weighted Bergman kernel is

(1.7) $$ K_\omega(z,\zeta) = -\frac{2n(\zeta)}{\omega(z)} \frac{\partial P_\omega(z,\zeta)}{\partial z}, \quad (z,\zeta) \in \Omega \times \partial\Omega, $$

where $n(\zeta)$ is interpreted as a complex number. In order to analyze the singularity of the weighted Bergman kernel, it is thus enough to study the regularity of the kernel P_ω.

Let f be a function defined on the boundary of Ω. The solution to the problem

(1.8) $$ \begin{cases} \bar{\partial}\omega^{-1}(z)\partial u(z) = 0, & z \in \Omega, \\ u(z) = f(z), & z \in \partial\Omega, \end{cases} $$

is given by the integral

$$
u(z) = P_\omega f(z) = \int_{\partial\Omega} P_\omega(z,\zeta) f(\zeta) \, ds(\zeta), \quad z \in \Omega,
$$

where $ds(\zeta)$ is the arc-length measure on $\partial\Omega$. Since $L_\omega = \bar{\partial}\omega^{-1}\partial$ is elliptic with a smooth symbol, the solution u is smooth away from the boundary and the local regularity of u near the boundary is completely determined by the local regularity of f. In Section 3, we study the boundary value problem (1.8) by a well-known method which transforms it to a pseudodifferential equation on the boundary (we recall some properties of pseudodifferential operators in Section 2). The idea is to extend u to a function u_0 which vanishes off Ω, and then apply the operator L_ω to u_0 to obtain a distribution which is supported on $\partial\Omega$. This distribution involves the boundary values f, the weight ω, and the inward normal derivative of u on the boundary, which a priori is unknown.

As an elliptic operator L_ω possesses a parametrix, that is, an operator $B_\omega : \mathcal{E}'(\mathbb{C}) \to \mathcal{D}'(\mathbb{C})$ such that $B_\omega \circ L_\omega - \mathrm{Id}$ and $L_\omega \circ B_\omega - \mathrm{Id}$ are smoothing operators (they map $\mathcal{E}'(\mathbb{C})$ into $C^\infty(\mathbb{C})$). Here Id is the identity operator, and $\mathcal{D}'(\mathbb{C})$ and $\mathcal{E}'(\mathbb{C})$ denote the set of distributions and distributions of compact support on $\mathbb{C}$, respectively.

From the properties of the parametrix B_ω, it follows that the distribution $u_0 - B_\omega(L_\omega u_0)$ can be represented by a function in $C^\infty(\mathbb{C})$. We point out that both u_0 and $B_\omega(L_\omega u_0)$ have a jump discontinuity of magnitude f at the boundary, so their difference is not defined pointwise on $\partial\Omega$. Since u_0 is identically zero outside Ω, the restriction of $B_\omega(L_\omega u_0)$ to the open set $\mathbb{C} \setminus \overline{\Omega}$ has an extension to a function in $C^\infty(\mathbb{C} \setminus \Omega)$. This gives the pseudodifferential equation, which holds modulo $C^\infty(\partial\Omega)$. By solving this equation modulo $C^\infty(\partial\Omega)$ and then tracing the steps backwards, we eventually obtain the singularity resolution of the weighted Bergman kernel. In our calculations, we use the same change of variables as in Theorem 1.1 to flatten out the boundary locally. This works well since the singular parts of P_ω and K_ω only depend on the local behavior of the weight near $\partial\Omega$.

1.4. Smooth extensions. In this paper, we will often extend functions smoothly to a larger domain subject to some side conditions. This can be achieved with the next result, or variants of it, due to Borel. The following version is from [15, p. 16].

THEOREM 1.3 (Borel). *For $j = 0, 1, \ldots$ let $f_j \in C^\infty(K)$, where K is a compact subset of $\mathbb{R}^n$, and let I be a compact neighborhood of 0 in $\mathbb{R}$. Then one can find $f \in C^\infty(K \times I)$ such that*

$$\left.\frac{\partial^j f(x,t)}{\partial t^j}\right|_{t=0} = f_j(x), \qquad j = 0, 1, \ldots.$$

The following terminology will be used. We say that a function $f \in C^\infty(X)$, $X \subset \mathbb{R}^n$ open, *vanishes to infinite order on* $K \subset X$ if f together with all its derivatives are zero on K.

2. Pseudodifferential operators

In this section, we shall present some properties and formulas for pseudodifferential operators which will be used in this paper. We restrict our attention to operators on function spaces over subset of $\mathbb{R}$ and $\mathbb{C}$. For more complete treatments, we refer to the textbooks on the subject, for example, [11, 14, 20].

The symbol class $S^m(I \times \mathbb{R})$, where $m \in \mathbb{N}$ and $I \subset \mathbb{R}$ is an interval, consists of all $a \in C^\infty(I \times \mathbb{R})$, such that for every compact set $K \subset I$, and pair of positive integers α and β, we have

$$|\partial_x^\alpha \partial_\xi^\beta a(x,\xi)| \leq C_{\alpha,\beta,K}(1 + |\xi|)^{m-\beta}, \qquad (x,\xi) \in K \times \mathbb{R},$$

for some constant $C_{\alpha,\beta,K}$. Suppose that the sequence $\{a_j\}_{j=0}^\infty$, $a_j \in S^{m_j}(I \times \mathbb{R})$, is such that $m_j \to -\infty$ monotonically as $j \to \infty$ and

$$a - \sum_{j=0}^n a_j \in S^{m_{n+1}}(I \times \mathbb{R})$$

for all integers $n \geq 0$, then the formal sum $\sum_{j=0}^\infty a_j$ is called an *asymptotic series* and a_0 the *principal symbol* for a. We write $a \sim \sum_{j=0}^\infty a_j$. The subclass $S_{\text{phg}}^m(I \times \mathbb{R})$ is the set of all $a \in S^m(I \times \mathbb{R})$ which has an asymptotic series $a \sim \sum_{j=0}^\infty a_j$ whose terms $a_j \in S^{m-j}(I \times \mathbb{R})$ are positively homogeneous of degree $m - j$ in ξ when $|\xi| > 1$.

An operator $A : C_0^\infty(I) \to \mathcal{D}'(I)$ belongs to the class $\Psi^m(I)$ ($\Psi_{\text{phg}}^m(I)$) of pseudodifferential operators if there exists a symbol $a \in S^m(I \times \mathbb{R})$ ($a \in S_{\text{phg}}^m(I \times \mathbb{R})$) such that

$$(2.1) \qquad Af(x) = \frac{1}{2\pi} \int_\mathbb{R} \int_I a(x,\xi) e^{i(x-y)\xi} f(y)\, dy d\xi, \qquad x \in \mathbb{R},$$

for all $f \in C_0^\infty(I)$. We shall often write σ_A for the symbol of the operator A. The space $\Psi_{\text{phg}}^m(I)$, $m \geq 0$, includes all differential operator of order less or equal to m on I with smooth coefficients. It is well-known that A has a continuous extension to $\mathcal{E}'(I) \to \mathcal{D}'(I)$, and that A does not increase the singular support. i.e., $\operatorname{sing\,supp} Af \subset \operatorname{sing\,supp} f$. The intersection of all $S^m(I \times \mathbb{R})$, $m \in \mathbb{N}$, is denoted by $S^{-\infty}(I \times \mathbb{R})$, and an operator in $\Psi^{-\infty}(I \times \mathbb{R})$ maps $\mathcal{E}'(I)$ into $C^\infty(I)$, and it is therefore called a *smoothing operator*. Furthermore, we say that an operator is *properly supported* if it and its adjoint maps compactly supported functions to compactly supported distributions. If A and B are pseudodifferential operators and at least one of them is properly supported, then the composition $A \circ B$ is defined and has the symbol

$$\sigma_{A \circ B}(x, \xi) = e^{-ix\xi} A(\sigma_B(\cdot, \xi)e^{i \cdot \xi}) = (2\pi)^{-1} \int_{\mathbb{R}} \int_{\mathbb{R}} a(x, \theta) b(t, \xi) e^{i(x-t)(\theta-\xi)} \, dt d\theta.$$

Moreover, the symbol $\sigma_{A \circ B}$ has the asymptotic expansion

$$(2.2) \qquad \sigma_{A \circ B}(x, \xi) \sim \sum_{\alpha=0}^{\infty} (-i)^\alpha \frac{\partial_\xi^\alpha \sigma_A(x, \xi) \partial_x^\alpha \sigma_B(x, \xi)}{\alpha!}.$$

The right hand side converges as an asymptotic series and is denoted by $\sigma_A \sharp \sigma_B(x, \xi)$. We call $\sharp$ the *sharp product*.

A pseudodifferential operator $A \in \Psi^m(I)$ is *elliptic* if for every compact set $K \subset I$ there exist constants C and $c > 0$ such that $|\sigma_A(x, \xi)| \geq c(1 + |\xi|)^m$ for all $x \in K$ and $|\xi| \geq C$. For every elliptic operator $A \in \Psi^m(I)$, there exists an operator $B \in \Psi^{-m}(I)$ such that $A \circ B - \operatorname{Id} \in \Psi^{-\infty}(I)$ and $B \circ A - \operatorname{Id} \in \Psi^{-\infty}(I)$, where Id is the identity operator. The operator B is called a *parametrix* for A, and if $A \in \Psi_{\mathrm{phg}}^m(I)$ then $B \in \Psi_{\mathrm{phg}}^{-m}(I)$. This property of being a parametrix is reflected in terms of the sharp product by

$$(2.3) \qquad \sigma_A \sharp \sigma_B(x, \xi) \sim \sigma_B \sharp \sigma_A(x, \xi) \sim 1.$$

In fact, in order to prove the existence of a parametrix, one solves the equation (2.3).

2.1. Calculation of parametrices for elliptic operators in $\Psi_{\mathrm{phg}}^{-1}(I)$. We shall now present formulas for the parametrix for an elliptic operator $P \in \Psi_{\mathrm{phg}}^{-1}(I)$ whose symbol has an asymptotic series of the form $p(x, \xi) \sim \sum_{k=1}^{\infty} \frac{p_k(x)}{\xi^k}$ for $|\xi| \geq 1$ (we assume that the singularity at $\xi = 0$ is taken care of by a suitable cut-off function). The ellipticity condition on P reduces to the requirement that $p_1(x) \neq 0$ everywhere.

PROPOSITION 2.1. *Suppose that $P \in \Psi_{\mathrm{phg}}^{-1}(I)$ is elliptic with symbol $p(x, \xi) \sim \sum_{k=1}^{\infty} \frac{p_k(x)}{\xi^k}$. If Q is a parametrix to P, then its symbol satisfies*

$$q(x, \xi) \sim \frac{\xi}{p_1(x)} + \sum_{n=0}^{\infty} \frac{q_n(x)}{\xi^n},$$

where $q_n(x) = \frac{r_{n+1, n+1}(x)}{p_1(x)}$, and the functions $r_{n,k}(x)$, $n \leq k$, are defined recursively by

$$r_{1,k}(x) = \frac{i\partial_x p_k(x) - p_{k+1}(x)}{p_1(x)} \quad \text{for } k = 1, 2, 3, \ldots,$$

$$r_{2,k} = \frac{p_1(x) r_{1,k}(x) - p_k(x) r_{1,1}(x)}{p_1(x)} \quad \text{for } k = 2, 3, 4, \ldots,$$

and for $n \geq 2$,

$$r_{n+1,k}(x) = r_{n,k}(x) - \frac{r_{n,n}(x)}{p_1(x)} \sum_{\alpha=0}^{k-n} i^\alpha \binom{n+\alpha-2}{\alpha} \partial_x^\alpha p_{k-n-\alpha+1}(x)$$

for $k = n+1, n+2, \ldots$.

Here, $\binom{n}{k} = \frac{n!}{k!(n-k)!}$ is the standard binomial coefficient.

PROOF. We shall find the terms in the asymptotic series inductively. In order to be a parametrix, the principal symbol for Q must be $\frac{\xi}{p_1(x)}$. Suppose that we have found the functions $q_0, q_1, \ldots, q_{n-2}$, $n \geq 1$, in the asymptotic series for q (if $n = 1$ we have found no functions). Let

$$Q_{n-2}(x,\xi) = \frac{\xi}{p_1(x)} + \sum_{k=0}^{n-2} \frac{q_k(x)}{\xi^k}.$$

It follows from the equation (2.3) for the parametrix and the sharp product that

(2.4)
$$\sum_{k=n}^{\infty} \frac{r_{n,k}(x)}{\xi^k} \sim 1 - Q_{n-2} \sharp P(x,\xi),$$

for some functions $\{r_{n,k}\}_{k=n}^{\infty}$. We now define

$$Q_{n-1}(x,\xi) = Q_{n-2}(x,\xi) + \frac{r_{n,n}(x)}{p_1(x)\xi^{n-1}}.$$

A calculation shows that $1 - Q_{n-1} \sharp P$ has order $-(n+1)$, and that

$$\sum_{k=n+1}^{\infty} \frac{r_{n+1,k}(x)}{\xi^k} \sim 1 - Q_{n-1} \sharp P(x,\xi),$$

where the functions $\{r_{n+1,k}\}_{k=n+1}^{\infty}$ can be computed in terms of the functions $\{r_{n,k}\}_{k=n}^{\infty}$ and $\{p_k\}_{k=-1}^{\infty}$. This implies that the function q_{n-1} in the asymptotic series for q is

$$q_{n-1}(x) = \frac{r_{n,n}(x)}{p_1(x)}.$$

The recursion formulas in the proposition follows from explicit calculation with the sharp product. □

For our application, we are concerned with a slight modification of the asymptotic series for P, namely $\operatorname{sgn}(\xi) \sum_{k=1}^{\infty} \frac{p_k(x)}{\xi^k}$. This case can be handled in the same way. We have the following corollary.

COROLLARY 2.2. *Proposition 2.1 holds also for $p(x,\xi) \sim \operatorname{sgn}(\xi) \sum_{k=1}^{\infty} \frac{p_k(x)}{\xi^k}$ if the asymptotic series for the parametrix is replaced by $\operatorname{sgn}(\xi) \left(\frac{\xi}{p_1(x)} + \sum_{k=0}^{\infty} \frac{q_k(x)}{\xi^k} \right)$.*

2.2. Composition of the Cauchy transform with a pseudodifferential operator. For a function h on $\mathbb{R}$, we define the analytic Cauchy transform of h by

$$(2.5) \qquad Jh(z) = \frac{1}{2\pi} \int_{\mathbb{R}} \frac{i}{z-t} h(t)\, dt, \qquad z \in \mathbb{C}_+.$$

whenever the right hand side exists. If $h \in L^2(\mathbb{R})$, then Jh is the orthogonal projection onto the Hardy space $H^2(\mathbb{C}_+)$. In this paper, the operator J appears in composition with proper pseudodifferential operators. A symbol $a \in S^m(\mathbb{R} \times \mathbb{R})$ can be extended to $S^m(\mathbb{C} \times \mathbb{R})$ such that a is almost analytic in z on the real line in the sense that $a(z, \xi) - a(t, \xi) = (z-t)R(z, t, \xi)$, $z \in \mathbb{C}$, $t \in \mathbb{R}$, for some $R \in S^m(\mathbb{C} \times \mathbb{R} \times \mathbb{R})$. This implies that $\bar{\partial}_z a(z, \xi)$ vanishes to infinite order at $\operatorname{Im} z = 0$ for every fixed $\xi \in \mathbb{R}$.

LEMMA 2.3. *Suppose that $A \in \Psi^m(\mathbb{R})$ is properly supported. Then for all $h \in \mathcal{E}'(\mathbb{R})$, we have*

$$J \circ Ah(z) \equiv A_J h(z) \quad \mathrm{mod}\ C^\infty(\overline{\mathbb{C}_+}),$$

where A_J is the operator

$$(2.6) \qquad A_J h(z) = \frac{1}{2\pi} \int_0^\infty \int_{\mathbb{R}} a(z, \xi) e^{i(z-t)\xi} h(t)\, dt d\xi, \qquad z \in \mathbb{C}_+.$$

The integral (2.6) is interpreted in the sense of distributions.

PROOF. The Cauchy transform can be written as a Fourier integral

$$Jh(z) = \frac{1}{2\pi} \int_{\mathbb{R}} \int_0^\infty e^{i(z-t)\xi} h(t)\, d\xi dt, \qquad z \in \mathbb{C}_+.$$

With R defined as above, we have

$$(2.7) \quad J \circ Ah(z) = \frac{1}{2\pi} \int_{\mathbb{R}} \int_{\mathbb{R}} \int_{\mathbb{R}} \frac{i}{z-t} a(z, \xi) e^{i(t-s)\xi} h(s)\, ds d\xi dt$$

$$+ \frac{i}{2\pi} \int_{\mathbb{R}} \int_{\mathbb{R}} \int_{\mathbb{R}} R(z, t, \xi) e^{i(t-s)\xi} h(s)\, ds d\xi dt.$$

It is easy to show that the first integral equals (2.6), so we only have to show that the second term represents a smooth function. The oscillatory integral

$$\frac{i}{2\pi} \int_{\mathbb{R}} \int_{\mathbb{R}} R(z, t, \xi) e^{i(t-s)\xi} h(s)\, ds d\xi$$

represents a distribution in $\mathcal{E}'(\mathbb{R})$ in the variable t which depends smoothly on $z \in \overline{\mathbb{C}_+}$. Therefore, when we apply it to the constant function 1, i.e., integrate on $\mathbb{R}$, we obtain a C^∞-smooth function in $z \in \overline{\mathbb{C}_+}$. The proposition is proved.   □

3. Boundary value problems for $\bar{\partial} \omega^{-1} \partial$

The purpose of this section is to find the solution kernel $P_\omega(z, \zeta)$ for the boundary value problem (1.8) modulo a C^∞-smooth function. By elliptic regularity, $P_\omega \in C^\infty(\overline{\Omega} \times \partial\Omega \setminus \triangle(\partial\Omega))$, so it suffices to study $P_\omega(z, \zeta)$ for z close to $\zeta \in \partial\Omega$. Let γ be a sub-arc of $\partial\Omega$, and let f be any function on $\partial\Omega$ which is compactly supported in γ. We shall investigate the local regularity near γ of the solution u to (1.8) with boundary values f. It is convenient to make the same change of variable as in Theorem 1.1 so we let N_γ be a neighborhood of γ, I an interval on $\mathbb{R}$, N a neighborhood of I, and $\psi : \mathbb{C}_+ \cap N \to \Omega \cap N_\gamma$ a conformal map which maps I onto γ. The function $v = (u \circ \psi) \cdot \overline{\psi'}$ solves the boundary value problem

(3.1)
$$\begin{cases} \bar{\partial} \omega_\psi^{-1}(z) \partial v(z) = 0, & z \in \mathbb{C}_+ \cap N, \\ v(z) = g(z), & z \in I, \end{cases}$$

where

$$\omega_\psi(z) = \omega(\psi(z)) |\psi'(z)|^2, \qquad z \in \overline{\mathbb{C}_+} \cap N,$$

and $g = (f \circ \psi) \cdot \overline{\psi'}$ on I. In order to apply the techniques described in the introduction we shall extend v to a function on the whole of $\mathbb{C}$. Let N' be a neighborhood of the support of f and take a cut-off function $\chi \in C_0^\infty(N)$ which is identically 1 on N'. We define v_0 to be $v\chi$ in $\mathbb{C}_+ \cap N$ and vanish elsewhere:

$$v_0(z) = v(z)\chi(z)H(y), \qquad z = x + iy \in \mathbb{C},$$

where $H(y)$ is the Heaviside function which is identically 1 for positive arguments and zero otherwise. Here we have extended ω_ψ to a strictly positive function in $C^\infty(\mathbb{C})$. We follow the standard procedure and apply the differential operator $L_{\omega_\psi} = \bar{\partial} \omega_\psi^{-1}(z) \partial$ to this extended function. Since v solves (3.1) and $\chi \in C_0^\infty(N)$, an explicit calculation yields

(3.2) $L_{\omega_\psi} v_0(z) = \dfrac{1}{4} \left(\dfrac{g(x)}{\omega_\psi(x)} \otimes \delta'(y) + h(x) \otimes \delta(y) \right) \quad \mod C^\infty(\mathbb{C}),$

$z = x + iy \in \mathbb{C}$, where we have put

(3.3)
$$h = \left(\frac{\partial_n v}{\omega_\psi} - g \frac{\partial \omega_\psi}{\omega_\psi^2} \right).$$

In general, if $u_0 = u 1_\Omega$, where u is a solution to (1.8) and 1_Ω the characteristic function of Ω, then $L_\omega u_0$ is a distribution with support on $\partial \Omega$. Let B_{ω_ψ} be a parametrix to L_{ω_ψ}, then

(3.4) $v_0(z) \equiv B_{\omega_\psi} \circ L_{\omega_\psi} v_0(z) \quad \mod C^\infty(\mathbb{C}).$

As a second order elliptic differential operator, L_{ω_ψ} possesses a parametrix B_{ω_ψ}, which is a pseudodifferential operator with symbol $b_{\omega_\psi} \in S_{\mathrm{phg}}^{-2}(\mathbb{R}^2 \times \mathbb{R}^2)$. Here, we have identified the complex plane $\mathbb{C}$ with $\mathbb{R}^2$. In the proposition below, we shall also use complex notation for the frequency variables. For simplicity, we drop the subscript ψ on the weight in the calculation of the parametrix.

PROPOSITION 3.1. *The symbol of the parametrix B_ω has the asymptotic series*

$$b_\omega(z, \xi) \sim -4 \sum_{n=0}^\infty \frac{2^n i^n \partial^n \omega(z)}{\xi \bar{\xi}^{n+1}}, \qquad \xi = \xi_1 + i\xi_2.$$

PROOF. Since the operators $\bar{\partial}$ and ∂ have the symbols $i\bar{\xi}/2$ and $i\xi/2$, the asymptotic series for their parametrices are simply $-2i/\bar{\xi}$ and $-2i/\xi$, respectively. From the factorization of the operator $L_\omega = \bar{\partial} \omega^{-1} \partial$, it follows that the symbol for the parametrix B_ω satisfies

(3.5) $b_\omega(z, \xi) \sim -4 \dfrac{1}{\bar{\xi}} \sharp \omega(z) \sharp \dfrac{1}{\xi}, \qquad \xi = \xi_1 + i\xi_2,$

where $\sharp$ is the sharp product introduced in equation (2.2). In $\mathbb{R}^2$, the sharp product between the symbols σ_A and σ_B can be written as

$$\sum_{\alpha=0}^{\infty}(-i)^{\alpha}\frac{(\partial_{x'}\partial_{\xi_1'}+\partial_{y'}\partial_{\xi_2'})^{\alpha}}{\alpha!}\left(\sigma_A(z,\xi')\sigma_B(z',\xi)\right)\bigg|_{z'=z,\xi'=\xi},$$

where $z' = x' + iy'$ and $\xi' = \xi_1' + i\xi_2'$. Since $(\partial_x\partial_{\xi_1}+\partial_y\partial_{\xi_2}) = 2(\partial_z\bar{\partial}_{\xi}+\bar{\partial}_z\partial_{\xi})$, and $\bar{\xi}^{-1}$ is anti-analytic away from the origin, we obtain

$$\frac{1}{\bar{\xi}}\sharp\omega(z)\sharp\frac{1}{\bar{\xi}} = \sum_{n=0}^{\infty}\frac{2^n i^n \partial^n\omega(z)}{\bar{\xi}\bar{\xi}^{n+1}}.$$

The result now follows from (3.5). $\square$

We see that all the terms in the asymptotic series for b_ω are rational functions in the frequency variables ξ_1 and ξ_2. This implies that B_ω satisfies the so called transmission condition [14, chapter 18], which in our case means that if $f_1, f_2 \in C_0^{\infty}(\mathbb{R})$, then the function

$$(3.6) \qquad w(z) = \frac{1}{4}B_\omega\left(f_1(x)\otimes\delta'(y)+f_2(x)\otimes\delta(y)\right)(z), \qquad z\in\mathbb{C}\setminus\mathbb{R},$$

can be extended to both $C^{\infty}(\overline{\mathbb{C}_+})$ and $C^{\infty}(\overline{\mathbb{C}_-})$ (but not to $C^{\infty}(\mathbb{C})$ since it has a jump discontinuity along the real line). In particular, w has boundary values on $\mathbb{R}$ from above and below and they are given by pseudodifferential operators applied to f_1 and f_2. In the next proposition, we calculate these pseudodifferential operators modulo smoothing operators.

PROPOSITION 3.2. *Let w be given by (3.6) and define*

$$w_\varepsilon(x) = w(x+i\varepsilon), \qquad x\in\mathbb{R}.$$

Then the limit $w_+ = \lim_{\varepsilon\to 0^+} w_\varepsilon$ exists in $C^{\infty}(\mathbb{R})$ and

$$w_+ = \frac{1}{2}\omega f_1 + \frac{1}{2}C_\omega(f_1) + \frac{1}{2}D_\omega(f_2),$$

where C_ω and D_ω are pseudodifferential operators on $\mathbb{R}$ whose symbols have the asymptotic expansions

$$c_\omega(x,\xi_1) \sim \mathrm{sgn}(\xi_1)\sum_{n=1}^{\infty}\frac{i^n\partial^n\omega(x)}{\xi_1^n},$$

and

$$d_\omega(x,\xi_1) \sim -\mathrm{sgn}(\xi_1)\sum_{n=1}^{\infty}\frac{i^{n-1}\partial^{n-1}\omega(x)}{\xi_1^n}.$$

Similarly, if $w_{-\varepsilon}(x) = w(x-i\varepsilon)$, then $w_- = \lim_{\varepsilon\to 0^+} w_{-\varepsilon}$ satisfies

$$w_- = -\frac{1}{2}\omega f_1 + \frac{1}{2}C_\omega(f_1) + \frac{1}{2}D_\omega(f_2).$$

For the proof, we need some inverse Fourier transforms of rational functions:

$$Q_n(y,\xi_1) = \frac{1}{2\pi}\int_{\mathbb{R}}\frac{1}{(\xi_1+i\xi_2)(\xi_1-i\xi_2)^n}e^{iy\xi_2}\,d\xi_2,$$

and
$$P_n(y, \xi_1) = \frac{1}{2\pi} \int_{\mathbb{R}} \frac{-i\xi_2}{(\xi_1 + i\xi_2)(\xi_1 - i\xi_2)^n} e^{iy\xi_2} \, d\xi_2,$$

for $n = 1, 2, 3, \ldots$. Note that $P_n(y, \xi_1) = -\partial_y Q_n(y, \xi_1)$.

LEMMA 3.3. *The following formulas hold for* $y, \xi_1 \in \mathbb{R} \setminus \{0\}$, *and* $n = 1, 2, 3, \ldots$

$$(3.7) \quad Q_n(y, \xi_1) =$$
$$\operatorname{sgn}(\xi_1) e^{-|y\xi_1|} \left(\frac{1}{(2\xi_1)^n} + \mathrm{H}(-y\xi_1) \sum_{k=0}^{n-2} \frac{1}{(n-k-1)!} \frac{(-y)^{n-k-1}}{(2\xi_1)^{k+1}} \right),$$

$$(3.8) \quad P_n(y, \xi_1) = \operatorname{sgn}(y) e^{-|y\xi_1|}$$
$$\times \left(\frac{1}{2} \frac{1}{(2\xi_1)^{n-1}} + \mathrm{H}(-y\xi_1) \sum_{k=0}^{n-2} \frac{1}{(n-k-1)!} \frac{(-y)^{n-k-1}}{(2\xi_1)^k} \right)$$
$$+ \operatorname{sgn}(\xi_1) e^{-|y\xi_1|} \mathrm{H}(-y\xi_1) \sum_{k=0}^{n-2} \frac{1}{(n-k-2)!} \frac{(-y)^{n-k-2}}{(2\xi_1)^{k+1}}.$$

We have the following limits as $y \to 0^{\pm}$

$$Q_n(0, \xi_1) = \operatorname{sgn}(\xi_1) \frac{1}{(2\xi_1)^n}, \qquad n = 1, 2, 3, \ldots,$$

$$P_1(0^+, \xi_1) = \frac{1}{2}, \qquad P_1(0^-, \xi_1) = -\frac{1}{2},$$

and

$$P_n(0, \xi_1) = \frac{1}{2} \operatorname{sgn}(\xi_1) \frac{1}{(2\xi_1)^{n-1}}, \qquad n = 1, 2, 3, \ldots.$$

PROOF. Define, for $n = 1, 2, 3, \ldots$, the functions

$$C_n(y, \xi_1) = \frac{1}{2\pi} \int_{\mathbb{R}} \frac{1}{(\xi_1 + i\xi_2)^n} e^{iy\xi_2} \, d\xi_2, \qquad y, \xi_1 \in \mathbb{R} \setminus \{0\}.$$

When $n = 1$, we interpret $C_n(y, \xi_1)$ as an oscillatory integral. By the Cauchy integral formula, we have

$$C_1(y, \xi_1) = \mathrm{H}(y\xi_1) \operatorname{sgn}(\xi) e^{-|y\xi_1|},$$

which, after differentiating $n - 1$ times with respect to ξ_1, yields

$$C_n(y, \xi_1) = \frac{y^{n-1}}{(n-1)!} \mathrm{H}(y\xi_1) \operatorname{sgn}(\xi) e^{-|y\xi_1|}.$$

From a partial fraction decomposition, it follows that

$$Q_1(y, \xi_1) = \frac{1}{(2\xi_1)^n} \left(C_1(y, \xi_1) + C_1(-y, \xi_1) \right) = \operatorname{sgn}(\xi_1) e^{-|y\xi_1|} \frac{1}{2\xi_1}.$$

The first formula can now be proved by induction over n, by using the relation

$$Q_n(y, \xi_1) = \frac{1}{2\xi_1} \left(Q_{n-1}(y, \xi_1) + C_n(-y, \xi_1) \right).$$

The second follows from the first after a differentiation with respect to y. $\square$

PROOF OF PROPOSITION 3.2. Let $\widehat{f_1}$ and $\widehat{f_2}$ be the Fourier transforms of f_1 and f_2, respectively. We can write w as the Fourier integral

$$(3.9) \quad w(z) = \frac{1}{(2\pi)^2} \int_{\mathbb{R}} \int_{\mathbb{R}} \frac{b_\omega(z,\xi)}{4} e^{i(x\xi_1 + y\xi_2)} \left(i\xi_2 \widehat{f_1}(\xi_1) + \widehat{f_2}(\xi_1) \right) d\xi_1 d\xi_2$$

$$= \frac{1}{4\pi} \int_{\mathbb{R}} \widetilde{c}_\omega(z,\xi_1) e^{ix\xi_1} \widehat{f_1}(\xi_1) \, d\xi_1 + \frac{1}{4\pi} \int_{\mathbb{R}} d_\omega(z,\xi_1) e^{ix\xi_1} \widehat{f_2}(\xi_1) \, d\xi_1,$$

where

$$\widetilde{c}_\omega(z,\xi_1) = \frac{1}{2\pi} \int_{\mathbb{R}} \frac{i\xi_2 b_\omega(z,\xi)}{2} e^{iy\xi_2} \, d\xi_2$$

and

$$d_\omega(z,\xi_1) = \frac{1}{2\pi} \int_{\mathbb{R}} \frac{b_\omega(z,\xi)}{2} e^{iy\xi_2} \, d\xi_2.$$

From the asymptotic series for b_ω together with Lemma 3.3, it follows that

$$\widetilde{c}_\omega(z,\xi_1) \sim 2\omega(z)P_1(y,\xi_1) + 2\sum_{n=1}^{\infty} 2^n i^n \partial^n \omega(z) P_{n+1}(y,\xi_1),$$

and

$$d_\omega(z,\xi_1) \sim -2\sum_{n=0}^{\infty} 2^n i^n \partial^n \omega(z) Q_{n+1}(y,\xi_1) = -\sum_{n=1}^{\infty} 2^n i^{n-1} \partial^{n-1} \omega(z) Q_n(y,\xi_1).$$

As $y \to 0^{\pm}$, we obtain the limits

$$\widetilde{c}_\omega(x \pm i0,\xi) \sim \pm\omega(x) + \mathrm{sgn}(\xi_1) \sum_{n=1}^{\infty} \frac{i^n \partial^n \omega(x)}{\xi_1^n}$$

and

$$d_\omega(x,\xi_1) \sim -\mathrm{sgn}(\xi_1) \sum_{n=1}^{\infty} \frac{i^{n-1} \partial^{n-1} \omega(x)}{\xi_1^n}.$$

Note that it is only the first term of $\widetilde{c}_\omega(z,\xi)$ which has different limits from above and below. Let the symbol c_ω be as $\widetilde{c}_\omega$ but without the first term. The convergence of w_ε now follows from the form of the factors P_n and Q_n and the continuity of the Fourier transform. $\qquad\square$

We now return to the boundary value problem (3.1). If we apply Proposition 3.2 to (3.2) and (3.4), we have for the function v_0

$$(3.10) \quad D_{\omega_\psi}(h) \equiv g - C_{\omega_\psi}(g/\omega_\psi) \mod C^\infty(\mathbb{R}),$$

where h is given by (3.3). This is the pseudodifferential equation mentioned in the introduction. Since D_{ω_ψ} is an elliptic pseudodifferential operator in $\Psi_{\mathrm{phg}}^{-1}(\mathbb{R})$, it has a parametrix $D_{\omega_\psi}^p \in \Psi_{\mathrm{phg}}^1(\mathbb{R})$. By applying $D_{\omega_\psi}^p$ to both sides of (3.10), we obtain

$$h \equiv D_{\omega_\psi}^p(g) - H_{\omega_\psi}(f) \mod C^\infty(\mathbb{R}),$$

where $M_{\omega_\psi^{-1}}$ is multiplication by ω_ψ^{-1} and $H_{\omega_\psi} = D_{\omega_\psi}^p \circ C_{\omega_\psi} \circ M_{\omega_\psi^{-1}}$.

Again by (3.4),

$$(3.11) \quad v_0(z) \equiv \frac{1}{4} B_{\omega_\psi}(g/\omega_\psi \otimes \delta'(y))(z)$$

$$+ \frac{1}{4} B_{\omega_\psi}\left(\left(D_{\omega_\psi}^p(g)(x) - H_{\omega_\psi}(g)(x) \right) \otimes \delta(y) \right)(z) \mod C^\infty(\mathbb{C}).$$

On the other hand, from the definition of v_0 and P_ω we have

$$(3.12) \qquad v_0(z) = \int_\gamma P_\omega(\psi(z), \zeta) f(\zeta) \, ds(\zeta) \overline{\psi'(z)}$$

$$(3.13) \qquad = \int_I P_\omega(\psi(z), \psi(\zeta)) g(\zeta) \frac{\psi'(\zeta)}{|\psi'(\zeta)|} \, ds(\zeta) \overline{\psi'(z)}.$$

Since the function f was arbitrary we obtain the singularity resolution for P_ω near γ by comparing the different expressions for v_0. In the next section, we calculate the asymptotic series for D_ω^p and H_ω, and use this to prove Theorem 1.1.

4. The analytic Bergman kernel

The purpose of this section is to find the singularity resolution of the Bergman kernel. For this we use the relations to boundary value problems for $\bar\partial \omega^{-1} \partial$, and we calculate the asymptotic series for the operators which appeared in the previous section. We also show that if $\log \omega$ is harmonic around a boundary point, then the expansion of K_ω at that point does not contain a logarithmic term.

According to (1.7), we can express the Bergman kernel K_ω near γ as

$$(4.1) \qquad K_\omega(\psi(z), \psi(\zeta)) = -\frac{2n(\psi(\zeta))}{\omega(\psi(z))} \partial_1 P_\omega(\psi(z), \psi(\zeta)), \qquad (z, \zeta) \in \mathbb{C}_+ \cap N \times I,$$

in terms of the conformal map ψ. Here, ∂_1 is the Wirtinger differential operator applied to the first argument and

$$n(\psi(\zeta)) = \frac{i\psi'(\zeta)}{|\psi'(\zeta)|}.$$

Since $\partial B_{\omega_\psi} \equiv M_{\omega_\psi} \circ \bar\partial^{-1}$ modulo a smoothing operator, we see from (3.11) that the function v_0 satisfies

$$(4.2) \qquad -\frac{2i}{\omega_\psi(z)} \partial v_0(z) \equiv -\frac{i}{2} \bar\partial^{-1}(g/\omega_\psi \otimes \delta'(y))(z)$$

$$-\frac{i}{2} \bar\partial^{-1}\Big(\big(D_{\omega_\psi}^p(g)(x) - H_{\omega_\psi}(g)(x) \big) \otimes \delta(y) \Big)(z) \quad \mod C^\infty(\mathbb{C}).$$

A combination of (3.12) and (4.1) gives

$$-\frac{2i}{\omega_\psi(z)} \partial v_0(z) = \int_I K_\omega(\psi(z), \psi(\zeta)) g(\zeta) \, ds(\zeta), \qquad z \in \mathbb{C}_+ \cap N'.$$

The singular part of the Bergman kernel is thus obtained by replacing g by a Dirac measure δ_ζ, $\zeta \in I$, in (4.2).

For the expansion we need the asymptotic series of the operators $D_{\omega_\psi}^p(z)$ and $H_{\omega_\psi}(z)$. Again we drop the subscript ψ in the calculations. It follows from the form of the asymptotic series for D_ω and C_ω, and Corollary 2.2 that

$$\sigma_{D_\omega^p}(x, \xi) \sim -\mathrm{sgn}(\xi) \left(\xi d_{-1}(x) + \sum_{n=0}^\infty \frac{d_n(x)}{\xi^n} \right),$$

and

$$\sigma_{H_\omega}(x, \xi) \sim \sum_{n=0}^\infty \frac{h_n(x)}{\xi^n},$$

where $\{d_n\}_{n=-1}^\infty$, and $\{h_n\}_{n=0}^\infty$ are sequences of functions defined on I. In the next two propositions we present formulas for these functions in terms of the weight ω.

PROPOSITION 4.1. *The functions* $\{d_n\}_{n=-1}^{\infty}$ *are given by*

$$d_{-1}(x) = \frac{1}{\omega(x)}, \quad \text{and } d_n(x) = \frac{r_{n+1,n+1}(x)}{\omega(x)}, \qquad n \in \mathbb{N},$$

where the functions $r_{n,k}$ *satisfy the recursion formulas*

$$(4.3) \qquad r_{1,k}(x) = \frac{i^k \partial^{k-1} \bar{\partial} \omega(x)}{\omega(x)}, \qquad k = 1, 2, 3, \ldots,$$

$$(4.4) \qquad r_{2,k}(x) = i^k \frac{\partial^{k-1} \bar{\partial} \omega(x) \omega(x) - \partial^{k-1} \omega(x) \bar{\partial} \omega(x)}{\omega(x)^2}, \qquad k = 2, 3, 4, \ldots,$$

and

$$(4.5) \qquad r_{n+1,k}(x) = r_{n,k}(x) - i^{k-n} \frac{r_{n,n}(x)}{\omega(x)} \sum_{\alpha=0}^{k-n} \binom{n+\alpha-2}{\alpha} \partial_x^\alpha \partial^{k-n-\alpha} \omega(x),$$

for $n \geq 2$ *and* $k \geq n+1$.

These formulas follow from Proposition 2.1 with $p_n(x) = i^{n-1} \partial^{n-1} \omega(x)$. The functions $\{h_n\}_{n=0}^{\infty}$ are obtained by a straightforward application of the sharp product (2.2). We omit the calculations.

PROPOSITION 4.2. *The functions* h_n *are given by the formulas*

$$(4.6) \qquad h_0(x) = \frac{\rho_0(x)}{\omega(x)} \text{ and } h_k(x) = \sum_{\alpha=0}^{k-1} (-i)^\alpha \binom{k-1}{\alpha} \rho_{k-\alpha}(x) \partial_x^\alpha \omega^{-1}(x),$$

where

$$\rho_0(x) = -i \frac{\partial \omega(x)}{\omega(x)},$$

$$\rho_1(x) = \frac{\partial \omega(x) \bar{\partial} \omega(x) - \partial \bar{\partial} \omega(x) \omega(x)}{\omega(x)^2} = -\frac{1}{4} \Delta \log \omega(z),$$

and

$$\rho_n(x) = i^{n+1} \frac{\partial^n \bar{\partial} \omega(x) \omega(x) - \partial^n \omega(x) \bar{\partial} \omega(x)}{\omega(x)^2}$$
$$- \sum_{k=1}^{n-1} \sum_{\alpha=0}^{n-k-1} i^{\alpha+k} \binom{n-k-1}{\alpha} \partial^k \partial_x^\alpha \omega(x) d_{n-k-\alpha}(x),$$

for $n = 2, 3, 4, \ldots$.

The functions ρ_n are defined such that the asymptotic series for $D_\omega^p \circ C_\omega$ is given by

$$\sum_{n=0}^{\infty} \frac{\rho_n(x)}{\xi^n}.$$

We are now ready for the proof of Theorem 1.1.

PROOF. To obtain $K_\omega(\psi(z), \psi(\zeta))$ modulo a C^∞-smooth kernel on $(z, \zeta) \in \overline{\mathbb{C}_+} \cap N \times I$, we replace g by the Dirac measure δ_ζ in (4.2). Since $D^p_{\omega_\psi}(\delta_\zeta)$ and $H_{\omega_\psi}(\delta_\zeta)$ are distributions on $\mathbb{R}$ we see that

$$K_\omega(\psi(z), \psi(\zeta)) \equiv \frac{1}{2\pi} \frac{1}{\omega_\psi(\zeta)} \frac{-1}{(z-\zeta)^2} - J \circ D^p_{\omega_\psi}(\delta_\zeta)(z)$$
$$+ J \circ H_{\omega_\psi}(\delta_\zeta)(z) \mod C^\infty(\overline{\mathbb{C}_+} \cap N \times I),$$

where J is the Cauchy transform defined in Section 2. From Lemma 2.3, we have

$$J \circ D^p_\omega(\delta_\zeta)(z) \equiv \frac{1}{2\pi} \int_0^\infty \sigma_{D^p_\omega}(z, \xi) e^{i(z-\zeta)\xi} \, d\xi \mod C^\infty(\overline{\mathbb{C}_+} \cap N \times I),$$

where the symbol $\sigma_{D^p_\omega}$ is extended as described there; the analogous relation holds for $J \circ H_\omega$. To treat this integral, we need a simple lemma which we state without proof.

LEMMA 4.3. _Fix_ $0 < m < M < \infty$. _Suppose_ $f(\xi) \in C^\infty([0, \infty[)$ _is identically_ 1 _for_ $\xi > M$, _and identically_ 0 _for_ $0 \le \xi < m$. _Then for_ $z \in \mathbb{C}_+$, _we have_

$$\int_0^\infty \frac{f(\xi) e^{iz\xi}}{\xi^k} \, d\xi = \begin{cases} -\frac{1}{z^2} + g_{-1}(z) & \text{for } k = -1, \\ \frac{i}{z} + g_0(z) & \text{for } k = 0, \\ \frac{i^{k-1} z^{k-1}}{(k-1)!} \log\left(\frac{i}{z}\right) + g_k(z) & \text{for } k = 1, 2, 3, \ldots, \end{cases}$$

for some entire functions g_k. _Moreover, if_ $a \in S^{-k}(\mathbb{C} \times \mathbb{R})$, $k \ge 2$, _then the function_

$$A(z) = \int_0^\infty g(z, \xi) e^{iz\xi} \, d\xi, \qquad z \in \overline{\mathbb{C}_+},$$

belongs to $C^{k-2}(\overline{\mathbb{C}_+})$.

From the form of the asymptotic series for the operators D^p_ω and H_ω and Lemma 4.3, it follows that

$$- J \circ D^p_\omega(\delta_\zeta)(z) \equiv \frac{d_{-1}(z)}{2\pi} \frac{-1}{(z-\zeta)^2} + \frac{d_0(z)}{2\pi} \frac{i}{z-\zeta}$$
$$+ \sum_{k=1}^n \frac{d_k(z)}{2\pi} \frac{i^{k-1}(z-\zeta)^{k-1}}{(k-1)!} \log\left(\frac{i}{z-\zeta}\right) \mod C^{n-1}(\overline{\mathbb{C}_+} \cap N \times I),$$

and

$$J \circ H_\omega(\delta_\zeta)(z) \equiv \frac{h_0(z)}{2\pi} \frac{i}{z-\zeta}$$
$$+ \sum_{k=1}^n \frac{h_k(z)}{2\pi} \frac{i^{k-1}(z-\zeta)^{k-1}}{(k-1)!} \log\left(\frac{i}{z-\zeta}\right) \mod C^{n-1}(\overline{\mathbb{C}_+} \cap N \times I),$$

for every integer $n = 1, 2, 3, \ldots$.

Since the Bergman kernel is anti-analytic in its second arguments, we can, by appealing to the Cauchy integral formula, extend the expansion to include $\zeta \in \overline{\mathbb{C}_+} \cap N$. Actually, we only need to substitute $\bar\zeta$ in the place of ζ. From the symmetry of K_ω, we also see that $\gamma_k(\zeta) = \overline{d_{k+1}(\zeta)} + \overline{h_{k+1}(\zeta)}$, $k = 0, 1, 2, \ldots$. An explicit calculation with the above formulas gives

$$
\begin{aligned}
&d_{-1}(x) = \tfrac{1}{\omega(x)}, \\
(4.7)\qquad &d_0(x) = i\frac{\partial\omega(x)}{\omega^2(x)}, \qquad\qquad h_0(x) = -i\frac{\partial\omega(x)}{\omega^2(x)}, \\
&d_1(x) = -\tfrac14\frac{\Delta\log\omega(x)}{\omega(x)}, \qquad h_1(x) = -\tfrac14\frac{\Delta\log\omega(x)}{\omega(x)},
\end{aligned}
$$

which yields the formula for γ_0. From this we also obtain the singular part of $\frac{\Psi_1(z,\zeta)}{(z-\zeta)^2}$, as stated in the theorem. $\qquad\square$

For two complex variables, L. Boutet de Monvel proved that the singularity resolution of the Bergman kernel does not contain a logarithmic part locally around a point at the boundary if and only if the intersection of a neighborhood of the point with the boundary is the image of a part of a sphere under a holomorphic mapping [5]. In our setting, we have the following corresponding result.

COROLLARY 4.4. *Suppose that $\Delta\log\omega(\zeta)$ vanishes at infinite order at the point $\zeta \in I$. Then the asymptotic expansion for the Bergman kernel does not contain a logarithmic term. This is in particular true when the weight locally equals the modulus of some zero-free analytic function.*

PROOF. Define ϕ such that $\omega = e^\phi$. We begin by showing that all the functions d_n, and h_n vanish at ζ for $n \geq 1$. By induction one can show that $\partial^k\omega(\zeta) = P_k(\partial^k\phi(\zeta), \partial^{k-1}\phi(\zeta), \ldots, \partial\phi(\zeta))\omega(\zeta)$, where P_k is a polynomial in $k-1$ variables. Since ϕ is harmonic at infinite order at ζ, it follows that

$$
\partial^{k-1}\bar\partial\omega(\zeta) = P_k(\partial^k\phi(\zeta), \partial^{k-1}\phi(\zeta), \ldots, \partial\phi(\zeta))\bar\partial\phi(\zeta)\omega(\zeta),
$$

and hence $\partial^{k-1}\bar\partial\omega(\zeta)\omega(\zeta) = \partial^{k-1}\omega(\zeta)\bar\partial\omega(\zeta)$. This shows that $r_{2,k}(\zeta) = 0$ for all $k = 2, 3, 4, \ldots$. The recursion formulas in Proposition 4.1 then shows that that $r_{k,l}(\zeta) = 0$ for all $k = 2, 3, 4, \ldots$, and $l = k, k+1, k+2, \ldots$. Therefore, the functions $d_n(\zeta)$ are zero for all $n \geq 1$. For similar reasons, the same holds for the functions $\{h_n\}_{n=1}^\infty$ as well. The result follows from the expansion of the singularity of the Bergman kernel. If ω is locally the modulus of some zero-free analytic function, then ϕ is harmonic, which yields the last assertion. $\qquad\square$

5. The harmonic Bergman kernel

If Ω is simply connected, then every harmonic function h on Ω can be decomposed into a sum of an analytic and an anti-analytic function on Ω. For a finitely connected domain Ω, the situation is slightly more complicated. Let n be the number of holes in Ω, and take n points, $\alpha_1, \alpha_2, \ldots, \alpha_n$, one from each hole. Then every harmonic function h on Ω can be written as

$$
h(z) = f_1(z) + \overline{f_2(z)} + \sum_{k=1}^n c_k \log|z - \alpha_k|, \qquad z \in \Omega,
$$

where $f_1, f_2 \in \mathcal{O}(\Omega)$ (the space of all analytic functions on Ω), and $c_1, c_2, \ldots, c_n$ are constants. Under an additional condition, we have a similar decomposition for the weighted harmonic Bergman space $HL^2(\Omega, \omega)$. Namely, the equality

$$
(5.1)\qquad HL^2(\Omega, \omega) = A^2(\Omega, \omega) + \overline{A^2(\Omega, \omega)} + \operatorname{span}\{\log|\cdot - \alpha_k| : k = 1, \ldots, n\}
$$

holds if and only if there exists a constant $\theta < 1$ such that

$$
(5.2)\qquad \left|\int_\Omega f^2(z)\omega(z)\, dA(z)\right| \leq \theta \int_\Omega |f(z)|^2\omega(z)\, dA(z),
$$

for all $f \in A^2(\Omega, \omega)$ such that

(5.3) $$\int_\Omega f(z)\omega(z) = 0.$$

This was proved in [**18**, Proposition 4.6] for the constant weight 1. Here, we have introduced the complex-conjugated space $\overline{A^2(\Omega, \omega)} = \{\bar{f} : f \in A^2(\Omega, \omega)\}$ and we let $A_0^2(\Omega, \omega)$ denote the space of functions $f \in A^2(\Omega, \omega)$ which satisfy (5.3). Note that the right hand side of (5.1) is always a subset of the left hand side. We call (5.2) the *weighted Friedrichs inequality*, and if it holds, then $A^2(\Omega, \omega) + \overline{A^2(\Omega, \omega)}$ is a closed subspace of $HL^2(\Omega, \omega)$ consisting of those harmonic functions which have well-defined harmonic conjugates in Ω. We show below how this enables us to develop the harmonic Bergman kernel in terms which involve the analytic and anti-analytic Bergman kernels. To prove that all terms in this sum, except for the analytic and anti-analytic Bergman kernels, are C^∞-smooth when Ω and ω satisfy the condition of Theorem 1.2, we use a quadrature identity for harmonic functions in combination with the singularity resolution of K_ω. We close the section with a discussion on the Friedrichs operator.

First, we need a result on orthogonal projections in Hilbert spaces due to A. Lenard [**17**].

THEOREM 5.1 (Lenard). *Suppose that E and F are two closed subspaces of a Hilbert space H, $E \cap F = \{0\}$, and that $\|P_E P_F\| < 1$ in operator norm, where P_E and P_F are the orthogonal projection onto E and F, respectively. Then the orthogonal projection onto $E \vee F$ is given by*

(5.4) $$P_{E \vee F} = (\text{Id} - P_F)(\text{Id} - P_E P_F)^{-1} P_E + (\text{Id} - P_E)(\text{Id} - P_F P_E)^{-1} P_F,$$

where $E \vee F$ is the smallest closed subspace which contain both E and F.

The operator $P_E P_F$ is called the *gap-operator* between the subspaces E and F.

Before we apply Lenard's formula, we introduce a relevant integral operator. Let T_ω be the linear operators defined for $f \in L^2(\Omega, \omega)$ by

(5.5) $$T_\omega f(z) = \int_\Omega T_\omega(z, \zeta) f(\zeta) \omega(\zeta) \, dA(\zeta), \qquad z \in \Omega,$$

with integral kernels

$$T_\omega(z, \zeta) = \int_\Omega K_\omega(z, \eta) K_\omega(\zeta, \eta) \omega(\eta) \, dA(\eta), \qquad z, \zeta \in \Omega,$$

We point out that all integral kernels considered here are with respect to the weighted measure $\omega \, dA$, and we shall in the sequel identify an integral operator on $L^2(\Omega, \omega)$ with its integral kernel. The symbols K_ω and Q_ω thus denote not only the analytic and harmonic Bergman kernel, but also the orthogonal projection onto the analytic and harmonic functions in $L^2(\Omega, \omega)$. Moreover, for an integral operator T on $L^2(\Omega, \omega)$ with integral kernel $T(z, \zeta)$, we write $\text{Re}\, T = \frac{1}{2}(T + \overline{T})$, where $\overline{T}$ is obtained from T by replacing the integral kernel by its complex conjugate, that is, for $f \in L^2(\Omega, \omega)$,

(5.6) $$\overline{T}f(z) = \int_\Omega \overline{T(z, \zeta)} f(\zeta) \omega(\zeta) \, dA(\zeta), \qquad z \in \Omega.$$

If n is an positive integer, then T^n denotes the composition of T with itself n times.

We shall now apply Lenard's theorem with $E = A_0^2(\Omega, \omega)$ and $F = \overline{A_0^2(\Omega, \omega)}$. The reproducing kernel for $A_0^2(\Omega, \omega)$ is

$$K_{\omega,0}(z, \zeta) = K_\omega(z, \zeta) - |\Omega|_\omega^{-1}, \qquad z, \zeta \in \Omega,$$

where $|\Omega|_\omega = \int_\Omega \omega \, dA$. In analogy with (5.5), we define the operator $T_{\omega,0}$ with the Bergman kernel $K_{\omega,0}$ instead. One can show that $A^2(\Omega, \omega)$ satisfies Friedrichs' inequality (5.2) if and only if $\|P_E P_F\| = \|T_{\omega,0}\| < 1$ in operator norm. From now on, we suppose that Friedrichs' inequality holds for $A^2(\Omega, \omega)$. Then $\widetilde{HL}_0^2(\Omega, \omega) = A_0^2(\Omega, \omega) + \overline{A_0^2(\Omega, \omega)}$ is a closed subspace of $HL^2(\Omega, \omega)$, and the orthogonal projection $\widetilde{Q}_{\omega,0}$ onto $\widetilde{HL}_0^2(\Omega, \omega)$ equals

(5.7) $(\mathrm{Id} - \overline{K_{\omega,0}})(\mathrm{Id} - T_{\omega,0})^{-1} K_{\omega,0} + (\mathrm{Id} - K_{\omega,0})(\mathrm{Id} - \overline{T_{\omega,0}})^{-1} \overline{K_{\omega,0}},$

according to (5.4). It is clear that

(5.8) $\widetilde{Q}_\omega(z, \zeta) = \widetilde{Q}_{\omega,0}(z, \zeta) + |\Omega|_\omega, \qquad z, \zeta \in \Omega,$

is the Bergman kernel for $\widetilde{HL}^2(\Omega, \omega) = A^2(\Omega, \omega) + \overline{A^2(\Omega, \omega)}$, which equals $HL^2(\Omega, \omega)$ in the simply connected case. If Ω is multiply connected with connectivity n, we choose n points $\alpha_1, \alpha_2, \ldots, \alpha_n$, one from each bounded component of the complement of Ω. Define the functions h_k, $k = 1, \ldots, n$, by

$$h_k(z) = \log|z - \alpha_k| - \widetilde{Q}_\omega(\log|\cdot - \alpha_k|)(z), \qquad z \in \Omega.$$

These functions are clearly orthogonal to $\widetilde{HL}^2(\Omega, \omega)$, and therefore, by applying the Gram–Schmidt orthogonalization procedure, we obtain an orthonormal basis $\{g_k\}_{k=1}^n$ for the orthogonal complement to $\widetilde{HL}^2(\Omega, \omega)$ in $HL^2(\Omega, \omega)$. The weighted harmonic Bergman kernel is then given by

(5.9) $\displaystyle Q_\omega(z, \zeta) = \widetilde{Q}_\omega(z, \zeta) + \sum_{k=1}^n g_k(z) g_k(\zeta), \qquad z, \zeta \in \Omega.$

For the rest of this section, we shall assume that Ω is a bounded domain in $\mathbb{C}$ with C^∞-smooth boundary, and that $\omega \in C^\infty(\overline{\Omega})$ is bounded from below. Before we prove Theorem 1.2, we need some preliminary results.

PROPOSITION 5.2. *If $g \in C^\infty(\overline{\Omega})$, then $K_\omega g \in C^\infty(\overline{\Omega})$.*

PROOF. By elliptic regularity, we know that the solution to the Dirichlet problem for $\bar{\partial} \omega^{-1} \partial$ in Ω belongs $C^\infty(\overline{\Omega})$ provided that the data f does. Thus,

$$z \mapsto \int_\Omega G_\omega(z, \zeta) f(\zeta) \omega(\zeta) \, dA(\zeta) \in C^\infty(\overline{\Omega})$$

for every $f \in C^\infty(\overline{\Omega})$. From the relation between the Bergman kernel K_ω and G_ω we have

$$\int_\Omega K_\omega(z, \zeta) g(\zeta) \omega(\zeta) \, dA(\zeta) = g(z) - \frac{1}{\omega(z)} \frac{\partial}{\partial z} \left(\int_\Omega G_\omega(z, \zeta) \bar{\partial} g(\zeta) \, dA(\zeta) \right).$$

The proposition now follows. $\square$

5.1. Quadrature identities. In [1], D. Aharonov and H. S. Shapiro made the following definition. A domain $\Omega \subset \mathbb{C}$ is a *quadrature domain* for analytic functions if there exists a distribution u with support on a finite set of points in Ω such that

$$u(f) = \int_\Omega f(z) \, dA(z)$$

for all $f \in A^2(\Omega)$. This can be generalized in two directions. Instead of distributions, we could consider integration of f with respect to a general measure with support in Ω on the left hand side, and we could also require that the identity should hold for all harmonic functions in $L^2(\Omega, \omega)$. If

$$(5.10) \qquad \int h(z) \, d\mu(z) = \int_\Omega h(z) \, dA(z)$$

holds for all bounded harmonic functions on Ω for some measure μ with compact support in Ω, we say that Ω is a *generalized quadrature domain* for harmonic functions. These ideas can of course be pursued in weighted Bergman spaces as well. The next proposition is a result in this direction.

PROPOSITION 5.3. *There exists a function $v \in C^\infty(\overline{\Omega})$ which vanishes to infinite order at the boundary such that*

$$(5.11) \qquad \int_\Omega h\omega \, dA = \int_\Omega hv \, dA,$$

holds for all area-integrable harmonic functions on Ω.

PROOF. Consider the boundary value problem

$$\begin{cases} \Delta\phi(z) = \omega(z), & z \in \Omega, \\ \phi(z) = 0, & z \in \partial\Omega. \end{cases}$$

Elliptic regularity theory shows that $\phi \in C^\infty(\overline{\Omega})$. Let the function h be harmonic on Ω and continuously differentiable on $\overline{\Omega}$. By Green's second formula, we have

$$\int_\Omega h\omega \, dA = \int_\Omega (h\Delta\phi - \phi\Delta h) \, dA = \int_{\partial\Omega} (h\partial_n\phi - \phi\partial_n h) \, ds = \int_{\partial\Omega} h\partial_n\phi \, ds.$$

According to Borel's theorem (Theorem 1.3), there exists a function $\nu \in C^\infty(\overline{\Omega})$ with boundary values zero and normal derivative equal to $\partial_n\phi$, such that $\Delta\nu$ vanishes to infinite order at the boundary. Again, by Green's second formula,

$$\int_{\partial\Omega} h\partial_n\phi \, ds = \int_{\partial\Omega} (h\partial_n\nu - \nu\partial_n h) \, ds = \int_\Omega (h\Delta\nu - \nu\Delta h) \, dA = \int_\Omega h\Delta\nu \, dA.$$

The function $v = \Delta\nu$ fulfills (5.11) for all harmonic functions in $C^1(\overline{\Omega})$. By an approximation argument, we can extend the result to all integrable harmonic functions on Ω.   □

The next result says that T_ω in (5.5) is a smoothing operator.

PROPOSITION 5.4. *The integral kernel for T_ω belongs to $C^\infty(\overline{\Omega} \times \overline{\Omega})$. As a consequence, T_ω is compact and maps $L^2(\Omega, \omega)$ continuously into $C^\infty(\overline{\Omega}) \cap \mathcal{O}(\Omega)$.*

PROOF. Let v be the function described in the previous proposition. For $z, \zeta \in \Omega$, we have

(5.12)
$$T_\omega(z, \zeta) = \int_\Omega K_\omega(z, \eta) K_\omega(\zeta, \eta) \omega(\eta) \, dA(\eta)$$
$$= \int_\Omega K_\omega(z, \eta) K_\omega(\zeta, \eta) v(\eta) \, dA(\eta),$$

since the integrand is anti-analytic in η. By the asymptotic expansion of the weighted Bergman kernel and the properties of v, it is clear that

$$\sup_{z, \zeta, \eta \in \Omega} \left| \partial_z^m K_\omega(z, \eta) \partial_\zeta^n K_\omega(\zeta, \eta) v(\eta) \right| < \infty,$$

for all positive integers m and n. Together with (5.12), this shows the result, since the kernel $T_\omega(z, \zeta)$ is analytic in both z and ζ. $\qquad\square$

We can now complete the proof of Theorem 1.2.

PROOF OF THEOREM 1.2. The first term in the representation (5.7) for $\widetilde{Q}_{\omega,0}$ can be written as

$$K_{\omega,0} + T_{\omega,0}(\text{Id} - T_{\omega,0})^{-1} K_{\omega,0} - \overline{K_{\omega,0}}(\text{Id} - T_{\omega,0})^{-1} K_{\omega,0}.$$

A simple argument, which uses that $(\text{Id} - T_{\omega,0})^{-1}$ is a bounded operator and $T_\omega(z, \zeta) \in C^\infty(\overline{\Omega} \times \overline{\Omega})$, shows that the integral kernels for the second and third of these terms belongs $C^\infty(\overline{\Omega} \times \overline{\Omega})$. Since the integral kernel for the second term of (5.7) is just the complex conjugate of the first, we have $\widetilde{Q}_\omega(z, \zeta) \equiv 2 \operatorname{Re} K_\omega(z, \zeta)$ modulo C^∞-smooth terms. This completes the proof in the simply connected case. In the multiply connected case, Proposition 5.2 implies that all the functions $\{g_k\}_{k=1}^n$ in (5.9) belong to $C^\infty(\overline{\Omega})$ and therefore

$$Q_\omega(z, \zeta) \equiv \widetilde{Q}_\omega(z, \zeta) \equiv 2 \operatorname{Re} K_\omega(z, \zeta). \qquad \mod C^\infty(\overline{\Omega} \times \overline{\Omega}),$$

$\qquad\square$

5.2. The Friedrichs operator. We close this section with some remarks on the eigenvalues of the Friedrichs operator. The *weighted Friedrichs operator* F_ω associated to the space $A^2(\Omega, \omega)$ is the conjugate-linear operator defined as $F_\omega = K_\omega C$, where C is the conjugation operator $Cf = \bar{f}$. Explicitly, we have, for $f \in A^2(\Omega, \omega)$,

$$F_\omega f(z) = \int_\Omega K_\omega(z, \zeta) \bar{f}(\zeta) \omega(\zeta) \, dA(\zeta), \qquad z \in \Omega.$$

Restricted to $A^2(\Omega, \omega)$, we have the relation $F_\omega = T_\omega C$. The square of the Friedrichs operator $S_\omega = F_\omega^2 = T_\omega^2$ is a self-adjoint linear operator and satisfies the operator inequality $0 \leq \langle S_\omega f, f \rangle_\omega \leq \|f\|_\omega^2$ for all $f \in A^2(\Omega, \omega)$. If the Friedrichs operator is compact, then, by the general theory of Hilbert spaces, there exists a complete orthonormal family of analytic functions $\{e_n\}_{n=0}^\infty$ in $A^2(\Omega, \omega)$, and a corresponding sequence of positive constants $\{\lambda_n\}_{n=0}^\infty$, which tends monotonically to zero as $n \to \infty$, with the property that for every function $f \in A^2(\Omega, \omega)$, we have

$$F_\omega f(z) = \sum_{n=0}^\infty \lambda_n \langle e_n, f \rangle_\omega e_n(z).$$

Since F_ω maps constants to its complex conjugate, it is clear that $\lambda_0 = 1$, and e_0 is a constant function with norm 1.

The singular numbers of a compact operator T are defined to be the eigenvalues of the positive operator $(T^*T)^{1/2}$. For a positive operator, the sequence of singular numbers coincides with the sequence of eigenvalues. It follows that $\{\lambda_n\}_{n=0}^\infty$ are the singular numbers for T_ω, and that $\{\lambda_n^2\}_{n=0}^\infty$ the eigenvalues for S_ω. An operator is said to belong to the Schatten p-class $\mathfrak{S}_p$ if its sequence of singular numbers is in l^p. Since the linear operator S_ω maps $L^2(\Omega, \omega) \to C^\infty(\overline{\Omega}) \cap \mathcal{O}(\Omega)$, it follows from the general theory of singular numbers [4] that S_ω lies in all Schatten classes with $p > 0$. This condition is equivalent to having the singular numbers λ_n decay faster than the reciprocal of any polynomial in n.

COROLLARY 5.5. *The operators T_ω and S_ω belong to $\mathfrak{S}_p$ for each $p > 0$.*

In [19], M. Putinar and H. S. Shapiro showed that if Ω admits a compactly supported quadrature measure, then the eigenvalues for the Friedrichs operator S satisfy $\lim_{n\to\infty} \lambda_n^{1/n} < 1$. Their proof extends to weighted Bergman spaces with compactly supported quadrature measures.

6. Final comments

Theorem 1.1 can also be obtained by a method which is close to Fefferman's idea. We shall describe this briefly when the domain is the unit disk $\mathbb{D}$. Let $\omega \in C^\infty(\overline{\mathbb{D}})$ be a strictly positive weight. We claim that it is enough to consider weights which have boundary values 1 everywhere on $\mathbb{T}$. Indeed, for any weight $\widetilde{\omega}$, there exists a non-zero analytic function $f \in C^\infty(\overline{\mathbb{D}})$ such that $|f|^2$ and $\widetilde{\omega}$ have the same boundary values. If we put $\omega = \widetilde{\omega}/|f|^2$, then the relation between the weighted Bergman kernels is

$$K_{\widetilde{\omega}}(z, \zeta) = \frac{1}{f(z)\overline{f(\zeta)}} K_\omega(z, \zeta), \qquad z, \zeta \in \mathbb{D}.$$

The Bergman kernel for the constant weight 1 is given by

$$K(z, \zeta) = \frac{1}{\pi} \frac{1}{(1 - z\bar\zeta)^2}, \qquad z, \zeta \in \mathbb{D}.$$

By using the reproducing property of the Bergman kernels, we obtain the following relation between K_ω and K:

(6.1) $$K_\omega(z, \zeta) - \int_{\mathbb{D}} K_\omega(z, \eta) E_\omega(\eta, \zeta)\, dA(\eta) = K(z, \zeta),$$

where

(6.2) $$E_\omega(z, \zeta) = \int_{\mathbb{D}} K(z, \eta)\, (1 - \omega(\eta))\, K(\eta, \zeta) f(\zeta)\, dA(\eta).$$

We associate the operator $E_\omega : L^2(\mathbb{D}) \to A^2(\mathbb{D})$ to the integral kernel $E_\omega(z, \zeta)$ by

$$E_\omega f(z) = \int_{\mathbb{D}} E_\omega(z, \zeta) f(\zeta)\, dA(\zeta), \qquad z \in \mathbb{D}.$$

Provided that the operator norm of E_ω is strictly less than 1, we can develop the Bergman kernel in a Neumann series

(6.3) $$K_\omega(z, \zeta) = K(z, \zeta) + \sum_{l=1}^\infty E_\omega^l(z, \zeta), \qquad z, \zeta \in \mathbb{D},$$

where $E_\omega^l(z, \zeta)$ is the integral kernel for E_ω composed with itself l times (this is a consequence of (6.1) interpreted as an operator relation). The singularity of the weighted Bergman kernel can now be captured by analyzing the Neumann series. One shows that the regularity of the terms increases as the power l increases, and each term, except for the first two, has an expansion

$$\psi_1(z, \zeta) \log\left(\frac{1}{1 - z\bar{\zeta}}\right) + \psi_2(z, \zeta)$$

for some $\psi_1, \psi_2 \in C^\infty(\overline{\mathbb{D}} \times \overline{\mathbb{D}})$. For the first term in the series in (6.3), we have

$$E_\omega(z, \zeta) = \frac{\phi(z, \zeta)}{(1 - z\bar{\zeta})^2} + \psi(z, \zeta) \log\left(\frac{1}{1 - z\bar{\zeta}}\right),$$

with $\phi, \psi \in C^\infty(\overline{\mathbb{D}} \times \overline{\mathbb{D}})$.

If E_ω does not have operator norm strictly less that 1, then a modified version of (6.3) can be used. Consider the subspace of functions in $A^2(\mathbb{D})$ which vanish at least to order n at the origin, where n is a positive integer. Let K_n be the reproducing kernel for this space:

$$K_n(z, \zeta) = \frac{1}{\pi} \frac{\left((n + 1)(z\bar{\zeta})^n + n(z\bar{\zeta})^{n+1}\right)}{(1 - z\bar{\zeta})^2}, \qquad z, \zeta \in \mathbb{D}.$$

If n is sufficiently large, then the operator $E_{\omega,n}$, obtained by replacing K with K_n in (6.2), has norm strictly less than one, and the analogue of (6.3) converges to the reproducing kernel for the corresponding subspace of $A^2(\mathbb{D}, \omega)$. The difference between this kernel and K_ω is a polynomial in z and $\bar{\zeta}$ of order at most $n - 1$, and it is therefore smooth on $\overline{\mathbb{D}} \times \overline{\mathbb{D}}$.

References

[1]   D. Aharonov and H. S. Shapiro, *Domains on which analytic functions satisfy quadrature identities*, J. Analyse Math. **30** (1976), 39–73.

[2]   M. Beals, C. Fefferman and R. Grossman, *Strictly pseudoconvex domains in $\mathbf{C}^n$*, Bull. Amer. Math. Soc. (N.S.) **8** (1983), no. 2, 125–322.

[3]   S. Bergman, *The Kernel Function and Conformal Mapping*, Mathematical Surveys, No. 5, American Mathematical Society, New York, N. Y., 1950.

[4]   M. Š. Birman and M. Z. Solomjak, *Estimates for the singular numbers of integral operators*, Uspehi Mat. Nauk **32** (1977), no. 1 (193), 17–84, 271.

[5]   L. Boutet de Monvel, *Le noyau de Bergman en dimension 2*, Séminaire sur les Équations aux Dérivées Partielles 1987–1988, École Polytech., Palaiseau, 1988, pp. Exp. No. XXII, 13.

[6]   L. Boutet de Monvel, *Singularity of the Bergman kernel*, Complex Geometry (Osaka, 1990), Dekker, New York, 1993, pp. 13–29.

[7]   L. Boutet de Monvel and J. Sjöstrand, *Sur la singularité des noyaux de Bergman et de Szegő*, Journées: Équations aux Dérivées Partielles de Rennes, Astérisque, No. 34–35 (1975) (1976), 123–164.

[8]   D. Catlin, *The Bergman kernel and a theorem of Tian*, Analysis and Geometry in Several-Complex Variables (Katata, 1997), Birkhäuser Boston, Boston, MA, 1999, pp. 1–23.

[9]   C. Fefferman, *The Bergman kernel and biholomorphic mappings of pseudoconvex domains*, Invent. Math. **26** (1974), 1–65.

[10]   P. R. Garabedian, *A partial differential equation arising in conformal mapping*, Pacific J. Math. **1** (1951), 485–524.

[11]   A. Grigis and J. Sjöstrand, *Microlocal Analysis for Differential Operators*, Cambridge University Press, Cambridge, 1994, An introduction.

[12]   H. Hedenmalm, S. Jakobsson and S. Shimorin, *A biharmonic maximum principle for hyperbolic surfaces*, J. Reine Angew. Math. **550** (2002), 25–75.

[13] H. Hedenmalm, B. Korenblum and K. Zhu, *Theory of Bergman spaces*, Graduate Texts in Mathematics, 199, Springer-Verlag, New York, 2000.

[14] L. Hörmander, *The Analysis of Linear Partial Differential Operators.* III, Springer-Verlag, Berlin, 1985.

[15] L. Hörmander, *The Analysis of Linear Partial Differential Operators.* I, second ed., Springer-Verlag, Berlin, 1990.

[16] S. Jakobsson, *Applications of semigroups to biharmonic Green functions*, J. Reine Angew. Math. **563** (2003), 53–82.

[17] A. Lenard, *The numerical range of a pair of projections*, J. Functional Analysis **10** (1972), 410–423.

[18] M. Putinar and H. S. Shapiro, *The Friedrichs operator of a planar domain* II, preprint, Royal Institute of Technology, Department of Mathematics, Stockholm, Sweden, 1999.

[19] M. Putinar and H. S. Shapiro, *The Friedrichs operator of a planar domain*, Operator Theory: Advances and Applications **113** (2000), 303–330.

[20] M. E. Taylor, *Pseudodifferential Operators*, Princeton University Press, Princeton, N.J., 1981.

SWEDISH DEFENCE RESEARCH AGENCY, FOI, SE-164 90 STOCKHOLM, SWEDEN
E-mail address: stefan.jakobsson@foi.se

Contemporary Mathematics
Volume **404**, 2006

Blaschke Sets for Bergman Spaces

Boris Korenblum

ABSTRACT. We characterize subsets S of the open unit disk $\mathbb{D}$ such that every zero sequence for a Bergman space A^p, $p > 0$, with elements in S is Blaschke.

1. Introduction

The following definition is an extension of the notion of a Blaschke set introduced by Krzysztof Bogdan [**B**].

DEFINITION. We call $S \subset \mathbb{D}$ a Blaschke set for a class X of analytic functions on $\mathbb{D} = \{z \in \mathbb{C} : |z| < 1\}$ if
(i) whenever $0 \not\equiv f \in X$, and $\{z_n\}_n$ are the zeros of f (counting multiplicities), with $z_n \in S$, the Blaschke condition holds:

$$(1) \qquad \sum_n (1 - |z_n|) < \infty \; ;$$

(ii) whenever $Z = \{z_n\}_n$ is a Blaschke sequence (i.e., (1) holds), with $z_n \in S$, there is an $f \in X$ whose zero sequence is Z.

REMARK. If X is made up of functions of bounded Nevanlinna characteristic, then this definition reduces to (ii). If $H^\infty \subset X$, it reduces to (i).

EXAMPLES.
1. Every subset of $\mathbb{D}$ is a Blaschke set for H^p, $0 < p < \infty$.
2. For analytic Lipschitz classes $Lip_\alpha(\mathbb{D})$, $\alpha > 0$, as well as for

$$A^\infty = \{f : f^{(n)} \in H^\infty, \forall n \geq 0\},$$

Blaschke sets are characterized by

$$(2) \qquad \int_0^{2\pi} \log \operatorname{dist}(e^{it}, S) dt > -\infty$$

where dist denotes the Euclidean distance. Note that for $Lip_\alpha(\mathbb{D})$ and A^∞, the zero sequences Z are characterized by (1) and (2), with S replaced by Z.

2000 *Mathematics Subject Classification.* Primary 30C15.

3. The Blaschke sets S for the class $\mathcal{D}$ of analytic functions with finite Dirichlet integral are characterized by (2) (see [**B**]). Note that $\mathcal{D}$-zero sequences cannot be described this way because there are $f \in \mathcal{D}$ whose zeros come arbitrarily close to every point of $\partial\mathbb{D}$ (see [**C**] and [**SS**]).

The purpose of this paper is to obtain a description of the Blaschke sets for Bergman spaces $A^p (p > 0)$ and growth spaces $A^{-\alpha} (\alpha > 0)$. Recall that A^p consists of functions f analytic on $\mathbb{D}$ such that

$$\|f\|_p^p = \int_{\mathbb{D}} |f(z)|^p \frac{dxdy}{\pi} < \infty \; ;$$

$A^{-\alpha}$ consists of analytic functions f with

$$\|f\|_{-\alpha} = \sup \left\{ (1 - |z|)^{\alpha} |f(z)| : z \in \mathbb{D} \right\} < \infty \; ;$$

We also consider the space

$$A^{-\infty} = \bigcup_{\alpha > 0} A^{-\alpha} = \bigcup_{p > 0} A^p.$$

We establish the following

THEOREM. *A set $S \subset \mathbb{D}$ is a Blaschke set for any of the spaces A^p, $A^{-\alpha}$, $A^{-\infty}$ if and only if* (2) *holds.*

To prove this theorem, we first reduce condition (2) to a form involving a collection of disjoint "tents" tightly surrounding S. The sufficiency of (2) then follows from the fact that "Stolz stars" $\mathcal{S}_F$ are $A^{-\infty}$-Blaschke sets if the entropy $\widehat{\kappa}(F)$ is finite (see (3) and [**HKZ**]). To prove the necessity of (2), we use some density concepts first introduced in [**K1**] and later refined in [**S**] and [**HKZ**].

ACKNOWLEDGMENTS. The author thanks Stefan Richter and Carl Sundberg for useful discussions. Special thanks are due to Carl Sundberg for bringing K. Bogdan's work [**B**] to the author's attention and to Alexander Borichev for sharing with the author an interesting observation (see the remark at the end of the paper).

2. An equivalent form of (2)

We assume that S contains a disk centered at 0 of radius $1/\sqrt{2}$.

We need some terminology.

A **tent** is an open subset h of $\mathbb{D}$ bounded by an arc $I \subset \partial\mathbb{D}$ of length less than $\pi/2$ and two straight lines through the endpoints of I forming with I an angle of $\pi/4$. The closed arc $\overline{I}$ will be called the **base** of the tent $h = h_I$. A tent h is said to **support** S if $h \cap S = \phi$ but $\overline{h} \cap \overline{S} \neq \phi$. A finite or countable collection of tents $\{h_n\}_n$ is a **belt** if h_n are pairwise disjoint and $\bigcup_n \overline{h}_n \supset \partial\mathbb{D}$. A collection of tents $\{h_n\}_n$ is an S-**belt** if h_n are pairwise disjoint, S-supporting, and $\bigcup_n \overline{h}_n \supset \partial\mathbb{D}\backslash S$. Note that an S-belt does not have to be a belt. If S is such that $\partial\mathbb{D}\backslash\overline{S} \neq \phi$, S-belts exist: to obtain one we start at an arbitrary point $\zeta_0 \in \partial\mathbb{D}\backslash\overline{S}$ and, moving counterclockwise, consecutively find points $\zeta_1, \zeta_2 \ldots$ such that the arcs between them are the bases of S-supporting tents; then we proceed similarly from ζ_0 in the opposite direction. We thus obtain a system of tents whose bases cover a component of $G = \partial\mathbb{D}\backslash\overline{S}$. Continuing this process for all the components, we obtain an S-belt.

An elementary computation shows that if $h = h_I$ is a tent supporting S, then

$$-|I| \log \frac{1}{|I|} - c|I| \leq \int_I \log \ \text{dist}(\zeta, S)|d\zeta| \leq -|I| \log \frac{1}{|I|} + c|I|$$

where c is a numerical constant. We thus obtain

LEMMA 1. *Let S be a subset of $\mathbb{D}$ such that $\partial\mathbb{D} \backslash \overline{S} \neq \phi$. Let $\{h_{I_n}\}_n$ be an S-belt. Then* (2) *holds if and only if*
(A) *the set $F_0 = \overline{S} \cap \partial\mathbb{D}$ has zero Lebesgue length;*
(B)

$$\sum_n \kappa(I_n) < \infty \ \text{where} \ \kappa(I) = |I| \log \frac{2\pi e}{|I|} \ .$$

($\kappa(I)$ *is called the κ-length of I).*

Note that (A) and (B) together are equivalent to

$$(3) \qquad \widehat{\kappa}(F) = \int_{\partial\mathbb{D}} \log \frac{2\pi}{d(\zeta, F)} |d\zeta| < \infty$$

where $F = F_0 \cup \Xi$ and Ξ consists of the endpoints of those bases such that $\overline{I}_n \subset G$; d denotes the angular distance.

The quantity $\widehat{\kappa}(F)$ is defined for all sets $F \subset \partial\mathbb{D}$ and is called the **entropy** of F. Closed sets with finite entropy are called **Beurling-Carleson sets.**

3. Sufficiency of (3)

Let $\Xi_1 \supset \Xi$ consist of **all** endpoints of the bases I_n (including those that are in F_0). Pick an increasing sequence $F_1 \subset F_2 \subset \ldots$ of finite subsets of Ξ_1 such that $\bigcup_n F_n = \Xi_1$. Then (3) implies

$$\lim_{n \to \infty} \widehat{\kappa}(F_n) = \widehat{\kappa}(F) \ .$$

Each F_n determines a belt whose tents are based on complementary arcs of F_n. Let H_n be the union of these tents. (Note that some of these tents are not S-supporting because they contain points from S.) The complement $\mathbb{D} \backslash H_n = \tau_n$ is a "Stolz Star", i.e., the union of Stolz angles with vertices in F_n and apertures of $\pi/2$.

Since $\widehat{\kappa}(F_n)$ are bounded, it follows that, whenever $0 \not\equiv f \in A^{-\infty}$, the Blaschke sums for those zeros of f lying in τ_n are bounded (see [**HKZ**, p. 118, Theorem 4.25]). We have $\sum_n \tau_n \supset S$ and $\tau_1 \subset \tau_2 \subset \ldots$, which implies that the Blaschke sum for the zeros of f lying in S is finite.

4. Necessity of (3)

Suppose now that $\widehat{\kappa}(F) = \infty$. Given an arbitrary fixed $p > 0$, we are going to construct a sequence $Z = \{z_n\}_n$, $z_n \in S$, such that Z is an A^p-zero sequence but

$$\sum_n (1 - |z_n|) = \infty.$$

In addition to the standard tools of $A^{-\infty}$-theory (density notions, premeasures, etc.), we will use some technical lemmas whose proofs are deferred to Section 5.

Recall that $F = F_0 \cup \Xi$ where $F_0 = \overline{S} \cap \partial\mathbb{D}$ and Ξ is a finite or countable set lying in $G = \partial\mathbb{D} \backslash F_0$. The cluster points (if any) of Ξ are in F_0.

We consider separately two cases depending on whether $\widehat{\kappa}(F_0)$ is infinite or finite.

Case 1. $\widehat{\kappa}(F_0) = \infty$. By Lemma 2 of Section 5, there is a sequence $\{\zeta_\nu\}_1^\infty$ of distinct points in F_0 such that the corresponding arcs $\{J_\nu\}_1^\infty$ between ζ_ν and $\zeta_{\nu+1}$ are pairwise disjoint, cover $\partial\mathbb{D}$, i.e., $\bigcup_\nu \overline{J}_\nu = \partial\mathbb{D}$, and $\widehat{\kappa}(\{\zeta_\nu\}) = \infty$, which is equivalent to

$$\sum_{\nu=1}^\infty \kappa(J_\nu) = \sum_{\nu=1}^\infty |J_\nu| \log \frac{2\pi e}{|J_\nu|} = \infty \ .$$

(Note that $\lim_{\nu\to\infty} \zeta_\nu = \zeta_1$.) Construct a premeasure (see [**K1**], [**K2**], [**HKZ**]) $d\mu = p|d\zeta| - d\sigma$ whose positive part has the density

$$p(\zeta) = \log \frac{2\pi}{d(\zeta, \{\zeta_\nu, \zeta_{\nu+1}\})}, \qquad \zeta \in J_\nu, \ \nu \geq 1 \ ,$$

and the negative singular part $-d\sigma$ puts on every point ζ_ν a Dirac mass equal to $-\kappa(J_\nu)$. Although both positive and negative parts are infinite on $\partial\mathbb{D}$, $d\mu$ is κ-bounded above, which means that there is a constant $c > 0$ such that for all arcs $I \subset \partial\mathbb{D}$

$$\mu(I) \leq c|I| \log \frac{2\pi e}{|I|} \ .$$

This enables us to consider a zero-free analytic function

$$f_\varepsilon(z) = \exp\left\{\varepsilon \int_{\partial\mathbb{D}} \frac{\zeta+z}{\zeta-z} d\mu(\zeta)\right\},$$

which is in A^p (or $A^{-\alpha}$) provided that $\varepsilon > 0$ is sufficiently small, and p (or α) are arbitrary but fixed positive numbers.

Now we use Lemma 3 of Section 5, to reduce all the singular masses at ζ_ν by a factor $1/2$ and compensate for that by extra zeros of high multiplicity at $z_\nu \in S$. We can ensure that the resulting function Φ is in A^p. The zeros z_ν of ϕ (counting multiplicities) form a non-Blaschke sequence of points from S (see Lemma 3 for details).

Case 2. $\widehat{\kappa}(F_0) < \infty$. Then we must have $\widehat{\kappa}(\Xi) = \infty$. Recall that Ξ includes all the endpoints of the base arcs of the S-belt that are in $G = \partial\mathbb{D}\backslash S$. Let $\{J_\nu\}_\nu$ be the sequence of these arcs arranged by decreasing lengths. Then $\widehat{\kappa}(F_0) < \infty$ together with $\widehat{\kappa}(\Xi) = \infty$ yield

$$\sum_{\nu=1}^\infty \kappa(J_\nu) = \sum_{\nu=1}^\infty |J_\nu| \log \frac{2\pi e}{|J_\nu|} = \infty \ .$$

It is always possible to find a decreasing sequence $1 > \lambda_1 > \lambda_2 > \cdots \to 0$ such that

$$\sum_{\nu=1}^\infty \lambda_\nu \kappa(J_\nu) = \infty \ .$$

Every $\overline{J}_\nu$ is the base of a tent h_{J_ν} that supports S; therefore there is at least one common point, say w_ν, in $\overline{S}$, $\overline{h}_{J_\nu}$ and $\mathbb{D}$. Take every w_ν and repeat it $[\frac{\lambda_\nu \kappa(J_\nu)}{1-|w_\nu|}]$ times. Let the resulting sequence be $Z = \{z_k\}_k$.

CLAIM. Z is a zero sequence for every A^p, $p > 0$. To prove the claim, we employ the notion of **upper asymptotic** κ**-density** of a sequence in $\mathbb{D}$. There are several equivalent definitions of this density. We will use the definition based on **radial stars** (see [**HKZ**]):

For an arbitrary finite set $M \subset \partial\mathbb{D}$, let r_M denote the union of radii from 0 to points in M. If $A = \{a_k\}_k$ is any sequence of points in $\mathbb{D}$, we form the **partial Blaschke sum** for A and r_M:

$$B(A, r_M) = \sum_k \{1 - |a_k| : a_k \in r_M\} ,$$

and define

(4) $$D^+(A) = \limsup_{\widehat{\kappa}(M) \to \infty} \frac{B(A, r_M)}{\widehat{\kappa}(M)}$$

where $\limsup$ is taken over all finite $M \subset \partial\mathbb{D}$.

The following result, although short of a full characterization of A^p-zero sets, is sharp enough for our purposes.

PROPOSITION. (See [**HKZ**, p. 130, Theorem 4.37].) *Let* $A = \{a_k\}_k$ *be a sequence of points in* $\mathbb{D}$ *and* $D^+(A)$ *be the upper asymptotic* κ*-density of* A. *If* $D^+(A) < \frac{1}{p}$, *then* A *is an* A^p*-zero sequence. If* $D^+(A) > \frac{1}{p}$, *then* A *is not an* A^p*-zero sequence.*

REMARK. This is a sharper version, due to Kristian Seip [**S**], of a similar but weaker result from [**K1**].

Now we can prove the claim by showing that $D^+(Z) = 0$. Let

$$Q = \left\{ q_\nu = \frac{w_\nu}{|w_\nu|} \right\}_\nu .$$

Every arc J_ν contains exactly one point from Q, namely q_ν. Obviously, for computing $D^+(Z)$, we can employ only those M that are finite subsets of Q. For such M, we have

$$B(Z, r_M) \leq \sum_\nu \left\{ \lambda_\nu \kappa(J_\nu) : q_\nu \in M \right\}$$

and

$$\widehat{\kappa}(M) \geq \sum_\nu \left\{ \kappa(J_\nu) : q_\nu \in M \right\}$$

(see Lemma 4 of Section 5). Therefore,

$$\frac{B(Z, r_M)}{\widehat{\kappa}(M)} \leq \sum_\nu \left\{ \lambda_\nu \kappa(J_\nu) : q_\nu \in M \right\} / \sum_\nu \left\{ \kappa(J_\nu) : q_\nu \in M \right\} .$$

It is convenient to use the following notations:

$$K(M) = \sum_\nu \left\{ \kappa(J_\nu) : q_\nu \in M \right\} ,$$

$$K_\lambda(M) = \sum_\nu \left\{ \lambda_\nu \kappa(J_\nu) : q_\nu \in M \right\} .$$

Let $\{M_n\}_n$ be a sequence of subsets of Q such that $\widehat{\kappa}(M_n) \to \infty$. Then we have

(5) $$\frac{B(Z, r_{M_n})}{\widehat{\kappa}(M_n)} \leq \frac{K_\lambda(M_n)}{K(M_n)} .$$

Suppose that $K(M_n) = O(1)(n \to \infty)$. Then obviously the left-hand side of (5) tends to 0. Also, if $K(M_n) \to \infty$, then the right-hand (and, with it, the left-hand) side of (5) tends to 0 because $\lambda_\nu \downarrow 0$. Therefore every sequence $\{M_n\}$, $M_n \subset Q$, $\widehat{\kappa}(M_n) \to \infty$, contains a subsequence $\{M_{n_k}\} = \{M'_k\}$, $n_1 < n_2 \ldots$, such that

$$\lim_{k\to 0} \frac{B(Z, r_{M'_k})}{\widehat{\kappa}(M'_k)} = 0 ,$$

which implies $D^+(Z) = 0$. Thus we have obtained a non-Blaschke A^p-zero sequence $\{z_k\}$ whose elements are in $\overline{S}$. Using now Lemma 5 of Section 5, we can replace z_k by nearby points $\tilde{z}_k$ from S so that the new sequence $\{\tilde{z}_k\}_k$ is still an A^p-zero sequence and non-Blaschke. This completes the proof of the Theorem.

5. Technical Lemmas

We give below the statement of the technical lemmas we used in proving the Theorem, together with a brief outline of their proofs.

DEFINITION. A sequence $\{\zeta_n\}_1^\infty$ of distinct points in $\partial\mathbb{D}$ is called $\mathbb{T}$-**monotone** if the open arcs I_n between ζ_n and ζ_{n+1} are pairwise disjoint and $\bigcup_n \overline{I} = \partial\mathbb{D}$. Note that it follows from this definition that $\lim_{n\to\infty} \zeta_n = \zeta_1$.

LEMMA 2. *Every closed set $F \subset \partial\mathbb{D}$ of infinite entropy contains a $\mathbb{T}$-monotone sequence $\{\zeta_n\}_n \subset F$ of infinite entropy.*

PROOF. We have

$$\widehat{\kappa}(F) = \int_{\partial\mathbb{D}} \log \frac{2\pi}{d(\zeta, F)} |d\zeta| = \infty .$$

By the Heine-Borel lemma, there is a point $\zeta_0 \in F$ such that every open arc J containing ζ_0 has the property

$$\int_J \log \frac{2\pi}{d(\zeta, F)} |d\zeta| = \infty .$$

Now we can find a nested system of open arcs such that $J_n \supset \overline{J}_{n+1}$, $\bigcap_n J_n = \{\zeta_0\}$, and a finite set $M_k \subset (J_n \backslash \overline{J}_{n+1}) \cap F$ such that

$$\int_{J_n \backslash \overline{J}_{n+1}} \log \frac{2\pi}{d(\zeta, M_n)} |d\zeta| \geq 1, \ n \geq 1 .$$

Taking the union $E = \bigcup_n M_n$ (or a suitable subset of E) and rearranging it in a sequence will prove the lemma. $\qquad\square$

LEMMA 3. *Let $f \in A^p (p > 0)$ have an "atomic singularity" at $z = 1$, i.e.,*

$$\limsup_{r\to 1^-} (1 - r) \log |f(r)| = -2m < 0 .$$

If $m_1 < m$, then
(i) $F(z) = e^{m_1 \frac{1+z}{1-z}} f(z)$ *is in A^p;*
(ii) *whenever $0 \neq \alpha_n \in \mathbb{D}$ and $\lim_{n\to\infty} \alpha_n = 1$, the function*

$$f_{\alpha_n}(z) = \left(\frac{\alpha_n - z}{1 - \overline{\alpha}_n z} \cdot \frac{|\alpha_n|}{\alpha_n} \right)^{N_n} F(z), \ where \ N_n = \left[\frac{m_1}{1 - |\alpha_n|} \right] ,$$

tends to f in the metric of A^p.

PROOF. (i) For any $r \in (0, \infty)$, the equation

$$\frac{1 - |z|^2}{|1 - z|^2} = r$$

defines a circle C_r internally tangent to $\partial \mathbb{D}$ at the point 1. Such circles are called **horocycles**. If f is in A^p and has atomic singularity m at 1, then the function

$$g(z) = e^{m \frac{1+z}{1-z}} f(z)$$

may not be in A^p; however, the integral

$$L(r) = \frac{1}{2\pi} \int_{C_r} |1 - \zeta|^2 |g(\zeta)|^p |d\zeta|$$

is finite and decreasing on $(0, \infty)$, and

$$\int_{\mathbb{D}} |f(z)|^p \frac{dxdy}{\pi} = \int_0^\infty e^{-mrp} L(r) dr < \infty .$$

This implies

$$\int_{\mathbb{D}} |F(z)|^p \frac{dxdy}{\pi} = \int_0^\infty e^{-(m-m_1)rp} L(r) dr < \infty .$$

(ii) then follows by the dominated convergence theorem. $\qquad \square$

LEMMA 4. *If $I \subset \partial \mathbb{D}$ is an arc, M is an arbitrary subset of $\partial \mathbb{D}$, and if at least one point from M is in $\overline{I}$, then*

$$\int_I \log \frac{2\pi}{d(\zeta, M)} |d\zeta| \geq \kappa(I) = |I| \log \frac{2\pi e}{|I|} .$$

PROOF. The minimum of the integral on the left for a given arc I is attained when M is a one-point set, and this point is one of the endpoints of I. A direct computation yields the required result. $\qquad \square$

LEMMA 5. *Let $f \in A^p$ have a zero at some point $a \in \mathbb{D}$. For arbitrary $\alpha \in \mathbb{D}$, define*

$$f_\alpha(z) = \frac{B_\alpha(z)}{B_a(z)} f(z)$$

where B_α, B_a are Blaschke factors

$$B_\alpha(z) = \frac{z - \alpha}{1 - \overline{\alpha}z}, \quad B_a(z) = \frac{z - a}{1 - \overline{a}z} .$$

Then f_α tends to f in the metric of A^p as $\alpha \to a$.

The proof is immediate and left to the reader.

REMARK. Alexander Borichev [**Bor**] found an alternative proof of the Theorem for Case 1 similar to our proof for Case 2. Instead of Lemma 3, Borichev's proof is based on the description of zero sets.

References

[B]   K. Bogdan, *On the zeros of functions with finite Dirichlet integral*, Kodai Math. J. **19** (1996), 7–16.

[Bor]   A. Borichev, *Private communication*, July 2004.

[C]   L. Carleson, *On the zeros of functions with bounded Dirichlet integral*, Math. Z. **56** (1952), 289–295.

[HKZ]   H. Hedenmalm, B. Korenblum and K. Zhu, *Theory of Bergman Spaces*, Graduate Texts in Mathematics, 199, Springer, 2000.

[K1]   B. Korenblum, *An extension of the Nevanlinna theory*, Acta Math. **135** (1975), 187–219.

[K2]   B. Korenblum, *A Beurling-type theorem*, Acta Math. **138** (1977), 265–293.

[S]   K. Seip, *Beurling type density theorems in the unit disk*, Invent. Math. **113** (1993), 21–39.

[SS]   H. S. Shapiro and A. L. Shields, *On the zeros of functions with finite Dirichlet integral and some related function spaces*, Math. Z. **80** (1962), 217–229.

DEPARTMENT OF MATHEMATICS AND STATISTICS, UNIVERSITY AT ALBANY, STATE UNIVERSITY OF NEW YORK, 1400 WASHINGTON AVENUE, ALBANY, NY 12222, USA

E-mail address: borisko@math.albany.edu

Contemporary Mathematics
Volume **404**, 2006

Phragmén-Lindelöf-type Problems for $A^{-\alpha}$

Xavier Massaneda and Pascal J. Thomas

ABSTRACT. We want to formalize problems of the following kind about holomorphic functions on the unit disk: given a size condition (that the function f belongs to a certain growth space), and a restriction on the growth of f on a subset S of the disk, what restriction do we get on the global growth of f? More explicitly, given $0 \leq \gamma \leq \beta \leq \alpha \leq \infty$, we want to know when the norm inequality $\sup_{z \in \mathbb{D}} (1 - |z|)^{\beta} |f(z)| \leq M \sup_{s \in S} (1 - |s|)^{\gamma} |f(s)|$ holds for any $f \in A^{-\alpha}$. We study the various implications between those properties for different values of the parameters, and give a necessary condition in an interesting special case.

1. Definitions and statement of results

Boris Korenblum often asks whether someone can find a characterization of the (discrete) subsets of the unit which have the Phragmén-Lindelöf property for the space $A^{-\infty}$, that is to say, such that any function in $A^{-\infty}$ that is bounded on the set must necessarily be bounded everywhere. We cannot answer this question, but we show that a certain generalization of the notion puts it into a common framework with a number of interesting problems in the area. The aim of this note is to take stock of the situation to date, and take a first step towards some further results. In particular, we give a (weak) necessary condition for the Phragmén-Lindelöf property.

1.1. Definitions, implications. Given $0 \leq \alpha < \infty$ and $S \subset \mathbb{D}$, let

$$A^{-\alpha}(S) = \left\{ f \in H(\mathbb{D}) : \|f\|_{\alpha, S} := \sup_{z \in S} (1 - |z|)^{\alpha} |f(z)| < \infty \right\},$$

where $H(\mathbb{D})$ stands for the set of all holomorphic functions in $\mathbb{D}$. In the case where $S = \mathbb{D}$, we write $\|f\|_{\alpha, \mathbb{D}} = \|f\|_{\alpha}$, and the spaces $A^{-\alpha}(\mathbb{D}) = A^{-\alpha}$ are the familiar growth spaces. Let also $A_+^{-\alpha} := \bigcap_{\beta > \alpha} A^{-\beta}$ and $A^{-\infty} = \bigcup_{\alpha > 0} A^{-\alpha}$.

2000 *Mathematics Subject Classification.* Primary 30C80, 30E99.

Key words and phrases. Phragmén-Lindelöf sequences, sampling sequence, density.

Both authors supported by a XTI program of the Comunitat de Treball dels Pirineus. First author also supported by the DGICYT grant MTM2005-08984-C02-02 and the CIRIT grant 2005-SGR00611.

We want to formalize problems of the following kind: given a size condition (that the function f belongs to a certain growth space), and a restriction on the growth of f on a subset S, what restriction do we get on the global growth of f?

DEFINITION. Let $0 \leq \gamma \leq \beta \leq \alpha \leq \infty$, with $\beta < \infty$. A set $S \subset \mathbb{D}$ is *of type* (α, β, γ), denoted $S \in \mathcal{T}(\alpha, \beta, \gamma)$, if and only if there exists a constant $M > 0$ such that for all $f \in A^{-\alpha}$, $\|f\|_\beta \leq M\|f\|_{\gamma, S}$.

Even though Banach space methods do not apply immediately, this turns out, where it makes sense, to be the same as the seemingly weaker inclusion property.

LEMMA 1. *Let* $0 \leq \gamma \leq \beta < \alpha \leq \infty$, *with* $\beta < \infty$. *A set* $S \subset \mathbb{D}$ *is of type* (α, β, γ) *if and only if*

$$A^{-\alpha} \cap A^{-\gamma}(S) \subset A^{-\beta}.$$

The proof is given in Section 2.

Special cases of the definition above coincide with previously studied properties:

- When $\alpha = \beta = \gamma > 0$, we have the $A^{-\alpha}$-*sampling* sets studied by Seip [**Se93b**]. The case $\alpha = \beta = \gamma = 0$ corresponds to the $H^\infty(\mathbb{D})$-*dominating* sets described earlier by Brown, Shields and Zeller [**BrShZe60**].
- Horowitz, Korenblum, and Pinchuk [**HoKoPi97**] studied $A^{-\infty}$-*sampling* sets, which in our terminology would be $\cap_{\gamma>0} \cap_{\varepsilon>0} \mathcal{T}(\infty, \gamma + \varepsilon, \gamma)$, or equivalently, sets such that $A^{-\infty} \cap A^{-\gamma}(S) \subset A_+^{-\gamma}$ for any $\gamma > 0$.
- *Weakly sufficient sets* are those for which the topology of $A^{-\infty}$ is determined by the restrictions of $f \in A^{-\infty}$ on the set, see [**KhTh98**], [**Sch74**]. This corresponds to the sets S such that for any $\gamma > 0$, there exists $\beta = \beta(\gamma)$ with $S \in \mathcal{T}(\infty, \beta, \gamma)$.
- Bonet and Domański [**BoDo03**] studied the sets of type (β, β, γ), which they termed (γ, β)-*sampling* sets, and of type (∞, β, γ), which they termed (γ, β)-*dominating* sets.

We introduce another piece of terminology.

DEFINITION. Let $0 \leq \beta < \alpha$. A set $S \subset \mathbb{D}$ is (α, β)-*Phragmén-Lindelöf* if and only if $S \in \mathcal{T}(\alpha, \beta, \beta)$.

Korenblum's Phragmén-Lindelöf sets correspond to our $(\infty, 0)$ case. An easy adaptation of [**Sch74**, Theorem 3.10] to our situation shows that any weakly sufficient set is $(\infty, 0)$-Phragmén-Lindelöf. Also, it is easy to see (taking powers) that any set of type $(\infty, \beta, 0)$ is really $(\infty, 0)$-Phragmén-Lindelöf.

We would like to clarify the relationships between those various classes of sets. First notice that it is trivial that $\mathcal{T}(\alpha, \beta, \gamma) \subset \mathcal{T}(\alpha', \beta', \gamma')$ when $\alpha \geq \alpha'$, $\beta \leq \beta'$, and $\gamma \geq \gamma'$, and readily seen by taking powers that $\mathcal{T}(\alpha, \beta, \gamma) \subset \mathcal{T}(\alpha/m, \beta/m, \gamma/m)$ whenever $m \in \mathbb{N}$.

THEOREM 2.

(a) *If* $\alpha \geq \alpha'$, $\beta \leq \beta' + \min(\alpha - \alpha', \gamma - \gamma')$, *and* $\gamma \geq \gamma'$, *then*

$$\mathcal{T}(\alpha, \beta, \gamma) \subset \mathcal{T}(\alpha', \beta', \gamma').$$

(b) *Conversely, if* $\alpha < \alpha'$; *or if* $\gamma < \gamma'$ *and* $\frac{\gamma}{\beta} < \frac{\gamma'}{\beta'}$, *then* $\mathcal{T}(\alpha, \beta, \gamma) \not\subset \mathcal{T}(\alpha', \beta', \gamma')$.

(c) *In the special situation of Phragmén-Lindelöf sets, we can give a more complete result:*
$\mathcal{T}(\alpha, \beta, \beta) \subset \mathcal{T}(\alpha', \beta', \beta')$ *if and only if* $\alpha \geq \alpha'$ *and* $\beta \geq \beta'$, *and the inclusion is proper when* $\alpha > \alpha'$ *or* $\beta > \beta'$.

The proof is given in Section 2. Notice that a number of cases remain open.

COROLLARY 3.

(a) *If S is $A^{-\alpha}$-sampling, then S is (α, β)-Phragmén-Lindelöf for all $\beta \leq \alpha$.*
(b) *If S is (α, β)-Phragmén-Lindelöf for some $\beta \leq \alpha$, then S is $A^{-\beta}$-sampling.*

1.2. A necessary condition. We now would like to give some conditions for Phragmén-Lindelöf sets in terms of densities.

For an arc $I \subset \mathbb{T}$, let $|I| = \sigma(I)$ stand for its normalized arc-length. We also use a normalized distance on $\mathbb{T}$:

$$d(e^{i\theta}, e^{i\psi}) = \frac{|\theta - \psi|}{\pi}, \quad d(\zeta, F) := \inf_{\xi \in F} d(\zeta, \xi) \quad \text{when } F \subset \mathbb{T}.$$

For a closed proper subset $F \subset \mathbb{T}$ such that $\mathbb{T} \setminus F = \bigcup_j I_j$, where the I_j are disjoint open arcs, the *Beurling-Carleson characteristic* of F is

$$\hat{\kappa}(F) := \sum_j |I_j| \left(1 + \log \frac{1}{|I_j|} \right) = \int_{\mathbb{T}} \log \frac{1}{d(\zeta, F)} d\sigma(\zeta).$$

The *density* of a set $S \subset \mathbb{D}$ is defined as

$$D(S) := \limsup_{\hat{\kappa}(F) \to \infty} \frac{\sum\limits_{s \in S \cap \Gamma(F)} \log \frac{1}{|s|}}{\hat{\kappa}(F)},$$

where $\Gamma(F)$ denotes the Korenblum star over F (see precise definition below).

S is a zero-set for $A_+^{-\alpha}$ if and only if $D(S) \leq \alpha$ ([**Se95**], [**HeKoZh00**, Theorem 4.15, p. 112]).

The Seip densities can be seen as versions of this, made uniform over automorphisms of the disk. More precisely, for $\frac{1}{2} < r < 1$, let

$$D(S, r) := \frac{\sum\limits_{s \in S \cap (\mathbb{D} \setminus \bar{D}(0, \frac{1}{2}))} \log \frac{1}{|s|}}{\log \frac{1}{1-r}}.$$

The *uniform upper and lower densities* are then defined respectively as

$$D_u^+(S) := \limsup_{r \to 1} \sup_{\varphi \in \text{Aut}(\mathbb{D})} D(\varphi(S), r)$$

$$D_u^-(S) := \liminf_{r \to 1} \inf_{\varphi \in \text{Aut}(\mathbb{D})} D(\varphi(S), r).$$

A set S is called *hyperbolically separated* if and only if there exists $\delta > 0$ such that for any $s \neq s' \in S$, $|\varphi_s(s')| \geq \delta$, where φ_s is an automorphism of $\mathbb{D}$ sending s to 0.

According to a well-known geometric characterization given by Seip, S is $A^{-\alpha}$-sampling if and only if it contains a hyperbolically separated subset $S' \subset S$ with $D_u^-(S') > \alpha$ ([**Se93b**], [**HeKoZh00**, Theorem 5.18, p. 153]).

A set of density less than α cannot be of type (α, β, γ) (see Lemma 5 below). From Corollary 3 we see also that an (α, β)-Phragmén-Lindelöf set must be of uniform lower density greater than β, and since this notion is invariant under automorphisms, *every* point of the boundary must be a non-tangential limit point of the set.

On the other hand, it follows from examples given in [**HoKoPi97**] and [**KhTh98**] that there exist sets of type (α, β, γ) with lower density equal to 0 as soon as $\beta > \gamma$. But it does follow from Theorem 2 that all of the sets we are studying are dominating sets for H^∞, and therefore that almost every point of the unit circle is a non-tangential limit point of the set.

We now define a more quantitative notion of non-tangential density, related to the above densities (see Lemma 8).

Given $z \in \mathbb{D}$, let

$$I_z^\gamma = \{\zeta \in \mathbb{T} : d(\zeta, z/|z|) < \gamma(1 - |z|)\}.$$

Also, for $\zeta \in \mathbb{T}$, $\gamma \in (0, 1/2]$, $t \in (0, 1)$ and $A \subset \mathbb{T}$ define:

$$\Gamma_\gamma(\zeta) = \{z \in \mathbb{D} : \zeta \in I_z^\gamma\} \qquad \Gamma_\gamma{}^t(\zeta) = \Gamma_\gamma(\zeta) \cap \bar{D}(1 - t)$$

$$\Gamma_\gamma(A) = \bigcup_{\zeta \in A} \Gamma_\gamma(\zeta) \qquad \Gamma_\gamma^t(A) = \Gamma_\gamma(A) \cap \bar{D}(1 - t),$$

where $D(t) = \{z \in \mathbb{D} : |z| < t\}$. Notice that $\Gamma_\gamma(\zeta)$ is contained in the half disk $\{z \in \mathbb{D} : |\arg z - \arg \zeta| < \pi/2\}$.

Given a sequence S in $\mathbb{D}$, consider the counting function

$$n_\gamma(\zeta, t; S) = \#(\Gamma_\gamma{}^t(\zeta) \cap S).$$

DEFINITION. Let S be a sequence in $\mathbb{D}$ and let $A \subset \mathbb{T}$. The *boundary density of S in A* is

$$BD_S^\gamma(A) := \lim_{t \searrow 0} \sup_{\substack{\zeta \in A \\ 0 < u \leq t}} \frac{n_\gamma(\zeta, u; S)}{\gamma |\log u|}.$$

THEOREM 4. *Let $S \in \mathcal{T}(\infty, 0, 0)$. Then, for all γ and all $I \subset \mathbb{T}$ interval*

$$BD_S^\gamma(I) = \infty.$$

The proof is given in Section 3. This theorem is quite far from being optimal. It only shows that any subset of $\mathbb{T}$ in which S has finite boundary density must be of empty interior. It would be natural to expect that the boundary density is infinite outside of a small exceptional set, in fact smaller than merely negligible for the 1-dimensional Lebesgue measure.

Indeed, as follows from the proofs in [**Hr78**, Section 9], or the more explicit version in [**BeCo86**] using [**Da77**] (see also [**BeCo88**] and [**Ko02**]), the correct measure should be the Hausdorff measure H_L associated to the function $L(x) := x|\log x|$. More precisely, any $f \in A^{-\infty}$ such that

$$\limsup_{r \to 1} |f(r\zeta)| < \infty, \quad \text{for all } \zeta \in \mathbb{T} \setminus E, \quad \text{with } H_L(E) = 0,$$

must belong to $H^\infty(\mathbb{D})$. And conversely, given a closed set $E \subset \mathbb{T}$ with $H_L(E) > 0$, one can find a positive measure μ supported on $E' \subset E$ such that $f(z) = \exp\{\int_\mathbb{T} \frac{\zeta + z}{\zeta - z} d\mu(\zeta)\}$ belongs to $A^{-\alpha} \setminus H^\infty(\mathbb{D})$ (in fact α can be chosen arbitrarily small) and is bounded on the set $\{z \in \mathbb{D} : 1 - |z|^2 \leq (d(z, E))^2\}$, which is as dense as one could possibly want near the points of $\mathbb{T} \setminus E$.

2. Inequalities and inclusions

In this section, we prove Lemma 1 and Theorem 2. The existence of regular uniformly distributed sequences will be useful.

PROPOSITION A. *For any* $t > 0$, *there exists a hyperbolically separated sequences* S_t *such that*

$$D_u^+(S_t) = D_u^-(S_t) = t .$$

For such sequences also $D(S_t) = t$ (see [**Se93b**, Proposition 3.1], [**HeKoZh00**, Section 5.4]).

We first need an auxiliary result which is of some interest in itself.

LEMMA 5. *Let* $\alpha > \beta \geq \gamma$. *Any set* S *such that* $A^{-\alpha} \cap A^{-\gamma}(S) \subset A^{-\beta}$ *is a uniqueness set for* $A^{-\alpha}$.

PROOF. Suppose that $f \in A^{-\alpha} \setminus \{0\}$ and $f|_S = 0$. Then $f \in A^{-\alpha} \cap A^{-\gamma}(S) \subset A^{-\beta}$.

Let $\delta := D(S) \leq \beta$ by the necessary condition on zero-sets. Since $\alpha > \beta$, we can pick $t \in (\beta - \delta, \alpha - \delta)$. Choose then S_t, disjoint from S, as in Proposition A. Then $D(S \cup S_t) = \delta + t \in (\beta, \alpha)$, and we can find $g \in A_+^{-(\delta+t)} \subset A^{-\alpha}$, g not identically zero, such that $g|_{S \cup S_t} = 0$. Since $D(S \cup S_t) > \beta$, we have $g \notin A^{-\beta}$, thus contradicting the inclusion. $\square$

The proof of Lemma 1 now proceeds verbatim as the proof of [**KhTh98**, Proposition 3, p. 439], where the roles of α, β, and γ are played respectively by $q + 2, q$, and p. What was denoted by $A^{-p}(S)$ in [**KhTh98**] would now be denoted by $A^{-p}(S) \cap A^{-\infty}$, and one should now choose $p_m^{\alpha-\beta} \geq 3^{2m}$ in [**KhTh98**, p. 440, l. 6] to take into account our slightly more general range of exponents.

PROOF OF THEOREM 2. (a) By the remarks before the statement of the theorem, there is no loss of generality to assume that $\beta' \leq \beta$. This proof is then an adaptation of [**BoDo03**, Theorem 1.4].

(c) Since $\mathcal{T}(\alpha, \beta', \beta') \subset \mathcal{T}(\alpha', \beta', \beta')$ trivially, it will be enough to prove $\mathcal{T}(\alpha, \beta, \beta) \subset \mathcal{T}(\alpha, \beta', \beta')$ when $\beta > \beta'$.

Let $f \in A^{-\alpha} \cap A^{-\beta'}(S)$ and define $f_n(z) = z^n f(z)$. For $z \in S$:

$$(1 - |z|)^\beta |f_n(z)| \leq |z|^n (1 - |z|)^{\beta-\beta'} \|f\|_{\beta',S} ,$$

and taking the supremum of the right hand side over z,

$$\|f_n\|_{\beta,S} \leq \frac{t^t n^n}{(t+n)^{t+n}} \|f\|_{\beta',S} \quad \text{where } t := \beta - \beta'.$$

By Lemma 1, $\|g\|_\beta \leq M\|g\|_{\beta,S}$, for any $g \in A^{-\alpha}$. Thus, for z with $|z| = r_{n,t} := n/(n+t)$, we find:

$$\frac{t^t n^n}{(t+n)^{t+n}} (1 - |z|)^{\beta'} |f(z)| = (1 - |z|)^\beta |f_n(z)| \leq \|f_n\|_\beta \leq M\|f_n\|_{\beta,S}$$

$$\leq M \frac{t^t n^n}{(t+n)^{t+n}} \|f\|_{\beta',S} ,$$

i.e., $(1 - |z|)^{\beta'} |f(z)| \leq M\|f\|_{\beta',S}$.

For arbitrary z there exists n such that $|z| \in [r_{n,t}, r_{n+1,t})$, and the maximum principle yields:

$$(1 - |z|)^{\beta'}|f(z)| \leq \frac{(1 - r_{n,t})^{\beta'}}{(1 - r_{n+1,t})^{\beta'}} M\|f\|_{\beta',S} .$$

This implies the existence of $C > 0$ such that $\|f\|_{\beta'} \leq CM\|f\|_{\beta',S}$, because $\lim\limits_{n \to \infty} \frac{1 - r_{n,t}}{1 - r_{n+1,t}} = 1$.

Given the previous results, to prove the converse and that the inclusions are proper when the inequalities are strict, it will be enough to see that:

  (i) For every $\alpha > \alpha'$, there exists $S \in \mathcal{T}(\alpha', \alpha', \alpha') \setminus \mathcal{T}(\alpha, 0, 0)$.
  (ii) If $\beta > \beta'$, there exists $S \in \mathcal{T}(\infty, \beta', \beta') \setminus \mathcal{T}(\beta, \beta, \beta)$.

PROOF OF (i). This is a special case of the first statement in (b), see proof below.

PROOF OF (ii). Let $t \in (\beta', \beta)$ and let S_t be as in Proposition A. Take also a sequence of radii $\{r_n\} \nearrow 1$ with $\lim\limits_{n \to \infty} \frac{1 - r_{n+1}}{1 - r_n} = 0$, and a sequence $\Lambda \subset \cup_{n \in \mathbb{N}}\{z : |z| = r_n\}$, as in [**HoKoPi97**, Example 2.5], such that

$$A^{-\infty} \cap A^{-\alpha}(\Lambda) \subset A_+^\alpha .$$

Notice that $D_u^-(S_0) = 0$ for all separated subsequence $S_0 \subset \Lambda$, hence Λ is not sampling for any $A^{-\alpha}$, $\alpha > 0$.

Define $S = S_t \cup \Lambda$. It is clear by Seip's characterization that S is not $A^{-\beta}$-sampling, since for every separated subsequence $S_0 \subset S$:

$$D_u^-(S_0) \leq D_u^+(S_0 \cap S_t) + D_u^-(S_0 \cap \Lambda) = D_u^+(S_0 \cap S_t) \leq D_u^+(S_t) = t < \beta .$$

To see that $S \in \mathcal{T}(\infty, \beta', \beta')$, let $f \in A^{-\beta'}(S)$. For any $\delta \in (\beta', t)$:

$$f \in A^{-\infty} \cap A^{-\beta'}(S) \subset A_+^{-\beta'} \subset A^{-\delta} .$$

S_t is $A^{-\delta}$-sampling, because it is hyperbolically separated and $D_u^-(S_t) = t > \delta$. Thus there exists $M = M(\delta) > 0$ such that

$$\|f\|_\delta \leq M(\delta)\|f\|_{\delta, S_t} \leq M(\delta)\|f\|_{A^{-\delta}(S)} .$$

Since the sampling constant $M(\delta)$ depends only on $D_u^-(S_t)$ and the separation of S_t, we can take the same constant M_0 for all δ close to β':

$$\|f\|_\delta \leq M(\delta)\|f\|_{\delta, S} \leq M_0\|f\|_{\delta, S} \qquad \text{for } \delta \searrow \beta' .$$

Letting δ tend to β' finishes the proof.

(b) Suppose that $\alpha' > \alpha$. Let $t \in (\alpha, \alpha')$ and let S_t be as in Proposition A. Thus S_t is $A^{-\alpha}$-sampling. Also, by [**Se95**], S_t is a zero set for $A^{-\alpha'}$, therefore by Lemma 5 it cannot be of type $\mathcal{T}(\alpha', \beta', \gamma')$.

Suppose now that $\gamma' > \gamma$, and $\gamma' > \gamma\frac{\beta'}{\beta}$. First we deal with the case when $\gamma' = \beta'$: by the proof of (c), above, we know that $\mathcal{T}(\alpha, \gamma, \gamma) \not\subset \mathcal{T}(\alpha, \gamma', \gamma')$, but trivially $\mathcal{T}(\alpha, \gamma, \gamma) \subset \mathcal{T}(\alpha, \beta, \gamma)$.

Assume then that $\beta' > \gamma'$. Set $M := \beta'/\gamma' \in (1, \beta/\gamma)$. By [**BoDo03**, Example 4.1] (itself a refinement of [**KhTh98**, Proposition 5]), there exists a countable union of circles centered at the origin E, such that for any $\delta, \delta' > 0$ and $t > \delta$

$$A^{-t} \cap A^{-\delta}(E) \subset A^{-\delta'} \text{ if and only if } \delta' > M\delta.$$

In particular, $E \in \mathcal{T}(\alpha, \beta, \gamma)$. But since $\beta' = M\gamma'$, we have $A^{-\alpha} \cap A^{-\gamma'}(E) \not\subset A^{-\beta'}$.

Note that E could be replaced by a discrete set (provided the points in each circle are close enough; we omit the details). $\qquad\square$

3. Density at the boundary

We may want to consider the true cones of vertex $\zeta \in \mathbb{T}$ and aperture $\pi\gamma$, $0 < \gamma < 1/\pi$, defined as the rotation $\tilde{\Gamma}_\gamma(\zeta) := \zeta \cdot \tilde{\Gamma}_\gamma(1)$ of

$$\tilde{\Gamma}_\gamma(1) = \{z \in \mathbb{D} : \operatorname{Re} z > 0 \; ; |\operatorname{Im} z| < \pi\gamma(1 - \operatorname{Re} z)\} \cup \{z \in \mathbb{D} : \operatorname{Re} z \le 0 \; ; |z| < \pi\gamma\} \; .$$

By construction, there exists $r_0 = r_0(\gamma)$ such that

(1) $\qquad \Gamma_\gamma(\zeta) \subset \tilde{\Gamma}_\gamma(\zeta) \qquad$ and $\qquad \tilde{\Gamma}_{\gamma/2}(\zeta) \cap (\mathbb{D} \setminus D(r_0)) \subset \Gamma_\gamma(\zeta) \; .$

Theorem 4 could also be stated using the density $B\tilde{D}_S^\gamma(I)$ defined as $BD_S^\gamma(I)$ but replacing the counting function $n_\gamma(\zeta, t; S)$ by $\tilde{n}_\gamma(\zeta, t; S) = \#(\tilde{\Gamma}_\gamma^t(\zeta) \cap S)$. It is immediate from (1) that for all γ and I:

(2) $$B\tilde{D}_S^{\gamma/2}(I) \le BD_S^\gamma(I) \le B\tilde{D}_S^\gamma(I) \; .$$

We first state some basic properties of the density.

LEMMA 6. *Given S and an interval $I \subset \mathbb{T}$, then*

$$BD_{S \cap \Gamma_\gamma(I)}^\gamma(\mathbb{T}) = BD_S^\gamma(I) \; .$$

PROOF. Let $S_I^\gamma = S \cap \Gamma_\gamma(I)$. By definition $BD_S^\gamma(I) = BD_{S_I^\gamma}^\gamma(I)$, thus it suffices to prove that $BD_{S_I^\gamma}^\gamma(I) = BD_{S_I^\gamma}^\gamma(\mathbb{T})$. It is clear that $BD_{S_I^\gamma}^\gamma(I) \le BD_{S_I^\gamma}^\gamma(\mathbb{T})$. In order to prove the reverse inequality, let $\zeta \notin I$ and take the closest end $\eta \in I$ to ζ. Then $n_\alpha(\zeta, t; S_I^\gamma) \le n_\alpha(\eta, t; S_I^\gamma)$; hence taking the supremum and the limit, we obtain the desired inequality. $\qquad\square$

LEMMA 7. *Let $S \subset \mathbb{D}$ and let $J = \{e^{i\theta} \in \mathbb{T} : \theta \in (-\frac{\pi}{2}, \frac{\pi}{2})\}$. Given γ and $I \subset \mathbb{T}$, there exists $\phi \in Aut(\mathbb{D})$ such that*

(a) *$\phi(I) = J$;*

(b) *$BD_{\phi(S)}^{\gamma/2}(J) \le 4\, BD_S^\gamma(I)$.*

PROOF. It will be more convenient to use the cones $\tilde{\Gamma}_\gamma$.

Apply first a rotation so that the interval is transformed into a symmetric interval $I = (e^{-i\alpha}, e^{i\alpha})$. Apply now the automorphism $\phi_a(z) = \frac{a+z}{1+az}$, where $a \in (-1, 1)$ is chosen so that (a) is satisfied. Explicitly, $a = -\frac{2\cos\alpha}{|1 - ie^{i\alpha}|^2}$.

Let $\zeta \in I$. Since ϕ_a is a Möbius transformation, the straight lines along the sides of the cone $\tilde{\Gamma}_\gamma(\zeta)$ are transformed into circles containing the points $\phi_a(\zeta)$, $\phi_a(\infty) = 1/a$, and cutting the unit circle at $\phi_a(\zeta)$ with angle γ. Hence

$$\tilde{\Gamma}_{\gamma/2}(\phi_a(\zeta)) \subset \phi_a\big(\tilde{\Gamma}_\gamma(\zeta)\big) \; .$$

The circle $|z| = \tau$ is transformed into a circle passing through $\phi_a(\tau) = \frac{a+\tau}{1+a\tau}$ and $\phi_a(-\tau) = \frac{a-\tau}{1-a\tau}$ and cutting the real line at these points with angle $\pi/2$. Thus, if $a \le 0$, that is, $|I| \le 1/2$, taking $t = 1 - \tau$ and $u = t\frac{1-a}{1+a-at}$, we have $\tilde{\Gamma}_{\gamma/2}^u(\phi_a(\zeta)) \subset \phi_a\big(\tilde{\Gamma}_\gamma^t(\zeta)\big)$, and

$$\frac{1}{\gamma/2} \frac{\tilde{n}_{\gamma/2}(\phi_a(\zeta), u; \phi_a(S))}{|\log u|} \le \frac{2}{\gamma} \frac{\tilde{n}_\gamma(\zeta, t; S)}{|\log u|} = \frac{2}{\gamma} \frac{\tilde{n}_\gamma(\zeta, t; S)}{\log \frac{1}{t}} \frac{\log \frac{1}{t}}{\log \frac{1}{t} - \log \frac{1-a}{1+a-at}} \; .$$

Hence

$$B\tilde{D}^{\gamma/2}_{\phi_a(S)}(\phi_a(I)) \le 2 \lim_{\substack{T\searrow 0 \\ \zeta\in I \\ t\le T}} \sup \frac{\tilde{n}_\gamma(\zeta,t;S)}{\gamma\log\frac{1}{t}} \frac{\log\frac{1}{t}}{\log\frac{1}{t}-\log\frac{1-a}{1+a-at}}$$

$$\le 2 B\tilde{D}^\gamma_S(I) \lim_{T\searrow 0} \frac{\log\frac{1}{T}}{\log\frac{1}{T}-\log\frac{1-a}{1+a-aT}} = 2B\tilde{D}^\gamma_S(I) \ .$$

The conclusion is then an immediate consequence of (2). In the case where $a > 0$, an analogous computation with $u = t\frac{1+a}{1-a+at}$ will yield the same result. $\qquad\square$

LEMMA 8. *Let $S \subset \mathbb{D}$. Then $D(S) \le BD^\gamma_S(\mathbb{T})$ for all $\gamma > 0$.*

PROOF. Let $S = (s_k)_k$ and define the measure $\mu = \sum_k \delta_{s_k}$. Then, for $E \subset \mathbb{D}$:

$$\gamma \sum_{s_k\in E} 1 - |s_k| = \int_E \int_{I^\gamma_z} d\sigma(\zeta)\, d\mu(z) = \int_\mathbb{T} \int_{E\cap\Gamma_\gamma(\zeta)} d\mu(z)\, d\sigma(\zeta)$$

$$= \int_\mathbb{T} \#(S \cap E \cap \Gamma_\gamma(\zeta))\, d\sigma(\zeta) \ .$$

Take $F \subset \mathbb{T}$ finite or countable, and $E = \Gamma_\gamma(F)$. By construction, there exists c_γ such that $\Gamma_\gamma(F) \cap \Gamma_\gamma(\zeta) \subset \Gamma^{c_\gamma d(\zeta,F)}_\gamma(\zeta)$. Thus

$$\gamma \sum_{s_k\in\Gamma_\gamma(F)} (1-|s_k|) \le \int_\mathbb{T} \#(S\cap\Gamma_\gamma(F)\cap\Gamma_\gamma(\zeta))\, d\sigma(\zeta) \le \int_\mathbb{T} n_\gamma(\zeta,c_\gamma d(\zeta,F);S)\, d\sigma(\zeta) \ .$$

By hypothesis, given $\varepsilon > 0$ there exists $t_\varepsilon > 0$ such that

$$n_\gamma(\zeta,t;S) \le \gamma(BD^\gamma_S(\mathbb{T})+\varepsilon)|\log t| \qquad \text{for all } 0 < t < t_\varepsilon \text{ and all } \zeta\in\mathbb{T}.$$

We separate the integral above into two terms. For ζ such that $c_\gamma d(\zeta,F) \ge t_\varepsilon$, we have $\Gamma^{c_\gamma d(\zeta,F)}_\gamma(\zeta) \subset D(1-c_\gamma d(\zeta,F)) \subset D(1-t_\varepsilon)$. Hence

$$\gamma \sum_{s_k\in\Gamma_\gamma(F)} (1 - |s_k|) = \int_\mathbb{T} \#(S \cap D(1-t_\varepsilon))\, d\sigma(\zeta) + \int_{\{\zeta:\, c_\gamma d(\zeta,F)\le t_\varepsilon\}} n_\gamma(\zeta,c_\gamma d(\zeta,F);S)\, d\sigma(\zeta)$$

$$\le C_1(\varepsilon,S) + \gamma(BD^\gamma_S(\mathbb{T})+\varepsilon)\int_\mathbb{T} \log\left(\frac{1}{c_\gamma d(\zeta,F)}\right) d\sigma(\zeta)$$

$$\le C_2(\varepsilon,S) + \gamma(BD^\gamma_S(\mathbb{T})+\varepsilon)\,\hat{\kappa}(F)$$

and, as desired,

$$\limsup_{\hat{\kappa}(F)\to\infty} \frac{\sum_{s_k\in\Gamma_\gamma(F)}(1-|s_k|)}{\hat{\kappa}(F)} \le BD^\gamma_S(\mathbb{T}) + \varepsilon \qquad \text{for all } \varepsilon > 0 \ .$$

$\qquad\square$

PROOF OF THEOREM 4. Assume there exists $I \subset \mathbb{T}$ with $BD^\gamma_S(I) < \infty$. We want to construct $H \in A^{-\infty} \setminus H^\infty$ such that $\sup_S |H| < \infty$, so that $S \notin \mathcal{T}(\infty,0,0)$.

By Lemma 7, we can assume that $I = J$. Take $\Omega = D(-1,\sqrt{2}) \cup \mathbb{D}$, so that the intersection angle between $\mathbb{T}$ and $\partial D(-1,\sqrt{2})$ at $\pm i$ is $\pi/4$. Consider then a conformal mapping $\psi : \Omega \longrightarrow \mathbb{D}$ with $\psi(J) = J$. Explicitly $\psi = \psi_3 \circ \psi_2 \circ \psi_1$, where $\psi_1(z) = \frac{i+z}{i-z}$, $\psi_2(z) = (iz)^{\frac{4}{5}}$ (the argument of iz running from 0 to $5\pi/4$),

$\psi_3(z) = i\frac{z-i}{z+i}$, from which we see that $\psi(i) = i, \psi(-i) = -i, \psi(1) = 1$ and there exists a constant $\eta > 0$ such that

(3) $$1 - |\psi(w)| \geq \eta(1 - |w|) \qquad \text{for all } w \in \mathbb{D}, \text{ and}$$

(4) $$\psi(\mathbb{D}) \subset \Gamma_{\gamma'}(J) \cup D(r_0) \qquad \text{for some } r_0 \text{ and } \gamma' \leq \frac{\pi}{2} - \frac{\pi}{5}.$$

Similarly to (4), we have $\psi(\mathbb{D}) \subset \tilde{\Gamma}_{\gamma'}(J)$.

LEMMA 9. *Let ψ be the conformal mapping above. Then*

$$BD^{\gamma/2}_{\psi(S) \cap \Gamma_{\gamma/2}(J)}(\mathbb{T}) \leq 2\, BD^{\gamma}_S(J)\ .$$

PROOF. By construction of ψ, there exists r_0 such that for any $\zeta \in J$

$$\Gamma^t_{\gamma/2}(\psi(\zeta)) \subset \psi(\Gamma^{\eta t}_\gamma(\zeta)) \cup D(r_0)\ .$$

Thus $\frac{n_{\gamma/2}(\psi(\zeta),t;\psi(S))}{\frac{\gamma}{2}|\log t|} \leq 2\, \frac{n_\gamma(\zeta,\eta t;S)+C_{r_0}}{\gamma|\log t|}$, and taking the supremum and the limit we obtain the desired inequality. $\qquad\qquad\square$

Using this and Lemma 6,

$$BD^{\gamma/2}_{\psi(S) \cap \Gamma_\gamma(J)}(\mathbb{T}) = BD^{\gamma/2}_{\psi(S)}(J) \leq 2\, BD^{\gamma}_S(J)\ .$$

Applying Lemma 8, we get $D(\psi(S) \cap \Gamma_{\gamma/2}(J)) \leq 2BD^{\gamma}_S(J)$; therefore there exists $f \in A^{-\alpha}_+ \setminus \{0\}$, $\alpha = 2BD^{\gamma}_S(J)$, such that $f \equiv 0$ on $\psi(S) \cap \Gamma_{\gamma/2}(J)$.

We want to multiply f by a function g with a suitable decrease on $\mathbb{D} \setminus \Gamma_{\gamma/2}(J)$. By (4), if $z \in \psi(S)$, there exists $\delta > 0$ such that $1 - |z| \geq \delta d(z/|z|, J)$.

Define, for a certain constant $c_1 > 0$ to be chosen later on, the function

$$g(z) = \exp\left\{-c_1 \int_{\mathbb{T}\setminus J} \frac{\zeta + z}{\zeta - z} \log \frac{1}{d(\zeta, J)}\, d\sigma(\zeta)\right\}\ .$$

Then

$$-\log|g(z)| = c_1 \int_{\mathbb{T}\setminus J} P_z(\zeta) \log \frac{1}{d(\zeta, J)}\, d\sigma(\zeta)\ ,$$

where $P_z(\zeta)$ stands for the Poisson kernel, and in particular $\|g\|_\infty \leq 1$.

For $z \notin \Gamma_{\gamma/2}(J)$, there exist $c_2, c_3 > 0$ such that

$$I_{c_2}(z) =: \{\zeta \in \mathbb{T} : d(z/|z|, \zeta) < c_2(1 - |z|)\} \subset \mathbb{T} \setminus J$$

and $d(I_{c_2}(z), J) \geq c_3(1 - |z|)$. Thus, for some $c_4 = c_4(c_2, \gamma)$:

$$\begin{aligned} -\log|g(z)| &\geq c_1 \int_{I_{c_2}(z)} P_z(\zeta) \log \frac{1}{d(\zeta, J)}\, d\sigma(\zeta) \\ &\geq c_1 c_3 \log \frac{1}{d(I_{c_2}(z), J)} \geq c_1 c_4 \log \frac{1}{(1 - |z|)}\ . \end{aligned}$$

Hence $|g(z)| \leq (1 - |z|)^{c_1 c_4}$ for $z \notin \Gamma_{\gamma/2}(J)$.

For any $e^{i\theta} \in J$,

$$\lim_{z \to e^{i\theta}, z \in \Gamma_{\gamma/2}(e^{i\theta})} |g(z)| = 1.$$

By Privalov's theorem, we can choose $e^{i\theta} \in J$ such that

$$\limsup_{z \to e^{i\theta}, z \in \Gamma_{\gamma/2}(e^{i\theta})} |f(z)| > 0.$$

Thus for any $k > 0$,

$$(5) \qquad \limsup_{z \to e^{i\theta}, z \in \Gamma_{\gamma/2}(e^{i\theta})} \frac{|f(z)g(z)|}{(1 - |z|)^k} = \infty .$$

Choose now $c_1 > 0$ such that $c_1 c_4 > 2BD_S^\gamma(J) + k$. Define $F(z) = \frac{f(z)g(z)}{(1-e^{-i\theta}z)^k}$ and $H = F \circ \psi$, where θ is as before.

Let us check the required properties for H:

(i) $\sup_S |H| < \infty$. If w is such that $\psi(w) \in \psi(S) \cap \Gamma_{\gamma/2}(J)$, we have then $f(\psi(w)) = 0$, and obviously $H(w) = 0$.

If w is such that $\psi(w) \in \psi(S) \cap (\mathbb{D} \setminus \Gamma_{\gamma/2}(J))$ and $z = \psi(w)$, then

$$|H(w)| = \frac{|f(z)g(z)|}{|1 - e^{-i\theta}z|^k} \leq \frac{|f(z)|(1 - |z|)^{c_1 c_4}}{(1 - |z|)^k} \leq \|f\|_{A^{-c_1 c_4 + k}} < \infty ,$$

since $c_1 c_4 - k > \alpha$.

(ii) $H \in A^{-\infty}$. It is clear that $F \in A_+^{-(\alpha+k)}$, since $f \in A_+^{-\alpha}$ and $g \in H^\infty$. The statement follows then from (3).

(iii) $H \notin H^\infty$. This follows from (5), since $|1 - e^{-i\theta}z| \simeq (1 - |z|)$ on $\Gamma_{\gamma/2}(e^{i\theta})$. $\square$

References

[BeCo86]   R. D. Berman and W. S. Cohn, *A radial Phragmén-Lindelöf Theorem for functions of slow growth*, Complex Variables Theory Appl. **6** (1986), 299–307.

[BeCo88]   R. D. Berman and W. S. Cohn, *Phragmén-Lindelöf theorems for subharmonic functions on the unit disk*, Math. Scand. **62** (1988), 269–293.

[BoDo03]   J. Bonet and P. Domański, *Sampling sets and sufficient sets for $A^{-\infty}$*, J. Math. Anal. Appl. **277** (2003), 651–669.

[BrShZe60]   L. Brown, A. Shields and K. Zeller, *On absolutely convergent exponential sums*, Trans. Amer. Math. Soc. **96** (1960), 162–183.

[Da77]   B. Dahlberg, *On the radial boundary values of subharmonic functions*, Math. Scand. **40** (1977), 301–317.

[HeKoZh00]   H. Hedenmalm, B. Korenblum and K. Zhu, *Theory of Bergman Spaces*, Springer, Graduate Texts in Mathematics, vol. 199 (2000).

[HoKoPi97]   C. A. Horowitz, B. Korenblum and B. Pinchuk, *Sampling sequences for $A^{-\infty}$*, Michigan Math. J. **44** (1997), 389–398.

[Hr78]   S. V. Hruščev, *The problem of simultaneous approximation and of removal of the singularities of Cauchy type integrals.* (Russian). Spectral theory of functions and operators, Trudy Mat. Inst. Steklov. **130** (1978), 124–195. (Translated in Proc. Steklov Inst. of Math. (1978), Issue 4, 133–202.)

[KhTh98]   Lê Hai Khôi and P. J. Thomas, *Weakly sufficient sets for $A^{-\infty}$*, Publ. Mat. **42** (1998), 435–448.

[Ko02]   B. Korenblum, *Asymptotic maximum principle*, Ann. Acad. Sci. Fenn. Math. **27** (2002), 249–255.

[Sch74]   D. M. Schneider, *Sufficient sets for some spaces of analytic functions*, Trans. Amer. Math. Soc. **197** (1974), 161–180.

[Se93a]   K. Seip, *Regular sets of sampling and interpolation for weighted Bergman spaces*, Proc. Amer. Math. Soc. **117** (1993), 213–219.

[Se93b]   K. Seip, *Beurling type density theorems in the unit disk*, Invent. Math. **113** (1993), 21–39.

[Se95]   K. Seip, *On Korenblum's density condition for the zero sequences of $A^{-\alpha}$*, J. Anal. Math. **67** (1995), 307–322.

DEPARTAMENT DE MATEMÀTICA APLICADA I ANÀLISI, UNIVERSITAT DE BARCELONA, GRAN VIA 585, 08071-BARCELONA, SPAIN
E-mail address: xavier@mat.ub.es

LABORATOIRE DE MATHÉMATIQUES EMILE PICARD, UMR CNRS 5580, UNIVERSITÉ PAUL SABATIER, 118 ROUTE DE NARBONNE, 31062 TOULOUSE CEDEX, FRANCE
E-mail address: pthomas@cict.fr

Contemporary Mathematics
Volume **404**, 2006

A Representation Formula for Reproducing Subharmonic Functions in the Unit Disc

Anders Olofsson

ABSTRACT. We present a representation formula of Poisson-Jensen type for subharmonic functions in the unit disc that are reproducing at the origin. The considerations apply to so-called Bergman inner functions.

0. Introduction

Let $\mathbb{D}$ be the unit disc and denote by $\mathbb{T} = \partial \mathbb{D}$ the unit circle. In this paper, we consider area integrable subharmonic functions u in $\mathbb{D}$ that are reproducing at the origin in the sense that the equality

$$h(0) = \int_{\mathbb{D}} h(z)u(z)dA(z)$$

holds for harmonic polynomials h. Here dA is the Lebesgue area measure normalized by a factor $1/\pi$. We also assume that u satisfies the growth assumption

$$\sup_{0 \leq r < 1} \frac{1}{2\pi} \int_{\mathbb{T}} u(re^{i\theta})d\theta < \infty.$$

The principal result of this paper is a representation formula of Poisson-Jensen type which is characteristic for this class of functions u (see Theorem 2.1 and Proposition 2.1). An immediate consequence of this representation formula is an inequality on the boundary for u: $u(e^{i\theta}) \geq 1$ for a.e. $e^{i\theta} \in \mathbb{T}$ (see Corollary 2.1 and Corollary 2.2). It is worth pointing out that using the above reproducing property of u, we establish the Poisson-Jensen formula with an absolutely continuous boundary measure for u assuming a less restrictive growth assumption than what is needed for general subharmonic functions.

The motivation for these considerations comes from Bergman space theory. For $0 < p < \infty$, we denote by $L_a^p(\mathbb{D})$ the standard Bergman space of all p-th power area integrable analytic functions in $\mathbb{D}$, that is, a function f is in $L_a^p(\mathbb{D})$ if and only if f is analytic in $\mathbb{D}$ and

$$\|f\|_p^p = \int_{\mathbb{D}} |f(z)|^p dA(z) < \infty.$$

2000 *Mathematics Subject Classification.* Primary 31A05; Secondary 30H05.

Key words and phrases. Bergman inner function, reproducing subharmonic function.

Research supported by the Göran Gustafsson Foundation.

A function G in $L_a^p(\mathbb{D})$ is said to be $L_a^p(\mathbb{D})$-inner if the equality

(0.1) $$h(0) = \int_{\mathbb{D}} h(z)|G(z)|^p dA(z)$$

holds for harmonic polynomials h. Note that in the above terminology, (0.1) means that the subharmonic function $u = |G|^p$ is reproducing at the origin.

The $L_a^p(\mathbb{D})$-inner functions has proved to be useful in the study of (z-)invariant subspaces of $L_a^p(\mathbb{D})$. A standard reference for recent developments is [8]. For instance, $L_a^p(\mathbb{D})$-inner functions can be used as contractive zero divisors (see [5, 7]). In the Hilbert space case $p = 2$, the defining property (0.1) has the operator theoretic interpretation that G is element in a wandering subspace for the Bergman shift operator (see [2]).

In the pioneering work [7], it was shown that the inequality $|G(e^{i\theta})| \geq 1$ for $e^{i\theta} \in \mathbb{T}$ holds when G is an $L_a^2(\mathbb{D})$-inner function which extends continuously to the boundary $\mathbb{T}$. By the results of the present paper, this inequality remains true in the refined sense that

(0.2) $$|G(e^{i\theta})|^p = 1 + P^*\big[(1 - |\cdot|^2)\Delta|G|^p\big](e^{i\theta}) \geq 1$$

for a.e. $e^{i\theta} \in \mathbb{T}$ when G is an $L_a^p(\mathbb{D})$-inner function which is also element in the Hardy space $H^p(\mathbb{D})$. Here $\Delta = \partial^2/\partial z \partial \bar{z}$ is the Laplacian and P^* denotes the balayage (sweeping out) operator for the Dirichlet problem in $\mathbb{D}$ (see Section 1 below). An anonymous referee has pointed out to us that a stronger local version of (0.2) is known (see [3, Theorem 5.8]).

We give two proofs of Theorem 2.1. The first proof uses the biharmonic Green function. The second proof relies on the F. Riesz representation formula for subharmonic functions.

1. Preliminaries

We use some standard notations from distribution theory. For instance, by $\mathcal{D}'(\mathbb{T})$ we denote the space of distributions on $\mathbb{T}$. A locally area integrable function u is identified with the Radon measure $u dA$.

We denote by $P(z, e^{i\theta}) = (1 - |z|^2)/|e^{i\theta} - z|^2$ the Poisson kernel for $\mathbb{D}$. The Green function for the Laplacian $\Delta = \partial^2/\partial z \partial \bar{z}$ in $\mathbb{D}$ is denoted by

$$G(z, \zeta) = \log\left|\frac{z - \zeta}{1 - \bar{\zeta}z}\right|^2, \quad (z, \zeta) \in \mathbb{D}^2,$$

and the Green potential by $G[\mu](z) = \int_{\mathbb{D}} G(z, \zeta) d\mu(\zeta)$.

The balayage (sweeping out) operator P^* is defined by

$$P^*[\mu](e^{i\theta}) = \int_{\mathbb{D}} P(z, e^{i\theta}) d\mu(z), \quad e^{i\theta} \in \mathbb{T}$$

(pointwise a.e.), for μ a complex regular Borel measure in $\mathbb{D}$. This definition coincides with the standard notion of balayage of measures. We have the duality formula

(1.1) $$\int_{\mathbb{D}} P[\varphi](z) d\mu(z) = \frac{1}{2\pi} \int_{\mathbb{T}} \varphi(e^{i\theta}) P^*[\mu](e^{i\theta}) d\theta,$$

where $P[\varphi](z) = \int_{\mathbb{T}} P(z, e^{i\theta})\varphi(e^{i\theta}) d\theta/2\pi$ is the Poisson integral of $\varphi \in C(\mathbb{T})$.

Let u be a function in $\mathbb{D}$ and $0 \leq r < 1$. We write u_r for the function defined by

$$u_r(e^{i\theta}) = u(re^{i\theta}), \quad e^{i\theta} \in \mathbb{T}.$$

A function u is said to be biharmonic in $\mathbb{D}$ if $\Delta^2 u = 0$ there. The Almansi representation asserts that a function u is biharmonic in $\mathbb{D}$ if and only if u has the form $u(z) = h_1(z) + |z|^2 h_2(z)$ for $z \in \mathbb{D}$, where h_j is harmonic in $\mathbb{D}$ $(j = 1, 2)$.

We have the following uniqueness result.

PROPOSITION 1.1. *Let u be a biharmonic function in $\mathbb{D}$ and assume that*

$$\lim_{r \to 1} u_r/(1 - r) = 0 \quad in \ \mathcal{D}'(\mathbb{T}).$$

Then $u(z) = 0$ for all $z \in \mathbb{D}$.

PROOF. By the Almansi representation, the function u has a convergent sum representation

$$u(z) = \sum_{k=-\infty}^{\infty} (a_k + b_k r^2) r^{|k|} e^{ik\theta}, \quad z = re^{i\theta} \in \mathbb{D}.$$

By assumption, we have that

$$\widehat{u}_r(k) = \frac{1}{2\pi} \int_{\mathbb{T}} u_r(e^{i\theta}) e^{-ik\theta} d\theta = (a_k + b_k r^2) r^{|k|} = o(1 - r)$$

as $r \to 1$. We conclude that $a_k = b_k = 0$ for all $k \in \mathbb{Z}$. □

A more elaborate version of Proposition 1.1 is planned to appear elsewhere.

We review some facts about biharmonic boundary value problems. The biharmonic Green function is the function Γ defined by

$$\Gamma(z, \zeta) = |z - \zeta|^2 \log \left| \frac{z - \zeta}{1 - \bar{\zeta}z} \right|^2 + (1 - |z|^2)(1 - |\zeta|^2), \quad (z, \zeta) \in \mathbb{D} \times \mathbb{D}.$$

If μ is a positive Radon measure in $\mathbb{D}$ such that $\int_{\mathbb{D}} (1 - |z|^2)^2 d\mu(z) < \infty$, then the biharmonic Green potential

$$\Gamma[\mu](z) = \int_{\mathbb{D}} \Gamma(z, \zeta) d\mu(\zeta), \quad z \in \mathbb{D},$$

exists and solves (in a weak sense) the boundary value problem

$$(1.2) \qquad \begin{cases} \Delta^2 \Gamma[\mu] & = \ \mu \quad \text{in } \mathbb{D}, \\ \Gamma[\mu], \ \partial_n \Gamma[\mu] & = \ 0 \quad \text{on } \mathbb{T}, \end{cases}$$

where ∂_n denotes differentiation in the inward normal direction. In (1.2), the first equation is satisfied in the sense of distribution theory and the last two boundary conditions holds in the sense that $\lim_{r \to 1} \Gamma[\mu]_r/(1 - r) = 0$ in $L^1(\mathbb{T})$ (see [**1**, Proposition 3.4]).

The function H defined by

$$H(z, \zeta) = (1 - |\zeta|^2) \frac{1 - |\zeta z|^2}{|1 - \bar{z}\zeta|^2}, \quad (z, \zeta) \in \bar{\mathbb{D}} \times \mathbb{D},$$

is known as the harmonic compensator. The functions Γ and H are related by the formula

$$(1.3) \qquad \Delta_z \Gamma(z, \zeta) = G(z, \zeta) + H(z, \zeta), \quad (z, \zeta) \in \mathbb{D}^2,$$

which can be verified by straightforward computation. Note that the function H is harmonic in its first argument.

2. The representation formula

For the proof of Theorem 2.1, we need some lemmas. The following lemma is well-known, but for the sake of completeness we include some details of proof.

LEMMA 2.1. *Let u be a subharmonic function in $\mathbb{D}$ and put $\Delta u = \mu$. Then*

$$\sup_{0 \leq r < 1} \frac{1}{2\pi} \int_{\mathbb{T}} u(re^{i\theta}) d\theta < \infty$$

if and only if $\int_{\mathbb{D}}(1 - |z|^2) d\mu(z) < \infty$.

PROOF. The convergence of the integral $\int_{\mathbb{D}}(1 - |z|^2) d\mu(z)$ is equivalent to the convergence of the Green potential $G[\mu]$. If $G[\mu]$ exists, then the function $h = u - G[\mu]$ is harmonic in $\mathbb{D}$ and majorizes u. Consequently, we have that $\int_{\mathbb{T}} u(re^{i\theta}) d\theta / 2\pi \leq h(0) = O(1)$.

We assume next that $\int_{\mathbb{T}} u(re^{i\theta}) d\theta / 2\pi = O(1)$, and consider the function $\tilde{u}$ defined by $\tilde{u}(z) = \int_{\mathbb{T}} u(ze^{-i\theta}) d\theta / 2\pi$ for $z \in \mathbb{D}$. The function $\tilde{u}$ is radial, subharmonic and bounded in $\mathbb{D}$. Let $\tilde{\mu} = \Delta \tilde{u}$. By the F. Riesz representation formula for subharmonic functions (see [**4**, Theorem 4.4.1] or [**9**, Theorem 3.3.6]), we have that $\int_{\mathbb{D}}(1 - |z|^2) d\mu(z) = \int_{\mathbb{D}}(1 - |z|^2) d\tilde{\mu}(z) < \infty$. □

For easy reference, we record the following lemma.

LEMMA 2.2. *Let μ be a positive Radon measure in $\mathbb{D}$ such that $\int_{\mathbb{D}}(1 - |z|^2) d\mu(z) < \infty$. Then*

$$\int_{\mathbb{D}} H(z, \zeta) d\mu(\zeta) = P\big[P^*\big[(1 - |\cdot|^2)\mu\big]\big](z), \quad z \in \mathbb{D}.$$

PROOF. Recall that the function H is harmonic in its first argument. By the Poisson integral formula, we have that

$$H(z, \zeta) = \frac{1}{2\pi} \int_{\mathbb{T}} P(z, e^{i\theta}) H(e^{i\theta}, \zeta) d\theta, \quad (z, \zeta) \in \mathbb{D}^2.$$

An integration using Fubini's theorem now shows that

$$\int_{\mathbb{D}} H(z, \zeta) d\mu(\zeta) = \frac{1}{2\pi} \int_{\mathbb{T}} P(z, e^{i\theta}) \int_{\mathbb{D}} \frac{1 - |\zeta|^2}{|e^{i\theta} - \zeta|^2}(1 - |\zeta|^2) d\mu(\zeta) \, d\theta,$$

which yields the conclusion of the lemma. □

We shall also need the following lemma about the normal derivative of a Green potential.

LEMMA 2.3. *Let μ be a complex regular Borel measure in $\mathbb{D}$. Then*

$$\lim_{r \to 1} G[\mu]_r / (1 - r) = -2P^*[\mu]$$

in the weak topology of $L^1(\mathbb{T})$.

PROOF. The assertion of the lemma means that

$$\lim_{r\to 1}\frac{1}{1-r}\int_{\mathbb{T}}G[\mu](re^{i\theta})\varphi(e^{i\theta})\frac{d\theta}{2\pi}=-2\int_{\mathbb{T}}P^*[\mu](e^{i\theta})\varphi(e^{i\theta})\frac{d\theta}{2\pi}$$

for every $\varphi\in L^\infty(\mathbb{T})$. By a change of integration, we have that

$$\frac{1}{1-r}\int_{\mathbb{T}}G[\mu](re^{i\theta})\varphi(e^{i\theta})\frac{d\theta}{2\pi}=\int_{\mathbb{D}}\Phi_r(\zeta)d\mu(\zeta),$$

where

$$\Phi_r(\zeta)=\frac{1}{1-r}\int_{\mathbb{T}}G(re^{i\theta},\zeta)\varphi(e^{i\theta})\frac{d\theta}{2\pi},\quad \zeta\in\mathbb{D}.$$

We have the asymptotic formula

(2.1) $$\frac{1}{1-r}G(re^{i\theta},\zeta)=-2\frac{1-|\zeta|^2}{|e^{i\theta}-\zeta|^2}+o(1)$$

as $r\to 1$, where the error term $o(1)$ is uniform in $e^{i\theta}\in\mathbb{T}$ and $\zeta\in K$; here $K\subset\mathbb{D}$ is a fixed compact set. By (2.1), we see that

$$\lim_{r\to 1}\Phi_r(\zeta)=-2\int_{\mathbb{T}}\frac{1-|\zeta|^2}{|e^{i\theta}-\zeta|^2}\varphi(e^{i\theta})\frac{d\theta}{2\pi},\quad \zeta\in\mathbb{D}.$$

Below we will show that the functions Φ_r are uniformly bounded as $r\to 1$. An application of the Lebesgue dominated convergence theorem now shows that

$$\lim_{r\to 1}\int_{\mathbb{D}}\Phi_r(\zeta)d\mu(\zeta)=-2\int_{\mathbb{T}}\int_{\mathbb{D}}\frac{1-|\zeta|^2}{|e^{i\theta}-\zeta|^2}d\mu(\zeta)\varphi(e^{i\theta})\frac{d\theta}{2\pi},$$

which yields the conclusion of the lemma.

We now prove (2.1). Taylor expanding the logarithm, we see that

$$G(z,\zeta)=\left|\frac{z-\zeta}{1-\bar{\zeta}z}\right|^2-1+O\left(\left(1-\left|\frac{z-\zeta}{1-\bar{\zeta}z}\right|^2\right)^2\right).$$

Recall the formula

$$1-\left|\frac{z-\zeta}{1-\bar{\zeta}z}\right|^2=\frac{(1-|z|^2)(1-|\zeta|^2)}{|1-\bar{\zeta}z|^2},\quad (z,\zeta)\in\mathbb{D}^2.$$

A computation now shows that

$$G(re^{i\theta},\zeta)=-(1-r^2)\frac{1-|\zeta|^2}{|1-\bar{\zeta}re^{i\theta}|^2}+O\left((1-r)^2\right)=-2(1-r)\frac{1-|\zeta|^2}{|e^{i\theta}-\zeta|^2}+o(1-r),$$

which yields formula (2.1).

We now show that the functions Φ_r are uniformly bounded as $r\to 1$. We have that

$$|\Phi_r(\zeta)|\le-\frac{1}{1-r}\int_{\mathbb{T}}G(re^{i\theta},\zeta)\frac{d\theta}{2\pi}\|\varphi\|_\infty,\quad \zeta\in\mathbb{D}.$$

The function $\zeta\mapsto\int_{\mathbb{T}}G(re^{i\theta},\zeta)d\theta/2\pi$ is subharmonic and radial in $\mathbb{D}$, harmonic for $|\zeta|\ne r$ and vanishes on $\mathbb{T}$. A straightforward computation shows that

(2.2) $$\frac{1}{2\pi}\int_{\mathbb{T}}G(re^{i\theta},\zeta)d\theta=\max\left(\log(r^2),\log|\zeta|^2\right),\quad \zeta\in\mathbb{D}.$$

For Φ_r, this gives the bound

$$|\Phi_r(\zeta)|\le 2\|\varphi\|_\infty\log(1/r)/(1-r),\quad \zeta\in\mathbb{D},$$

which is uniform as $r\to 1$. $\square$

We now state the formula for a reproducing subharmonic function.

THEOREM 2.1. *Let u be a subharmonic function in $\mathbb{D}$ such that*

$$(2.3) \qquad \sup_{0 \le r < 1} \frac{1}{2\pi} \int_{\mathbb{T}} u(re^{i\theta}) d\theta < \infty,$$

and assume that u is area integrable in $\mathbb{D}$ and reproducing at the origin in the sense that the equality

$$(2.4) \qquad h(0) = \int_{\mathbb{D}} h(z)u(z)dA(z)$$

holds for harmonic polynomials h. Then the function u admits the representation

$$(2.5) \qquad u(z) = 1 + P\big[P^*\big[(1 - |\cdot|^2)\Delta u\big]\big](z) + G[\Delta u](z), \quad z \in \mathbb{D}.$$

FIRST PROOF OF THEOREM 2.1. Let us write $\mu = \Delta u$ for the Riesz measure of u. We consider the Green potential

$$\Phi(z) = G[u - 1](z) = \int_{\mathbb{D}} G(z, \zeta)\big(u(\zeta) - 1\big)dA(\zeta), \quad z \in \mathbb{D}.$$

Note that $\Delta^2 \Phi = \mu$ in $\mathcal{D}'(\mathbb{D})$. By Lemma 2.3, we know that $\lim_{r \to 1} \Phi_r/(1 - r) = -2P^*[u - 1]$ in $\mathcal{D}'(\mathbb{T})$. By the reproducing property of u and (1.1), we have that

$$\frac{1}{2\pi} \int_{\mathbb{T}} e^{ik\theta} P^*[u - 1](e^{i\theta}) d\theta = \int_{\mathbb{D}} z^k \big(u(z) - 1\big) dA(z) = 0$$

for $k \ge 0$ and similarly for $k < 0$. Thus the Fourier coefficients of the function $P^*[u - 1]$ all vanish and we conclude that $P^*[u - 1](e^{i\theta}) = 0$ for a.e. $e^{i\theta} \in \mathbb{T}$. By Proposition 1.1, we now conclude that the function Φ has the representation

$$(2.6) \qquad \Phi(z) = \Gamma[\mu](z) = \int_{\mathbb{D}} \Gamma(z, \zeta) d\mu(\zeta), \quad z \in \mathbb{D},$$

where Γ is the biharmonic Green function. Indeed, the function $\Phi - \Gamma[\mu]$ is biharmonic in $\mathbb{D}$ (Weyl's lemma) and has vanishing boundary values in the sense that $\lim_{r \to 1} (\Phi - \Gamma[\mu])_r/(1 - r) = 0$ in $\mathcal{D}'(\mathbb{T})$.

Recall formula (1.3) and note that $\int_{\mathbb{D}} (1 - |z|^2) d\mu(z) < \infty$ by Lemma 2.1. Applying a Laplacian to the representation (2.6), we obtain that

$$(2.7) \quad u(z) - 1 = \int_{\mathbb{D}} \Delta_z \Gamma(z, \zeta) d\mu(\zeta) = \int_{\mathbb{D}} G(z, \zeta) d\mu(\zeta) + \int_{\mathbb{D}} H(z, \zeta) d\mu(\zeta), \quad z \in \mathbb{D}.$$

(Formally speaking, we compute the distributional Laplacian.) By Lemma 2.2, this last formula (2.7) yields (2.5). $\qquad \square$

SECOND PROOF OF THEOREM 2.1. Let $\mu = \Delta u$ and note that $\int_{\mathbb{D}} (1 - |z|^2) d\mu(z) < \infty$ by Lemma 2.1. By the F. Riesz theorem (see [4, Theorem 4.4.1] or [9, Theorem 3.3.6]), the function u has the representation

$$(2.8) \qquad u(z) = G[\mu](z) + 1 + h(z), \quad z \in \mathbb{D},$$

where h is harmonic. We now compute h to show that this representation (2.8) reduces to (2.5).

We first note that $\int_{\mathbb{D}} |G[\mu]| dA = -\int_{\mathbb{D}} G[\mu] dA = \int_{\mathbb{D}} (1 - |z|^2) d\mu(z) < \infty$, so that h is area integrable in $\mathbb{D}$. By the reproducing property of u, we have that

$$(2.9) \qquad 0 = \int_{\mathbb{D}} z^k \big(u(z) - 1\big) dA(z) = \int_{\mathbb{D}} z^k G[\mu](z) dA(z) + \int_{\mathbb{D}} z^k h(z) dA(z).$$

for $k \geq 0$. The first integral on the right hand side in (2.9) is computed by

$$\int_{\mathbb{D}} z^k G[\mu](z) dA(z) = \int_{\mathbb{D}} \int_{\mathbb{D}} z^k G(z, \zeta) dA(z) d\mu(\zeta) = -\frac{1}{k+1} \int_{\mathbb{D}} \zeta^k (1 - |\zeta|^2) d\mu(\zeta).$$

We also have that

$$\int_{\mathbb{D}} z^k h(z) dA(z) = \widehat{h}(-k)/(k+1)$$

for $k \geq 0$, where $h(z) = \sum \widehat{h}(k) r^{|k|} e^{ik\theta}$ for $z = re^{i\theta} \in \mathbb{D}$ is the series expansion of h. By (2.9), we conclude that

$$\widehat{h}(-k) = \int_{\mathbb{D}} \zeta^k (1 - |\zeta|^2) d\mu(\zeta)$$

for $k \geq 0$. By complex conjugation, we also have that $\widehat{h}(k) = \int_{\mathbb{D}} \bar{\zeta}^k (1 - |\zeta|^2) d\mu(\zeta)$ for $k > 0$. A straightforward computation shows that

$$h(z) = \sum_{k=-\infty}^{\infty} \widehat{h}(k) r^{|k|} e^{ik\theta} = \int_{\mathbb{D}} \frac{1 - |z\bar{\zeta}|^2}{|1 - z\bar{\zeta}|^2} (1 - |\zeta|^2) d\mu(\zeta) = \int_{\mathbb{D}} H(z, \zeta) d\mu(\zeta),$$

where $z = re^{i\theta} \in \mathbb{D}$ and H is the harmonic compensator function. The conclusion of the theorem now follows by (2.8) and Lemma 2.2. $\qquad \square$

In the context of Theorem 2.1, the assumption of area integrability of u is needed for the integrals in (2.4) to make sense. Similarly, the growth assumption (2.3) is needed for the Green potential $G[\Delta u]$ to exist (see Lemma 2.1).

In Theorem 2.1, we have established the Poisson-Jensen formula for u with an absolutely continuous boundary measure using the reproducing property (2.4). Let us comment on the case of general subharmonic functions in $\mathbb{D}$. The standard assumption for the F. Riesz formula $u = G[\mu] + h$ (h a harmonic function in $\mathbb{D}$) to hold is (2.3). The standard assumption for the Poisson-Jensen formula $u = P[\sigma] + G[\mu]$ (σ a complex regular Borel measure on $\mathbb{T}$) to hold is that

$$\frac{1}{2\pi} \int_{\mathbb{T}} |u(re^{i\theta})| d\theta = O(1) \quad \text{as } r \to 1.$$

This last condition of bounded L^1-means as $r \to 1$ is sometimes formulated as $\int_{\mathbb{T}} u(re^{i\theta})^+ d\theta/2\pi = O(1)$ as $r \to 1$; here $u^+ = \max(u, 0)$. For the boundary measure σ to be absolutely continuous, a slightly stronger assumption is needed. For full details, we refer to [9, Section 3.3] and [6].

We now turn to some corollaries of Theorem 2.1.

COROLLARY 2.1. *Let u be as in Theorem 2.1. Then*

$$u(e^{i\theta}) = \lim_{r \to 1} u(re^{i\theta}) = 1 + P^*\left[(1 - |\cdot|^2)\Delta u\right](e^{i\theta})$$

for a.e. $e^{i\theta} \in \mathbb{T}$.

PROOF. A classical result of Littlewood asserts that the Green potential $G[\mu]$ has radial limit 0 a.e. on $\mathbb{T}$ (see [4, Theorem 4.6.4]). It is well-known that a similar statement holds for the Poisson integral (see [4, Theorem 4.6.6]). The corollary follows by taking radial limits. $\qquad \square$

COROLLARY 2.2. *Let u be as in Theorem 2.1. Then*

$$\lim_{r \to 1} u_r = 1 + P^*\big[(1 - |\cdot|^2)\Delta u\big] \quad in \ L^1(\mathbb{T});$$

here $u_r(e^{i\theta}) = u(re^{i\theta})$ for $e^{i\theta} \in \mathbb{T}$ and $0 \le r < 1$. In particular, we have that

$$\lim_{r \to 1} \frac{1}{2\pi} \int_{\mathbb{T}} |u(re^{i\theta}) - 1| d\theta = \frac{1}{2\pi} \int_{\mathbb{T}} \big(u(e^{i\theta}) - 1\big) d\theta,$$

where $u(e^{i\theta})$ is as in Corollary 2.1.

PROOF. It is clearly sufficient to prove the first assertion of the corollary. Let $\mu = \Delta u$ and recall that $\int_{\mathbb{D}} (1 - |z|^2) d\mu(z) < \infty$. By Theorem 2.1, we have that

$$u_r = 1 + P\big[P^*\big[(1 - |\cdot|^2)\mu\big]\big]_r + G[\mu]_r \quad in \ L^1(\mathbb{T}).$$

Using formula (2.2), we see that

$$\|G[\mu]_r\|_{L^1(\mathbb{T})} = -\int_{\mathbb{D}} \max\big(\log(r^2), \log|\zeta|^2\big) d\mu(\zeta)$$

which tends to 0 by, say, monotone convergence. Thus $G[\mu]_r \to 0$ in $L^1(\mathbb{T})$. By a standard regularization procedure, we know that $P[P^*[(1-|\cdot|^2)\mu]]_r \to P^*[(1-|\cdot|^2)\mu]$ in $L^1(\mathbb{T})$. $\qquad\square$

For u a subharmonic function reproducing at the origin, the (weaker) condition of upper bounded integral means implies the (stronger) condition of bounded L^1-means as $r \to 1$.

We now give a converse of Theorem 2.1.

PROPOSITION 2.1. *Let u be a subharmonic function in $\mathbb{D}$ satisfying the growth condition (2.3) and assume that the representation formula (2.5) holds. Then u has bounded L^1-means as $r \to 1$ and is reproducing at the origin in the sense that the equality*

$$h(0) = \int_{\mathbb{D}} h(z)u(z)dA(z)$$

holds for harmonic polynomials h.

PROOF. Let $\mu = \Delta u$. Recall that $\int_{\mathbb{D}}(1 - |z|^2) d\mu(z) < \infty$ by Lemma 2.1, so that formula (2.5) makes sense. As in the proof of Corollary 2.2, we see that the L^1-means of u are bounded as $r \to 1$. In fact, the assertion of Corollary 2.2 holds. By Lemma 2.2, formula (2.5) yields (2.7). By formula (2.7), we have that

$$\int_{\mathbb{D}} h(z)u(z)dA(z) = h(0) + \int_{\mathbb{D}}\int_{\mathbb{D}} h(z)\big(G(z,\zeta) + H(z,\zeta)\big)d\mu(\zeta)dA(z) =$$

$$= h(0) + \int_{\mathbb{D}}\int_{\mathbb{D}} h(z)\Delta_z\Gamma(z,\zeta)dA(z)d\mu(\zeta),$$

where in the last step we have used Fubini's theorem. An application of Green's formula (Green's second identity) shows that

$$\int_{\mathbb{D}} h(z)\Delta_z\Gamma(z,\zeta)dA(z) = 0,$$

which concludes the proof of the proposition. $\qquad\square$

References

[1] A. Abkar and H. Hedenmalm, *A Riesz representation formula for super-biharmonic functions*, Ann. Acad. Sci. Fenn. Math. **26** (2001), 305–324.

[2] A. Aleman, S. Richter and C. Sundberg, *Beurling's theorem for the Bergman space*, Acta Math. **177** (1996), 275–310.

[3] A. Aleman, S. Richter and C. Sundberg, *The majorization function and the index of invariant subspaces in the Bergman spaces*, J. Anal. Math. **86** (2002), 139–182.

[4] D. H. Armitage and S. J. Gardiner, *Classical Potential Theory*, Springer, 2001.

[5] P. Duren, D. Khavinson, H. S. Shapiro and C. Sundberg, *Contractive zero-divisors in Bergman spaces*, Pacific J. Math. **157** (1993), 37–56.

[6] L. Gårding and L. Hörmander, *Strongly subharmonic functions*, Math. Scand. **15** (1964), 93–96.

[7] H. Hedenmalm, *A factorization theorem for square area integrable analytic functions*, J. Reine Angew. Math. **422** (1991), 45–68.

[8] H. Hedenmalm, B. Korenblum and K. Zhu, *Theory of Bergman Spaces*, Springer, 2000.

[9] L. Hörmander, *Notions of Convexity*, Birkhäuser, 1994.

FALUGATAN 22 1TR, SE-113 32 STOCKHOLM, SWEDEN
E-mail address: anderso@math.su.se

Contemporary Mathematics
Volume **404**, 2006

On Dominating Sets for Bergman Spaces

Fernando Pérez-González and Julio C. Ramos

To Professor Boris Korenblum on the occasion of his 80th birthday

ABSTRACT. Let $\varepsilon > 0$ and consider the sector $\Sigma_\varepsilon = \{z \in \mathbb{C} : |\arg(z)| < \varepsilon\}$. We prove that there is a $\delta > 0$ such that

$$(0.1) \qquad \int_{f^{-1}(\Sigma_\varepsilon)} |f(z)|^p dA(z) \geq \delta \int_{\mathbb{D}} |f(z)|^p dA(z),$$

for f in certain subclasses of the weighted Bergman space A_α^p. Our results are in line with a theorem due to D. Marshall and W. Smith [**MS99**], where it was shown that the estimate in (0.1) holds for any univalent function in A^1 satisfying $f(0) = 0$. We also construct sets $S_f(\rho, \lambda) \subset \mathbb{D}$ for which the estimate (0.1), with $f^{-1}(\Sigma_\varepsilon)$ replaced by $S_f(\rho, \lambda)$, holds for all $f \in A_\alpha^p$.

1. Introduction

Let $\mathbb{D}$ be the open unit disk of the complex plane $\mathbb{C}$. For $p > 0$ and $\alpha > -1$, the weighted Bergman space $A_\alpha^p := A_\alpha^p(\mathbb{D})$ is the space of all analytic functions on $\mathbb{D}$ such that

$$\|f\|_{\alpha,p} := \left(\int_{\mathbb{D}} |f(z)|^p dA_\alpha(z) \right)^{\frac{1}{p}} < \infty.$$

The normalized two-dimensional Lebesgue measure on $\mathbb{D}$ is denoted by $dA(z) = \frac{1}{\pi} dx dy = \frac{1}{\pi} r dr d\theta$ and $dA_\alpha(z) := (\alpha + 1)(1 - |z|^2)^\alpha dA(z)$. For $\alpha = 0$, we put A^p, the usual Bergman space.

In [**Lu81**], D. Luecking gave necessary and sufficient conditions on a subset G of $\mathbb{D}$ in order that the operation $f \to f|_G$ from A^p to $L^p(G)$ have closed range, i.e., if G has positive measure, then

$$\int_{\mathbb{D}} |f(z)|^p dA(z) \leq const. \int_G |f(z)|^p dA(z), \qquad f \in A^p.$$

2000 *Mathematics Subject Classification*. Primary 30H05.

The first author has been partially supported by a grant of Dirección General de Investigación-MCyT, Spain, Proyecto no. BFM2002-02098 and by EU Training Research Network Contract No. HPRN-CT-2000-00116. The second author received partial financial support from Universidad de Oriente, Venezuela, to cover several visits to Universidad de La Laguna. He is also very grateful to Departamento de Análisis Matemático de la Universidad de La Laguna for the hospitality provided him.

We say a subset G of $\mathbb{D}$ is a *dominating set for* A_α^p if $\int_G |f(z)|^p dA_\alpha(z)$ is comparable to $\|f\|_{\alpha,p}^p$ for any $f \in A_\alpha^p$. In fact, in [**Lu81**], D. Luecking described dominating sets for A^p in analytic and geometric terms, and it was showed in [**PR01**] that such characterizations also hold for weighted Bergman spaces A_α^p, $\alpha > -1$ and $p \geq 1$ as is collected in Theorem A below.

For $a \in \mathbb{D}$ and $\eta \in (0,1)$, we will put $D_\eta(a) := \{z \in \mathbb{D} : |z - a| < \eta(1 - |a|)\}$, and $\Delta(a, r) = \{z \in \mathbb{D} : |\varphi_a(z)| < r\}$ is the pseudo-hyperbolic disk, where φ_a is the the Möbius transformation $\varphi_a(z) = \frac{a-z}{1-\bar{a}z}$.

THEOREM A. *Let G be a measurable subset of $\mathbb{D}$, $\alpha > -1$ and $p \geq 1$. The following statements are equivalent:*

(1) *G is a dominating set for A_α^p.*
(2) *There is a $\delta > 0$ such that $A(G \cap D) > \delta A(\mathbb{D} \cap D)$ for any disk D centred in the unit circle.*
(3) *There are constants $\delta_0 > 0$ and $\eta \in (0,1)$ such that $A(G \cap D_\eta(a)) > \delta_0 A(D_\eta(a))$, for all $a \in \mathbb{D}$.*
(4) *There are constants $\delta_1 > 0$ and $\eta_1 \in (0,1)$ such that $A(G \cap \Delta(\alpha, \eta_1)) > \delta_1 A(\Delta(\alpha, \eta_1))$, for all $a \in \mathbb{D}$.*

A measurable function f induces a Borel measure on $\mathbb{C}$ defined by

$$\mu_f(E) = \int_{f^{-1}(E)} |f| \, dA.$$

In [**MS99**], D. Marshall and W. Smith studied the problem of the angular distribution of mass by such a measure in the sense that if one considers the sector $\Sigma_\varepsilon = \{w \in \mathbb{C} : |arg(w)| < \varepsilon\}$, $\varepsilon > 0$, they proved the following result.

THEOREM B. *For every $\varepsilon > 0$, there exists a $\delta > 0$ such that if $f \in A^1$ is univalent and $f(0) = 0$, then*

$$(1.1) \qquad \int_{f^{-1}(\Sigma_\varepsilon)} |f| dA \geq \delta \|f\|_1.$$

Theorem B says that the measure μ_f cannot be too asymmetric, and that inverse images of sectors Σ_ε by certain functions in A^1 behave like *uniformly* dominating sets for such subclass in A^1.

Moreover, it was also implicitly stated in [**MS99**] that if it is not assumed that the function to be conformal, then (1.1) holds for sectors Σ_ε, whenever $\varepsilon > \varepsilon_0$, with $\varepsilon_0 < \frac{\pi}{2}$. Theorem B can not be extended to A^p for $p > 1$ as is shown in [**MS99**, Example 4.3]. Indeed, for $p > 1$, Marshall and Smith showed that there is an angle $\varepsilon_1 = \varepsilon_1(p) > 0$ such that the inequality (1.1) is not true, in general, for $\varepsilon \in (0, \varepsilon_1)$. In fact, it follows from a counterexample of Marshall and Smith in [**MS99**] that $\varepsilon_1 \to \pi$ as $p \to \infty$. It is an open problem to prove (1.1) without the restriction that f be univalent. This is equivalent to a conjecture involving quasi-conformal mappings made by M. Ortel and W. Smith in [**OS88**] (see [**MS99**, p. 94]).

In Section 2, we give two results showing that there exist an $\varepsilon_0 \geq 0$ so that the estimate

$$(1.2) \qquad \int_{f^{-1}(\Sigma_\varepsilon)} |f(z)|^p \, dA_\alpha(z) \geq \delta \int_{\mathbb{D}} |f(z)|^p \, dA_\alpha(z)$$

holds for each $\varepsilon \geq \varepsilon_0$, and for each function f in some subclasses of A_α^p, $\delta > 0$ being a constant independent of f. We argue as in [**Lu81**] to construct, in Section 3, other subsets S_f, depending on some parameters, so that estimate (1.2) also holds for any function in A_α^p with $\delta > 0$ independent of f (see Theorem 3.3).

ACKNOWLEGMENTS. We would like to thank Donald Marshall for many useful discussions. We also appreciate very much the comments and suggestions of the referee.

2. Inverse Images of Sectors

Using ideas from Marshall and Smith in [MS99], we showed in [PR] that, for $\alpha > -1$ and $p \in (0, 1]$, every univalent function in A_α^p with $f(0) = 0$ satisfies (1.2) for all $\varepsilon > 0$. Moreover, for $\alpha \in (-1, 0)$ and $p = 1$, then (1.2) fails for some non-univalent functions f with $\varepsilon \in (0, \varepsilon_2)$, where $\varepsilon_2 = \varepsilon_2(\alpha) > 0$. However, for any function in A_α^1 fixing the origin, we have the following result.

THEOREM 2.1. *Suppose that $\alpha > -1$, then there are constants $\varepsilon_0 = \varepsilon_0(\alpha, p) \in (0, \frac{\pi}{2})$ and $\delta = \delta(\alpha, p) > 1$ such that*

$$\int_{\mathbb{D}} |f(z)| \, dA_\alpha(z) \leq \delta \int_{f^{-1}(\Sigma_{\varepsilon_0})} |f(z)| \, dA_\alpha(z),$$

for any $f \in A_\alpha^1$ satisfying $f(0) = 0$.

PROOF. The theorem follows from a simple modification of reasonings in [OS88] and [MS99]. However, the argument we give, which is a bit different, might be of interest since it is more direct.

We start by recalling that there is a positive constant K (depending only on α), such that any real harmonic function u in $\mathbb{D}$ satisfies

$$(2.1) \qquad \int_{\mathbb{D}} |v(z)| \, dA_\alpha(z) < K \int_{\mathbb{D}} |u(z)| \, dA_\alpha(z),$$

where v is the conjugate harmonic function of u with $v(0) = 0$. (See [HKZ00, p. 26] or [DS88, Theorem 2.3]). In particular, if $f \in A_\alpha^1$ with $f(0) = 0$, we can write

$$(2.2) \qquad \int_{\mathbb{D}} |f(z)| \, dA_\alpha(z) < K \int_{\mathbb{D}} |\operatorname{Re} f(z)| \, dA_\alpha(z).$$

Since $0 = \int_0^{2\pi} \operatorname{Re} f(re^{i\theta}) d\theta$, $r < 1$, it follows that

$$\int_{\{re^{i\theta} \,:\, \operatorname{Re} f(re^{i\theta}) > 0\}} \operatorname{Re} f(re^{i\theta}) d\theta = - \int_{\{re^{i\theta} \,:\,]\operatorname{Re} f(re^{i\theta}) < 0\}} \operatorname{Re} f(re^{i\theta}) d\theta.$$

Therefore, $\int_0^{2\pi} |\operatorname{Re} f(re^{i\theta})| d\theta \leq 2 \int_{\{re^{i\theta} \,:\, \operatorname{Re} f(re^{i\theta}) > 0\}} \operatorname{Re} f(re^{i\theta}) d\theta$, and another integration gives

$$(2.3) \qquad \int_{\mathbb{D}} |\operatorname{Re} f(re^{i\theta})| dA_\alpha(z) \leq 2 \int_{\{z \in \mathbb{D} \,:\, \operatorname{Re} f(z) > 0\}} \operatorname{Re} f(z) dA_\alpha(z).$$

Now, we split the set $\{z \in \mathbb{D} : \operatorname{Re} f(z) > 0\}$ into the sets $A = \{z \in \mathbb{D} : \operatorname{Re} f(z) > 0, \; f(z) \in \Pi\}$ and $B = \{z \in \mathbb{D} : \operatorname{Re} f(z) > 0, \; f(z) \in \Sigma_{\frac{\pi}{2} - \eta}\}$, where $\eta > 0$ and Π represents the right-half plane minus the sector $\Sigma_{\frac{\pi}{2} - \eta}$ symmetric with respect to the positive real axis. Since $\operatorname{Re} f(z) \leq \tan(\eta) |\operatorname{Im} f(z)|$ for $z \in A$, by (2.3), we obtain that

$$\frac{1}{2}\int_{\mathbb{D}} |\operatorname{Re} f(z)|dA_\alpha(z) \;\leq\; \int_A \operatorname{Re} f(z)dA_\alpha(z) + \int_B \operatorname{Re} f(z)dA_\alpha(z)$$

$$\leq\; \tan(\eta)\int_{\mathbb{D}} |\operatorname{Im} f(z)|dA_\alpha(z) + \int_{f^{-1}(\Sigma_{\frac{\pi}{2}-\eta})} |f(z)dA_\alpha(z).$$

Hence, choosing $\eta > 0$ so that $1 - 2K\tan(\eta) > 0$ and using (2.1) and (2.2), we get

$$\int_{\mathbb{D}} |f(z)|\,dA_\alpha(z) \leq \frac{2K}{1 - 2K\tan(\eta)} \int_{f^{-1}\left(\Sigma_{\frac{\pi}{2}-\eta}\right)} |f(z)|\,dA_\alpha(z),$$

and the theorem is proved with $\varepsilon_0 = \frac{\pi}{2} - \eta$ and $\delta = \frac{2K}{1-2K\tan(\eta)}$. $\qquad\square$

Theorem 2.1 says that there is an angle $\varepsilon_3 = \varepsilon_3(\alpha) < \frac{\pi}{2}$ such that the estimate (1.2) is true for all $\varepsilon \geq \varepsilon_3$, and for *any* function in A_α^1, $\alpha > -1$, with $f(0) = 0$. Recalling the comments at the beginning of this section, it is an open question what can be said for $\varepsilon \in (\varepsilon_2, \varepsilon_3)$ whenever $\alpha < 0$ and, even, to find the best values of ε_2 and ε_3.

We modify an example in [**MS99**] to show that Theorem 2.1 cannot be extended for all $\varepsilon > 0$ and $p > 1$.

For $n = 1, 2, \ldots$, put

$$(2.4)\qquad\qquad \varepsilon_n = \frac{p-1}{p}\pi - \frac{\alpha}{2p}\pi + \frac{1}{n},$$

and suppose α and p have been chosen in such a way that $\varepsilon_n > \frac{\pi}{2}$, $n \geq 1$. Without loss of generality, we can take n large enough so that $\varepsilon_n < \pi$ for all n. We mention that the ε_n's in (2.4) were used for similar tasks in [**PR**] (for $p = 1$), and for $\alpha = 0$ in [**MS99**].

For each n, consider the Riemann application f_n that sends conformally the open disk $\mathbb{D}$ onto the domain $\Omega_n := \mathbb{C} \setminus (1 + \Sigma_{\varepsilon_n})$, normalized so that $f_n(0) = 0$ and $f_n'(0) > 0$. Let $\varepsilon_\infty = \lim_{n\to\infty} \varepsilon_n = \frac{p-1}{p}\pi - \frac{\alpha}{2p}\pi$ and $\Omega_\infty = \mathbb{C} \setminus (1 + \Sigma_\infty)$. By composition of standard conformal maps, one can reach an explicit representation for these functions:

$$f_n(z) = 1 - \left(\frac{1-z}{1+z}\right)^{2\left(1-\frac{\varepsilon_n}{\pi}\right)}.$$

We claim that $f_n \in A_\alpha^p$ for all n. To check such an assertion, it is enough to prove that $(1+z)^{-2\left(1-\frac{\varepsilon_n}{\pi}\right)} \in A_\alpha^p$. But this follows from Theorem 1.7 in [**HKZ00**, p. 7]) if we set $c = -\frac{2p}{n\pi} < 0$. Moreover, since

$$\|(1+z)^{-2\left(1-\frac{\varepsilon_n}{\pi}\right)}\|_{\alpha,p} = \int_{\mathbb{D}} \frac{(1-|z|^2)^\alpha}{|1+z|^{2+\alpha+c}}dA(z) \sim 1, \qquad \text{as } z \to -1,$$

we obtain that $\|f_n\|_{\alpha,p} < \infty$ for any $n = 1, 2, \cdots$. On the other hand, note that

$$f_n(z) \to f(z) = 1 - \left(\frac{1-z}{1+z}\right)^{\frac{1}{p}(2+\alpha)}, \qquad z \in \mathbb{D},$$

as $n \to \infty$, and $f \notin A_\alpha^p$ (cf. $c = 0$ in [**HKZ00**, p. 7]). An application of Fatou's lemma yields

$$(2.5)\qquad\qquad \|f_n\|_{\alpha,p} \to \infty, \qquad \text{as } n \to \infty.$$

Let now $\varepsilon_c = \frac{\pi}{2}$. With simple trigonometric calculations, it is easy to see that

$$\Sigma_{\varepsilon_c} \cap \Omega_n \subset \Sigma_{\varepsilon_c} \cap \Omega_\infty = \Sigma_{\varepsilon_c} \cap (\mathbb{C} \setminus (1 + \Sigma_{\varepsilon_\infty})) \subset D(0, R),$$

where $R = |\tan(\varepsilon_\infty)|$. In this situation, for each $\varepsilon \in (0, \varepsilon_c)$ we have

$$(2.6) \qquad \int_{f_n^{-1}(\Sigma_\varepsilon)} |f_n(z)|^p \, dA_\alpha(z) \leq \int_{f_n^{-1}(\Sigma_{\varepsilon_c})} |f_n(z)|^p \, dA_\alpha(z) \leq R^p \int_{\mathbb{D}} dA_\alpha(z) = R^p,$$

for any n. But, if δ were a constant such that

$$\|g\|_{\alpha,p} < \delta \int_{g^{-1}(\Sigma_{\varepsilon_0})} |g(z)|^p dA_\alpha(z)$$

for *any* univalent function $g \in A_\alpha^p$ fixing the origin, we would obtain that

$$\|f_n\|_{\alpha,p} < \delta \int_{f_n^{-1}(\Sigma_{\varepsilon_0})} |f_n(z)|^p dA_\alpha(z)$$

for all n, which produces a contradiction between (2.5) and (2.6). Consequently, we can assert that Theorem 2.1 cannot be true for *any* function $f \in A_\alpha^p$ with $f(0) = 0$, and $\varepsilon_0 < \frac{\pi}{2}$.

A deep analysis of the proof of Theorem B in [**MS99**] shows that the condition $f'(0) \neq 0$ plays a fundamental role. This circumstance is exploited in Lemma 2.2, which can be of independent interest, and which we use to see that inverse images of sectors by some functions in A_α^p become dominating sets.

LEMMA 2.2. *Assume $\varepsilon > 0$, and let $p \geq 1$.*

(a) *There is a universal constant K such that: for each function g analytic on $\mathbb{D}$, $g \not\equiv 0$, with $g(0) = 0$, we have*

$$(2.7) \qquad \int_{g^{-1}(\Sigma_\varepsilon)} |g(z)|^p dA(z) \geq \varepsilon K \frac{|g'(0)|^{2p+4}}{\|g\|_p^{p+4}}.$$

(b) *Moreover, if $\alpha > -1$, then there is a constant $K(\alpha)$ (depending only on α) such that for any g analytic on $\mathbb{D}$, $g \not\equiv 0$, with $g(0) = 0$, we have*

$$(2.8) \qquad \int_{g^{-1}(\Sigma_\varepsilon)} |g(z)|^p dA_\alpha(z) \geq \varepsilon K(\alpha) \frac{|g'(0)|^{2p+4}}{\|g\|_{\alpha,p}^{p+4}}.$$

PROOF. Without loss of generality, we can assume that $\|g\|_p = 1$, i.e., $f = \frac{g}{\|g\|_p}$. By the Cauchy integral formula,

$$f'(z) = \frac{1}{2\pi i} \int_{|\zeta|=r} \frac{f(\zeta)}{(\zeta - z)^2} d\zeta, \qquad |z| < r < 1,$$

and an integration from $r = 3/4$ to $r = 1$ gives

$$|f'(z)| \frac{1}{4} \leq \int_{3/4}^1 \int_0^{2\pi} \frac{|f(re^{i\theta})|}{(\frac{1}{2})^2} \frac{r \, dr \, d\theta}{2\pi}.$$

It follows by the Hölder inequality that

$$(2.9) \qquad |f'(z)| \leq 8 \int_{\mathbb{D}} |f| dA \leq 8\|f\|_p = 8,$$

and, by Schwarz's lemma,

$$|f'(z) - f'(0)| \leq 64|z|, \qquad |z| \leq 1/4.$$

So, if we put

(2.10)
$$R := \frac{|f'(0)|}{64},$$

then

$$|f'(z) - f'(0)| < |f'(0)|, \qquad |z| < R.$$

In particular, for $z_1, z_2 \in D(0, R)$, we have

$$|f(z_2) - f(z_1) - f'(0)(z_2 - z_1)| \leq \int_{z_1}^{z_2} |f'(z) - f'(0)||dz| < |z_2 - z_1||f'(0)|.$$

Therefore, $f(z_1) \neq f(z_2)$ so that the function $\frac{f(Rz)}{Rf'(0)}$ is one to one in $\mathbb{D}$. Then, by the 1/4-Koebe theorem $\frac{f(R\mathbb{D})}{Rf'(0)} \supset D(0, \frac{1}{4})$, and we have

$$D\left(0, \frac{|f'(0)|^2}{256}\right) \subset f(R\mathbb{D}).$$

Now, from (2.9),

$$\int_{f^{-1}(\Sigma_\varepsilon)} |f(z)|^p dA \geq \int_{f^{-1}(\Sigma_\varepsilon \cap \{R/2 < |w| < R\})} |f(z)|^p dA$$

$$\geq \left(\frac{R}{2}\right)^p \frac{1}{8^2} \int_{f^{-1}(\Sigma_\varepsilon \cap \{R/2 < |w| < R\})} |f'(z)|^2 dA.$$

Hence, using (2.10), we obtain

$$\int_{f^{-1}(\Sigma_\varepsilon)} |f(z)|^p dA \geq R^p R^2 \varepsilon C = \varepsilon K |f'(0)|^{2p+4}.$$

(2.7) follows because $f = g/\|g\|_p$ and $f^{-1}(\Sigma_\varepsilon) = g^{-1}(\Sigma_\varepsilon)$.

To see (b), we can argue in the same way and observe that $(1 - |z|)^\alpha \geq C_1(\alpha)$, a constant that depends only on α, whenever $z \in f^{-1}(\{R/2 < |w| < R\})$. We omit the details. $\qquad \square$

It should be observed that Lemma 2.2 is really about the image of a fixed ball centered at 0, not the range of g on the whole disk, and for this reason it might be useful in trying to generalize the main result in [**MS99**]. In particular, while $|g'(0)|/\|g\|_p$ has always an upper bound, part (a) in Lemma 2.2 gives (0.1) for those functions with a lower bound for $|g'(0)|/\|g\|_p$.

We are ready to see that inverse images of sectors are dominating sets for uniformly bounded and locally univalent in the origin functions of A_α^p.

THEOREM 2.3. *Let* $p \geq 1$, $\alpha > -1$, $M > 0$, *and* $a \in \mathbb{C}$, $a \neq 0$. *Then, given any* $\varepsilon > 0$, *there exists a constant* $\delta = \delta(\alpha, p, \varepsilon) > 0$ *such that*

(2.11)
$$\int_{\mathbb{D}} |f(z)|^p \, dA_\alpha(z) \leq \delta \int_{f^{-1}(\Sigma_\varepsilon)} |f(z)|^p \, dA_\alpha(z)$$

for any function $f \in A_\alpha^p$ *satisfying* $f(0) = 0 = f'(0) - a$ *and* $\|f\|_{\alpha,p} \leq M$.

PROOF. It is well-known that any ball in A_α^p is a normal family. We will assume that the statement fails to get a contradiction. In this case, for some $\varepsilon > 0$, we can find a sequence of functions f_n holomorphic on $\mathbb{D}$ such that

(2.12)
$$n \int_{f_n^{-1}(\Sigma_\varepsilon)} |f_n(z)|^p \, dA_\alpha(z) \leq \|f_n\|_{\alpha,p} \leq M, \qquad n = 1, 2, \ldots$$

with $f_n(0) = f'_n(0) - a = 0$, for all n. By normality, taking a subsequence if necessary, f_n converges to f uniformly on compact subsets. In particular, $f(0) = 0$, $f'(0) = a$, and, by Fatou's lemma, $\|f\|_{\alpha,p} \leq M$. Moreover, from (2.12) it is easy to see that

$$\int_{f^{-1}(\Sigma_\varepsilon)} |f(z)|^p dA_\alpha(z) = 0,$$

which is in contradiction with Lemma 2.2. □

REMARK. The referee has pointed out to us that Theorem 2.3 can also be deduced immediately from Lemma 2.2 without the normal families argument. Indeed, since $\frac{|f'(0)|}{\|f\|_{\alpha,p}} \geq \frac{|a|}{M}$, the integral on the left side of (2.8) is bounded below by $\varepsilon K\left(\frac{a}{M}\right)^{2p+4}$ which is what is claimed in Theorem 2.3. In fact, this argument also shows that a more general result is true: If we just assume a lower bound on $\frac{|f'(0)|}{\|f\|_{\alpha,p}}$ instead $f'(0) = a$ and $\|f\|_{\alpha,p} < M$, we can state

THEOREM 2.4. *Let $p \geq 1$ and $\alpha > -1$. Then, given any $\varepsilon > 0$, there is a constant $\delta = \delta(\alpha, p, \varepsilon) > 0$ such that*

$$(2.13) \qquad \int_{\mathbb{D}} |f(z)|^p \, dA_\alpha(z) \leq \delta \int_{f^{-1}(\Sigma_\varepsilon)} |f(z)|^p \, dA_\alpha(z)$$

for every function $f \in A^p_\alpha$, $f(0) = 0$ such that $\frac{|f'(0)|}{\|f\|_{\alpha,p}}$ is bigger than some positive constant N.

3. Other class of dominating sets

While Theorem A describes dominating sets for the classes A^p_α, in this section we use the arguments of Luecking in [**Lu81**] to construct other sets S_f, that despite their depending on some parameters, behave like dominating sets for $f \in A^p_\alpha$.

We will assume $\varepsilon \in (0, 1)$ is fixed. It is easy to see that there exist a constant $K_1(\varepsilon, \alpha) > 0$ such that

$$(3.1) \qquad \left(1 - |z|^2\right)^\alpha > K_1(\varepsilon, \alpha) \left(1 - |a|^2\right)^\alpha$$

for any $z \in D_\varepsilon(a) = \{z \in \mathbb{D} : |z - a| < \varepsilon(1 - |a|)\}$ and all $a \in \mathbb{D}$. Given $\lambda \in (0, 1)$ and a function f holomorphic in $\mathbb{D}$, we put

$$E(\lambda, f)(a) := E_\lambda(a) = \{z \in D_\varepsilon(a) : |f(z)| > \lambda|f(a)|\}.$$

Whenever the function f is clear in the context, we just write $E(\lambda, f)(a) := E_\lambda(a)$. It is easy to see that for $\lambda_1 < \lambda_2 < 1$, $E(\lambda_2, f)(a) \subset E(\lambda_1, f)(a)$, $a \in \mathbb{D}$.

Given a positive, translation invariant doubling measure μ and an analytic function f on $\mathbb{D}$, we define the set

$$(3.2) \qquad S_f(\rho, \lambda) = \{z \in \mathbb{D} : \mu(E_\lambda(z)) > \rho\mu(D_\varepsilon(z))\},$$

for each $\rho, \lambda \in (0, 1)$. Note that for λ fixed, if $\rho_1 < \rho_2 < 1$, then $S_f(\rho_2, \lambda) \subset S_f(\rho_1, \lambda)$. Sets $S_f(\rho, \lambda)$ behave as dominating sets for A^p_α as we will show in Theorem 3.3 below whose proof rests upon two lemmas.

LEMMA 3.1. *Given $\varepsilon \in (0, 1)$, there exist a constant $C(\varepsilon) > 0$ such that for any z in $\mathbb{D}$ we have*

$$(3.3) \qquad I := \int_{\mathbb{D}} \frac{1}{\mu(D_\varepsilon(a))} \mathbf{1}_{D_\varepsilon(a)}(z) d\mu(a) \leq C(\varepsilon),$$

where $\mathbf{1}_{D_\varepsilon(a)}$ denotes the characteristic function of the set $D_\varepsilon(a)$.

PROOF. We observe that $D_\varepsilon(a) \subset \Delta(a,\varepsilon)$ and, since $z \in \Delta(a,\varepsilon)$ if and only if $a \in \Delta(z,\varepsilon)$, then $\mathbf{1}_{D_\varepsilon(a)}(z) \leq \mathbf{1}_{\Delta(a,\varepsilon)}(z) = \mathbf{1}_{\Delta(z,\varepsilon)}(a)$. Therefore,

$$(3.4) \qquad I \leq \int_{\Delta(z,\varepsilon)} \frac{1}{\mu(D_\varepsilon(a))} d\mu(a);$$

moreover, since $\Delta(z,\varepsilon)$ is an Euclidean disk of radius R, we get that $A(\Delta(z,\varepsilon)) \leq \frac{4\pi\varepsilon^2}{(1-\varepsilon^2)^2}(1-|z|)^2$. Hence there is a constant $C(\varepsilon) > 0$ such that $R = C(\varepsilon)\text{radius}(D_\varepsilon(a))$ and $\mu(\Delta(z,\varepsilon)) \leq C(\varepsilon)\mu(D_\varepsilon(a))$. Here we have used that μ is a translation-invariant doubling measure. This, with (3.4), yields $I \leq C(\varepsilon)$, and (3.3) follows. $\qquad\square$

Fix $r_0 \in (0,1)$. For each $f \in A_\alpha^p$, we consider the set

$$(3.5) \qquad A = \left\{ a \in \mathbb{D} : |f(a)|^p \leq \frac{r_0}{\mu(D_\varepsilon(a))} \int_{D_\varepsilon(a)} |f(z)|^p d\mu(z) \right\}.$$

Note that, by Lemma 3.1, there exists a constant $C_2(\varepsilon) > 0$ such that

$$(3.6) \qquad \int_A |f(z)|^p d\mu(z) \leq r_0 C_2(\varepsilon) \int_{\mathbb{D}} |f(z)|^p d\mu(z).$$

We also define the operator

$$T_\lambda f(a) = \frac{1}{A(E_\lambda(a))} \int_{E_\lambda(a)} |f(z)|^p dA(z),$$

where $\lambda \in (0,1)$. We split the unit disk into two sets G and $M = \mathbb{D} \setminus G$, where

$$G = \left\{ a \in \mathbb{D} : |f(a)|^p > r_0^3 T_\lambda f(a) \right\}.$$

LEMMA 3.2. *Suppose r_0 and ρ are given in $(0,1)$, and let $f \in A_\alpha^p$. If $\lambda < r_0^{\frac{3}{p}\left(\frac{\rho}{1-\rho}\right)}$, then $G \subset S_f(\rho,\lambda)$.*

PROOF. Since $\log|f|$ is concave, for any $a \in G$ we have that

$$A(D_\varepsilon(a))\log|f(a)| \leq \int_{D_\varepsilon(a)} \log|f(z)| dA(z)$$

$$\leq (A(D_\varepsilon(a)) - A(E_\lambda(a)))\log\lambda|f(a)| + \frac{1}{p}A(E_\lambda(a))\log T_\lambda f(a)$$

$$< \left\{ \frac{3}{p}\left(\frac{\rho}{1-\rho}\right)\log(r_0) + \log|f(a)| \right\} A(D_\varepsilon(a))$$

$$\qquad - \frac{3}{p(1-\rho)}\log(r_0) A(E_\lambda(a)),$$

from which, a simple calculation leads to $A(E_\lambda(a)) > \rho A(D_\varepsilon(a))$. Consequently, $G \subset S_f(\rho,\lambda)$, as desired. $\qquad\square$

We are in position to state the main result of this section.

THEOREM 3.3. *Let $f \in A_\alpha^p$, $\alpha > -1$ and $p \geq 1$. Given ε and ρ in $(0,1)$, there are positive constants λ and δ in $(0,1)$, not depending on f, satisfying*

$$(3.7) \qquad \int_{S_f(\rho,\lambda)} |f(z)|^p dA_\alpha(z) > \delta \int_{\mathbb{D}} |f(z)|^p dA_\alpha(z).$$

PROOF. Suppose $r_0 \in (0,1)$ is fix, and take λ as in Lemma 3.2. Let us decompose the corresponding set M into two sets $M_A = M \cap A$ and $M_B = M \setminus A$, where A is as in (3.5) with μ replaced by the Lebesgue measure. Then, by (3.6) and (3.1), we can write

$$\int_{M_A} |f(z)|^p dA_\alpha(z) \le r_0 C_2(\varepsilon, \alpha) \int_{\mathbb{D}} |f(z)|^p dA_\alpha(z).$$

CLAIM. For $a \in M_B$ and $\lambda \le 1/2$, there is an absolute constant such that

$$(3.8) \qquad A(E_\lambda(a)) > C r_0^2 A(D_\varepsilon(a)).$$

To see this, take any point $a \in M_B$ and set, say, $R = \frac{1}{32} r_0 \varepsilon (1 - |a|)$. We will show that

$$(3.9) \qquad D(a; R) \subset E_\lambda(a).$$

To see (3.9), let $r = \frac{\varepsilon}{2}(1 - |a|)$ and take $z \in D(a, R)$. By the Cauchy integral formula, we can write

$$(3.10) \qquad |f(z) - f(a)| \le \frac{1}{8} r_0 M_r,$$

where $M_r := \sup\{|f(t)| : |t - a| = r\}$. Since $|f|$ is subharmonic, $D(t, r) \subset D_\varepsilon(a)$, using the Hölder inequality, and that $a \in M_B$, one has

$$M_r \le \frac{4}{A(D_\varepsilon(a))} \int_{D_\varepsilon(a)} |f(w)| dA(w) < \frac{4}{r_0} |f(a)|.$$

Thus, a substitution in (3.10) yields $|f(z) - f(a)| < \frac{1}{2}|f(a)|$ for all $z \in D(a, R)$. By the triangle inequality, one gets $|f(z)| > \frac{1}{2}|f(a)| \ge \lambda|f(a)|$ for any $z \in D(a, R)$ and (3.9) holds. From this, an easy calculation leads to (3.8) with $C = \frac{1}{32^2}$.

Now, since $M_B \subset M$, using Fubini's theorem and estimate in (3.1), we have

$$\int_{M_B} |f(a)|^p dA_\alpha(a)$$

$$\le \frac{r_0^3}{K_1(\varepsilon, \alpha)} \int_{\mathbb{D}} |f(z)|^p \left(\int_{M_B} \frac{1}{A(E_\lambda(a))} \mathbf{1}_{E_\lambda(a)}(z) dA(a) \right) dA_\alpha(z)$$

$$\le C_3(\varepsilon, \alpha) r_0 \int_{\mathbb{D}} \left(\int_{M_B} \frac{1}{A(D_\varepsilon(a))} \mathbf{1}_{D_\varepsilon(a)}(z) dA(a) \right) |f(z)|^p dA_\alpha(z)$$

$$\le C_4(\varepsilon, \alpha) r_0 \int_{\mathbb{D}} |f(z)|^p dA_\alpha(z).$$

Note that we have used (3.8) and Lemma 3.1. Hence, if we select $r_0 = r_0(\varepsilon, \alpha) < \frac{1}{C_2(\varepsilon, \alpha) + C_4(\varepsilon, \alpha)}$, choose λ in such a way that $\lambda < \min\left\{ \frac{1}{2}, r_0^{\frac{3}{p}\left(\frac{\rho}{1-\rho}\right)} \right\}$, and according to the inclusion in Lemma 3.2 we get

$$\int_{\mathbb{D}} |f(z)|^p dA_\alpha(z) \le \int_{S_f(\rho,\lambda)} |f(z)|^p dA_\alpha(z) + \int_M |f(z)|^p dA_\alpha(z)$$

$$< \int_{S_f(\rho,\lambda)} |f(z)|^p dA_\alpha(z)$$

$$+ r_0 \left(C_2(\varepsilon, \alpha) + C_4(\varepsilon, \alpha) \right) \int_{\mathbb{D}} |f(z)|^p dA_\alpha(z),$$

and the proof is complete.   □

REMARK. It should be noted that Theorem 3.3 also holds, with minor modifications in its proof, for analytic functions in $L^p(\mu)$ where μ is a translation-invariant doubling measure.

3.1. An Application. Theorem A asserts that dominating sets for the class A_α^p must have considerable mass near the unit circle. However, we can apply Theorem 3.3 to give an example of a function f such that a certain dominating set for such f is contained in a disk $D(0,r)$, $r < 1$.

Consider the function $f(z) = z^n$, and let $\rho = \frac{7}{8}$. According Theorem 3.3, we can find a $\lambda_0 > 0$, independent of f, such that $S_f = S_f\left(\frac{7}{8}, \lambda_0\right)$ is a dominating set for such a f. Now, let $\eta \in (0,1)$ and $r = \frac{\eta}{1+\eta}$. We will show that if we take n large enough, then $S_f \subset D(0,r)$. Indeed, take $a \notin D(0,r)$ so that $|a| > \frac{\eta}{1+\eta}$, $|a| - \eta(1 - |a|) > 0$ and $0 \notin D_\eta(a)$. Moreover, note that we have

$$E_{\lambda_0}(a) = \left\{ z \in D_\eta(a) : |z| > \sqrt[n]{\lambda_0}|a| \right\}.$$

Since $\sqrt[n]{\lambda_0} \to 1$ as $n \to \infty$, we can choose n big enough so that

$$A\left(D_\eta(a) \setminus E_{\lambda_0}(a)\right) \geq \frac{1}{2} A\left(D_\eta(a) \cap D(0;|a|)\right) \geq \frac{1}{8} A\left(D_\eta(a)\right).$$

This is possible because $D_\eta(a)$ is centered at the boundary of $D(0;|a|)$. Then

$$A\left(E_{\lambda_0}(a)\right) \leq A\left(D_\eta(a)\right) - A\left(D_\eta(a) \setminus E_{\lambda_0}(a)\right) \leq \frac{7}{8} A\left(D_\eta(a)\right),$$

and so $a \notin S_f = S_f\left(\frac{7}{8}, \lambda_0\right) = \left\{ z \in \mathbb{D} : A\left(E_{\lambda_0}(z)\right) > \frac{7}{8} A\left(D_\eta(z)\right) \right\}$. Thus, we can conclude that if n is big and $f(z) = z^n$, then $S_f \subset D(0,r)$ and

$$\int_{D(0,r)} |f(z)|^p \, dA_\alpha(z) \geq \int_{S_f} |f(z)|^p \, dA_\alpha(z) \geq \delta \int_{\mathbb{D}} |f(z)|^p \, dA_\alpha(z),$$

where δ is given by Theorem 3.3.

References

[Ah76]   L. V. Ahlfors, *Complex Analysis*, second ed., McGraw-Hill, New York, 1976.

[Co78]   J. Conway, *Functions of One Complex Variable*, Springer Verlag, New York, 1978.

[DS88]   S. Djrbashian and F. Shamoian, *Topics in the Theory of A_α^p Spaces*, Teubner-Texte Zur Mathematic, Leipzig, 1988.

[Du70]   P. Duren, *Theory of H^p Spaces*, Academic Press, New York, 1970.

[HKZ00] H. Hedenmalm, B. Korenblum and K. Zhu, *Theory of Bergman Spaces*, Springer-Verlag, New York, 2000.

[Lu81]   D. Luecking, *Inequalities on Bergman spaces*, Ill. J. Math. **25** (1981), 1–11.

[MS99]   D. Marshall and W. Smith, *The angular distribution of mass by Bergman functions*, Rev. Mat. Iberoamericana **15** (1999), 93–116.

[OS88]   M. Ortel and W. Smith, *The argument of an extremal dilatation*, Proc. Amer. Math. Soc. **104** (1988), 498–502.

[PR01]   F. Pérez-González and J. C. Ramos, *Conjuntos dominantes en espacios de Bergman con peso*, in Margarita Matematica (L. Español and J.L. Varona, eds.), Sdo. Publicaciones, Universidad de La Rioja, Logroño, 2001, pp. 97–109.

[PR]   F. Pérez-González and J. C. Ramos, *The angular distribution of mass by weighted Bergman functions*, Divulgaciones Matem. **12** (2004), 65-86.

[Po92]   Ch. Pommerenke, *Boundary Behaviour of Conformal Maps*, Springer Verlag, Berlin, 1992.

[Ra95]   T. Ransford, *Potential Theory in the Complex Plane*, Cambridge Univesity Press, London Math. Society, Student Texts, no. 28, Cambridge, 1995.

[Zh90]   K. Zhu, *Operator Theory in Function Spaces*, Marcel Dekker, New York-Basel, 1990.

DEPARTAMENTO DE ANÁLISIS MATEMÁTICO, UNIVERSIDAD DE LA LAGUNA, 38271 LA LAGUNA, TENERIFE, SPAIN
E-mail address: fernando.perez.gonzalez@ull.es

DEPARTAMENTO DE MATEMÁTICAS, UNIVERSIDAD DE ORIENTE, CUMANÁ, EDO. SUCRE, VENEZUELA
E-mail address: jramos@sucre.udo.edu.ve

Contemporary Mathematics
Volume **404**, 2006

Trigonometric Obstacle Problem and Weak Factorization

Serguei Shimorin

ABSTRACT. Using an argument related to an obstacle problem for trigonometric polynomials, we prove that any real valued polynomial is the difference of two positive polynomials of the same degree with the control of L^1-norm. This fact has an application to weak factorization of polynomials.

1. Introduction

Let $\omega(x_1, \ldots, x_d)$ be a positive bounded function defined in $\mathbb{R}^d$ and 2π-periodic in each of variables $x_1, \ldots, x_d$. Given a multiindex

$$\alpha = (\alpha_1, \ldots, \alpha_d) \in \mathbb{Z}_+^d,$$

we consider a set $T_{\omega,\alpha}$ of trigonometric polynomials

$$(1.1) \qquad p(x_1, \ldots, x_d) = \sum_{\substack{-\alpha_1 \leqslant \nu_1 \leqslant \alpha_1 \\ \vdots \\ -\alpha_d \leqslant \nu_d \leqslant \alpha_d}} \hat{p}(\nu_1, \ldots, \nu_d) e^{i(\nu_1 x_1 + \cdots + \nu_d x_d)}$$

satisfying

$$p(x_1, \ldots, x_d) \geqslant \omega(x_1, \ldots, x_d), \quad x_l \in \mathbb{R}, \, l = 1, \ldots, d.$$

For a trigonometric polynomial p of the form (1.1), we write

$$\beta = \deg p$$

if $\beta = (\beta_1, \ldots, \beta_d) \in \mathbb{Z}_+^d$ is such a multiindex that β_ν is the degree of p with respect to the variable x_ν for each $\nu = 1, \ldots, d$. In other words, the set $T_{\omega,\alpha}$ consists of trigonometric polynomials p in variables $x_1, \ldots, x_d$ that satisfy

$$\deg p \leqslant \alpha \qquad \text{and} \qquad p \geqslant \omega.$$

The aim of the first part of the present paper is the study of the following quantity:

$$D_{\omega,\alpha} := \inf\{\|p\|_{L^1} : p \in T_{\omega,\alpha}\},$$

where the norm of p is taken in $L^1((-\pi, \pi)^d)$ with respect to the Lebesgue measure. It turns out that $D_{\omega,\alpha}$ can be characterized modulo absolute constants by the

---

2000 *Mathematics Subject Classification.* Primary 42A05; Secondary 47A68.

"obstacle function" ω. The following kind of maximal function appears in such a characterization:

$$(1.2) \qquad M_\alpha[\omega](x_1, \ldots, x_d) := \sup_{Q_\alpha(x)} \omega(t_1, \ldots, t_d),$$

the supremum being taken over the set $Q_\alpha(x)$ of those $t = (t_1, \ldots, t_d)$ that

$$|t_\nu - x_\nu| \leqslant (\alpha_\nu + 1)^{-1}, \quad \nu = 1, \ldots, d.$$

Our first result is

THEOREM 1.1. *There exist constants $c(d)$ and $C(d)$ depending on the dimension d but not on α such that for any multiindex $\alpha \in \mathbb{Z}_+^d$,*

$$(1.3) \qquad c(d)\|M_\alpha[\omega]\|_{L^1} \leqslant D_{\omega,\alpha} \leqslant C(d)\|M_\alpha[\omega]\|_{L^1}.$$

This theorem was originally inspired by the following question: if p is a real-valued trigonometric polynomial of certain degree α, is it possible to represent it as a difference of two positive trigonometric polynomials of the same degree with the control of L^1-norm? The answer turned out to be positive as shows the next theorem which will be proved by a method similar to that of the proof of Theorem 1.1.

THEOREM 1.2. *There exists a constant $A(d)$ depending only on the dimension d but not on α such that for any real-valued trigonometric polynomial p of degree α, there exist positive trigonometric polynomials p^+ and p^- satisfying*

    (i) $\deg p^+ \leqslant \alpha$, $\deg p^- \leqslant \alpha$;

    (ii) $p = p^+ - p^-$;

    (iii) $\|p^+\|_{L^1} \leqslant A(d)\|p\|_{L^1}$; $\quad \|p^-\|_{L^1} \leqslant A(d)\|p\|_{L^1}$.

It should be mentioned that this theorem remains valid even in the operator-valued context, i.e., for polynomials taking values in the space $\mathcal{B}(\mathcal{H})$ of bounded linear operators in some Hilbert space $\mathcal{H}$ (of course, "real-valued" should be replaced by "self-adjoint-valued" in this context).

In the one-dimensional situation, the last theorem can be combined with Fejér-Riesz' theorem saying that a positive trigonometric polynomial is a square of modulus of an analytic polynomial of the same degree (both considered as defined on the unit circle $\mathbb{T}$). This observation, together with an application of a shift operator, implies

THEOREM 1.3. *Let $P(z)$ be an analytic polynomial of a complex variable z of degree $2n$. Then there exist analytic polynomials q_ν and r_ν, $\nu = 1, 2, 3, 4$, such that*

    (i) $\deg q_\nu \leqslant n$, $\deg r_\nu \leqslant n$;

    (ii) $P(z) = \displaystyle\sum_{\nu=1}^{4} q_\nu(z) r_\nu(z);$ *and*

    (iii) $\displaystyle\sum_{\nu=1}^{4} \|q_\nu\|_{L^2(\mathbb{T})} \|r_\nu\|_{L^2(\mathbb{T})} \leqslant A \|P\|_{L^1(\mathbb{T})}.$

Here, A is an absolute constant.

This theorem was also obtained independently by A. Volberg (private communication).

Theorem 1.3 remains valid in the operator-valued context as well (with absolutely the same proof).

A two-dimensional version of Theorem 1.3 is also true in the scalar case with more addends instead of four. But the proof is quite different. We shall derive it in Section 3 from the recent result of Ferguson and Lacey [1] on weak factorization of functions from H^1 in the bidisk.

Theorems 1.1 and 1.2 will be proved in Section 2. We consider only the case $d = 2$ there. The proofs in the general case can be obtained by obvious modifications. Section 3 is devoted to Theorem 1.3 and its two-dimensional counterpart.

2. An obstacle problem

In this section, we prove Theorems 1.1 and 1.2 in the case $d = 2$. Let ω and $\alpha = (\alpha_1, \alpha_2)$ be fixed.

We prove first the inequality

$$(2.1) \qquad D_{\omega, \alpha} \leqslant C\|M_\alpha[\omega]\|_{L^1}.$$

Let

$$(2.2) \qquad \Phi_n(t) = \sum_{k=-n}^{n} \left(1 - \frac{|k|}{n+1}\right) e^{ikt} = \frac{1}{n+1} \frac{\sin^2 \frac{n+1}{2} t}{\sin^2 \frac{t}{2}}$$

denote the Fejér kernel. We define

$$q(x_1, x_2) := \int_{-\pi}^{\pi} \int_{-\pi}^{\pi} M_\alpha[\omega](x_1 - t_1, x_2 - t_2) \Phi_{\alpha_1}(t_1) \Phi_{\alpha_2}(t_2) \frac{dt_1}{2\pi} \frac{dt_2}{2\pi}$$

or, in brief notation

$$q = M_\alpha[\omega] * (\Phi_{\alpha_1} \otimes \Phi_{\alpha_2}).$$

Clearly, q is a trigonometric polynomial satisfying $\deg q \leqslant \alpha$ and $\|q\|_{L^1} \leqslant \|M_\alpha[\omega]\|_{L^1}$. On the other hand,

$$M_\alpha[\omega](x_1 - t_1, x_2 - t_2) \geqslant \omega(x_1, x_2)$$

if $|t_1| \leqslant (\alpha_1 + 1)^{-1}$, $|t_2| \leqslant (\alpha_2 + 1)^{-1}$, which implies

$$q(x_1, x_2) \geqslant \int_{-(\alpha_1+1)^{-1}}^{(\alpha_1+1)^{-1}} \int_{-(\alpha_2+1)^{-1}}^{(\alpha_2+1)^{-1}} \omega(x_1, x_2) \Phi_{\alpha_1}(t_1) \Phi_{\alpha_2}(t_2) \frac{dt_1}{2\pi} \frac{dt_2}{2\pi}.$$

Since

$$\int_{-(n+1)^{-1}}^{(n+1)^{-1}} \Phi_n(t)\, dt \geqslant a,$$

where a is an absolute constant, we obtain that there is an absolute constant $C > 0$ such that

$$q(x_1, x_2) \geqslant C^{-1}\omega(x_1, x_2).$$

Taking finally $p = Cq$, we obtain a trigonometric polynomial from the set $T_{\omega, \alpha}$ satisfying $\|p\|_{L^1} \leqslant C\|M_\alpha[\omega]\|_{L^1}$ which proves (2.1).

To prove the estimate

$$c\|M_\alpha[\omega]\|_{L^1} \leqslant D_{\omega, \alpha},$$

we assume that p is an arbitrary trigonometric polynomial from the set $T_{\omega,\alpha}$. Let

$$(2.3) \quad r(x_1, x_2) := p(x_1, x_2) + \int_{-(\alpha_1+1)^{-1}}^{(\alpha_1+1)^{-1}} \left| \frac{\partial p}{\partial x_1}(x_1 - t_1, x_2) \right| dt_1$$

$$+ \int_{-(\alpha_2+1)^{-1}}^{(\alpha_2+1)^{-1}} \left| \frac{\partial p}{\partial x_2}(x_1, x_2 - t_2) \right| dt_2$$

$$+ \int_{-(\alpha_1+1)^{-1}}^{(\alpha_1+1)^{-1}} \int_{-(\alpha_2+1)^{-1}}^{(\alpha_2+1)^{-1}} \left| \frac{\partial^2 p}{\partial x_1 \partial x_2}(x_1 - t_1, x_2 - t_2) \right| dt_1 dt_2$$

or, in brief notation,

$$r = p + 2\pi \left| \frac{\partial p}{\partial x_1} \right| * \left(\chi_{Q_\alpha^1} \otimes \delta_0 \right) + 2\pi \left| \frac{\partial p}{\partial x_2} \right| * \left(\delta_0 \otimes \chi_{Q_\alpha^2} \right) + 4\pi^2 \left| \frac{\partial^2 p}{\partial x_1 \partial x_2} \right| * \chi_{Q_\alpha},$$

where χ_E means the characteristic function of a set E,

$$Q_\alpha^l = [-(\alpha_l + 1)^{-1}, (\alpha_l + 1)^{-1}], \quad l = 1, 2,$$

and

$$Q_\alpha = Q_\alpha(0) = Q_\alpha^1 \times Q_\alpha^2 = \{(t_1, t_2) : |t_l| \leqslant (\alpha_l + 1)^{-1}, \quad l = 1, 2\}.$$

Since

$$\omega(x_1 - t_1, x_2 - t_2) \leqslant p(x_1 - t_1, x_2 - t_2)$$

and

$$(2.4) \quad p(x_1 - t_1, x_2 - t_2) = p(x_1, x_2) - \int_0^{t_1} \frac{\partial p}{\partial x_1}(x_1 - \tau_1, x_2) d\tau_1$$

$$- \int_0^{t_2} \frac{\partial p}{\partial x_2}(x_1, x_2 - \tau_2) d\tau_2 + \int_0^{t_1} \int_0^{t_2} \frac{\partial^2 p}{\partial x_1 \partial x_2}(x_1 - \tau_1, x_2 - \tau_2) d\tau_1 d\tau_2,$$

we obtain

$$\omega(x_1 - t_1, x_2 - t_2) \leqslant r(x_1, x_2)$$

if $(t_1, t_2) \in Q_\alpha$ and hence $M_\alpha[\omega] \leqslant r$. On the other hand, Bernstein's inequality implies that

$$\left\| \frac{\partial p}{\partial x_l} \right\|_{L^1} \leqslant \alpha_l \|p\|_{L^1}, \quad l = 1, 2$$

and

$$\left\| \frac{\partial^2 p}{\partial x_1 \partial x_2} \right\|_{L^1} \leqslant \alpha_1 \alpha_2 \|p\|_{L^1}$$

which gives us

$$(2.5) \quad \|r\|_{L^1} \leqslant A \|p\|_{L^1}$$

with an absolute constant A. Therefore,

$$\|M_\alpha[\omega]\|_{L^1} \leqslant A \|p\|_{L^1}$$

and it remains to take the infimum over all $p \in T_{\omega,\alpha}$.

Now, we turn to the proof of Theorem 1.2. Let p be an arbitrary real-valued trigonometric polynomial of degree $\alpha = (\alpha_1, \alpha_2)$. We define a function r by a formula similar to (2.3) with $p(x_1, x_2)$ replaced by $|p(x_1, x_2)|$. Using representation (2.4), we obtain

$$p(x_1 - t_1, x_2 - t_2) \leqslant r(x_1, x_2)$$

if $(t_1, t_2) \in Q_\alpha$. In addition, $\|r\|_{L^1} \leqslant A\|p\|_{L^1}$. We define then a trigonometric polynomial p^+ as

$$p^+(x_1, x_2) := C_\alpha r * (\Phi_{\alpha_1} \otimes \Phi_{\alpha_2})(x_1, x_2)$$

$$= C_\alpha \int_{-\pi}^{\pi} \int_{-\pi}^{\pi} r(x_1 - t_1, x_2 - t_2) \Phi_{\alpha_1}(t_1) \Phi_{\alpha_2}(t_2) \frac{dt_1}{2\pi} \frac{dt_2}{2\pi},$$

where the constant C_α is defined by

$$C_\alpha^{-1} = \int_{-(\alpha_1+1)^{-1}}^{(\alpha_1+1)^{-1}} \int_{-(\alpha_2+1)^{-1}}^{(\alpha_2+1)^{-1}} \Phi_{\alpha_1}(t_1) \Phi_{\alpha_2}(t_2) \frac{dt_1}{2\pi} \frac{dt_2}{2\pi}.$$

Clearly, all constants C_α are bounded by an absolute constant. In view of the above properties of r, p^+ satisfies

$$deg\, p^+ \leqslant \alpha;$$

$$\|p^+\|_{L^1} \leqslant C_\alpha \|r\|_{L^1} \leqslant C_\alpha A \|p\|_{L^1};$$

and

$$p^+(x_1, x_2) \geqslant C_\alpha \int_{-(\alpha_1+1)^{-1}}^{(\alpha_1+1)^{-1}} \int_{-(\alpha_2+1)^{-1}}^{(\alpha_2+1)^{-1}} r(x_1 - t_1, x_2 - t_2) \Phi_{\alpha_1}(t_1) \Phi_{\alpha_2}(t_2) \frac{dt_1}{2\pi} \frac{dt_2}{2\pi}$$

$$\geqslant C_\alpha p(x_1, x_2) \cdot \int_{-(\alpha_1+1)^{-1}}^{(\alpha_1+1)^{-1}} \int_{-(\alpha_2+1)^{-1}}^{(\alpha_2+1)^{-1}} \Phi_{\alpha_1}(t_1) \Phi_{\alpha_2}(t_2) \frac{dt_1}{2\pi} \frac{dt_2}{2\pi} = p(x_1, x_2),$$

Letting now $p^- := p^+ - p$, we obtain the desired decomposition $p = p^+ - p^-$ from Theorem 1.2.

REMARK 2.1. The above argument of the proof of Theorem 1.2 works without changes for operator-valued trigonometric polynomials p with self-adjoint values.

3. Weak factorization

In this section, all trigonometric polynomials will be considered as defined on the unit circle $\mathbb{T}$ in one-dimensional case or on $\mathbb{T}^2$ in the case $d = 2$.

We recall first Fejér-Riesz' theorem saying that a one-dimensional positive trigonometric polynomial of degree n (written in complex form)

$$p(z) = \sum_{k=-n}^{n} \hat{p}(k) z^k, \qquad z \in \mathbb{T}$$

has the form

$$p(z) = |q(z)|^2,$$

where

$$q(z) = \sum_{k=0}^{n} \hat{q}(k) z^k$$

is an analytic polynomial of the same degree. An operator-valued version of this theorem was proved by Rosenblum and Rovnyak [3].

Let now

$$P(z) = \sum_{k=0}^{2n} \widehat{P}(k) z^k$$

be an arbitrary analytic polynomial of degree $2n$. Considering a trigonometric polynomial

$$p(z) = \bar{z}^n P(z) = \sum_{k=-n}^{n} \widehat{P}(k+n)z^k,$$

one can decompose it using Theorem 1.2 as a sum

$$p(z) = \sum_{\varepsilon = \pm 1, \pm i} \varepsilon p_\varepsilon(z),$$

where p_ε are positive trigonometric polynomials of degree at most n satisfying $\|p_\varepsilon\|_{L^1(\mathbb{T})} \leqslant C\|p\|_{L^1(\mathbb{T})}$. Writing $p_\varepsilon = |q_\varepsilon|^2$ and letting $r_\varepsilon(z) = \varepsilon z^n \overline{q_\varepsilon(z)}$, one arrives at the desired weak factorization

$$P = \sum_{\varepsilon = \pm 1, \pm i} q_\varepsilon r_\varepsilon$$

from Theorem 1.3.

Clearly, there are no problems with operator-valued polynomials as well.

Unfortunately, the above argument cannot be applied to obtain weak factorization of trigonometric polynomials in several variables. The reason for this is that the Fejér-Riesz theorem fails in several variables even in weak form: there are positive trigonometric polynomials

$$p(z_1, \ldots, z_d) = \sum_{\substack{-\alpha_1 \leqslant k_1 \leqslant \alpha_1 \\ \vdots \\ -\alpha_d \leqslant k_d \leqslant \alpha_d}} \hat{p}(k_1, \ldots, k_d)z_1^{k_1} \ldots z_d^{k_d}, \qquad z_l \in \mathbb{T}, \; l = 1, \ldots, d$$

not representable as sums of squares of moduli of analytic polynomials

$$q(z_1, \ldots, z_d) = \sum_{\substack{0 \leqslant k_1 \leqslant \alpha_1 \\ \vdots \\ 0 \leqslant k_d \leqslant \alpha_d}} \hat{q}(k_1, \ldots, k_d)z_1^{k_1} \ldots z_d^{k_d}.$$

The existence of such polynomials p was proved by Rudin [4], and an explicit example of such a polynomial is

$$p(e^{ix_1}, e^{ix_2}) = (1 + \cos x_1)^2 \sin^2 x_2 + \sin^2 x_1 (1 + \cos x_2)^2$$
$$+ (1 - \cos x_1)^2 (1 - \cos x_2)^2 - 3\sin^2 x_1 \sin^2 x_2$$

which can be obtained by Rudin's method from a known Motzkin's example (see, e.g., [2])

$$M(x, y) = x^4 y^2 + x^2 y^4 + 1 - 3x^2 y^2, \quad (x, y) \in \mathbb{R}^2$$

of a positive algebraic polynomial not representable as a sum of squares.

However, in the case $d = 2$, a weak factorization theorem analogous to Theorem 1.3 can be derived from a recent result of Ferguson and Lacey on weak factorization of functions from the Hardy space H^1 in the bidisk.

Given an analytic polynomial $r(z_1, z_2)$ in two variables, we shall say that $n = (n_1, n_2) \in \mathbb{Z}_+^2$ is the degree of r if n_l is the degree of r with respect to the variable z_l, $l = 1, 2$.

THEOREM 3.1. *Let*

$$P(z_1, z_2) = \sum_{\substack{0 \leqslant k_1 \leqslant 2n_1 \\ 0 \leqslant k_2 \leqslant 2n_2}} \widehat{P}(k_1, k_2) z_1^{k_1} z_2^{k_2}$$

be an analytic scalar polynomial in variables z_1, z_2 of degree $(2n_1, 2n_2)$. Then there exist analytic polynomials q_ν and r_ν, $\nu = 1, \ldots, N$, such that

(i) $\deg q_\nu \leqslant (n_1, n_2)$; $\deg r_\nu \leqslant (n_1, n_2)$, $\nu = 1, \ldots, N$;

(ii) $P = \sum_{\nu=1}^{N} q_\nu r_\nu$;

(iii)

(3.1) $$\sum_{\nu=1}^{N} \|q_\nu\|_{L^2(\mathbb{T}^2)} \|r_\nu\|_{L^2(\mathbb{T}^2)} \leqslant A \|P\|_{L^1(\mathbb{T}^2)}.$$

Here, $N = 4(2n_1 + 1)(2n_2 + 1)$ and A is an absolute constant.

To prove this theorem, we reduce first this weak factorization problem from analytic to trigonometric polynomials as we did in the proof of Theorem 1.3. We define so

$$p(z_1, z_2) := \bar{z}_1^{n_1} \bar{z}_2^{n_2} P(z_1, z_2), \qquad (z_1, z_2) \in \mathbb{T}^2.$$

As before, it is enough to obtain a weak factorization

$$p = \sum_{\nu=1}^{N} q_\nu \overline{r_\nu}$$

with analytic polynomials q_ν, r_ν of degree (n_1, n_2). We shall say that such a weak factorization is L^1-norm-controlled if

(3.2) $$\sum_{\nu} \|q_\nu\|_{L^2} \|r_\nu\|_{L^2} \leqslant A \|p\|_{L^1}$$

with an absolute constant A.

The first lemma shows that if p admits an L^1-norm-controlled weak factorization with some number of addends, then there exists another L^1-norm-controlled weak factorization with at most $4(2n_1 + 1)(2n_2 + 1)$ addends.

LEMMA 3.1. *Assume that a trigonometric polynomial p of degree (n_1, n_2) can be weakly factorized as*

$$p = \sum_{\nu \geqslant 1} q_\nu \overline{r_\nu},$$

where q_ν, r_ν are analytic polynomials of degree $\leqslant (n_1, n_2)$ satisfying (3.2). Then there exist analytic polynomials $\tilde{q}_\nu, \tilde{r}_\nu$, $\nu = 1, \ldots, 4(2n_1 + 1)(2n_2 + 1) =: N$ of degree at most (n_1, n_2) such that

$$p = \sum_{\nu=1}^{N} \tilde{p}_\nu \overline{\tilde{r}_\nu}$$

and

$$\sum_{\nu=1}^{N} \|\tilde{q}_\nu\|_{L^2} \|\tilde{r}_\nu\|_{L^2} \leqslant 4A \|p\|_{L^1}.$$

PROOF. Without loss of generality, we may assume that $\|q_\nu\|_{L^2} = \|r_\nu\|_{L^2}$ for each $\nu \geqslant 1$. Using polarization, we obtain

$$p = \frac{1}{4} \sum_{\varepsilon = \pm 1, \pm i} \sum_{\nu \geqslant 1} \varepsilon |q_\nu + \varepsilon r_\nu|^2 = \frac{1}{4} \sum_{\varepsilon = \pm 1, \pm i} \varepsilon P_\varepsilon,$$

where

$$P_\varepsilon = \sum_{\nu \geqslant 1} |q_\nu + \varepsilon r_\nu|^2$$

is a positive trigonometric polynomial representable as a sum of squares of analytic polynomials. Moreover,

$$\|P_\varepsilon\|_{L^1} = \sum_{\nu \geqslant 1} \|q_\nu + \varepsilon r_\nu\|_{L^2}^2$$

$$\leqslant 2 \sum_{\nu \geqslant 1} \left(\|q_\nu\|_{L^2}^2 + \|r_\nu\|_{L^2}^2 \right) = 4 \sum_{\nu \geqslant 1} \|q_\nu\|_{L^2} \|r_\nu\|_{L^2} \leqslant 4A \|p\|_{L^1}.$$

An easy algebraic argument (see, e.g., [4], the proof of Lemma 1.3) implies now that each P_ε is representable as a sum of at most $(2n_1 + 1)(2n_2 + 1)$ squares

$$P_\varepsilon = \sum_{\nu=1}^{(2n_1+1)(2n_2+1)} |\tilde{q}_{\varepsilon,\nu}|^2,$$

and we have

$$\sum_{\nu=1}^{(2n_1+1)(2n_2+1)} \|\tilde{q}_{\varepsilon,\nu}\|_{L^2}^2 = \|P_\varepsilon\|_{L^1} \leqslant 4A \|p\|_{L^1}.$$

After summation over $\varepsilon = \pm 1, \pm i$, we are done. $\square$

REMARK 3.1. A similar result holds as well if p admits originally an integral weak factorization

$$p = \int_a^b q_t \overline{r_t} \, dt$$

with estimate

$$\int_a^b \|q_t\|_{L^2} \|r_t\|_{L^2} \, dt \leqslant A \|p\|_{L^1}.$$

Now, we assume without loss of generality that both n_1 and n_2 are even, i.e., $n_1 = 2m_1$ and $n_2 = 2m_2$. Obvious modifications needed in the case where some of n_l (or both) are odd are left to the reader.

The next lemma is a principal ingredient of the whole proof. It is a direct consequence of the Ferguson-Lacey weak factorization theorem for the space $H^1(\mathbb{D}^2)$.

Let R denote a rectangle $\{(k_1, k_2) : -2m_l \leqslant k_l \leqslant 2m_l, \, l = 1, 2\}$. We denote by R_{--} its corner $\{(k_1, k_2) : -2m_l \leqslant k_l \leqslant -m_l, \, l = 1, 2\}$. The obvious notations R_{+-}, R_{-+}, and R_{++} stand for three remaining similar corners.

LEMMA 3.2. *For any trigonometric polynomial p of degree $(2m_1, 2m_2)$, there exist analytic polynomials p_ν, r_ν of degree at most (m_1, m_2) such that*

(i) *A trigonometric polynomial*

$$p_{--}(z_1, z_2) := \sum_{\nu \geqslant 1} q_\nu \cdot \overline{\left(z_1^{m_1} z_2^{m_2} r_\nu\right)}$$

satisfies

$$\widehat{p}_{--}(k_1, k_2) = \widehat{p}(k_1, k_2) \qquad \text{if} \qquad (k_1, k_2) \in R_{--};$$

(ii)

$$\sum_{\nu \geqslant 1} \|q_\nu\|_{L^2} \|r_\nu\|_{L^2} \leqslant A\|p\|_{L^1}.$$

Since, clearly, $\text{supp}(\widehat{p}_{--}) \subset \{(k_1, k_2) \: : \: k_l \leqslant 0, l = 1, 2\}$, this Lemma implies that one can find a weak factorization of a part of p corresponding to the corner R_{--} without changing $\widehat{p}$ at other corners R_{+-}, R_{-+}, and R_{++}.

PROOF. Let

$$P(z_1, z_2) = z_1^{2m_1} z_2^{2m_2} p(z_1, z_2).$$

This is an analytic polynomial which can be considered as an element of $H^1(\mathbb{D}^2)$. By the theorem of Ferguson-Lacey [1], there exist functions f_ν, g_ν from $H^2(\mathbb{D}^2)$ such that

$$P = \sum_{\nu \geqslant 1} f_\nu g_\nu \qquad \text{in} \quad \mathbb{D}^2$$

and

$$\sum_{\nu \geqslant 1} \|f_\nu\|_{L^2} \|g_\nu\|_{L^2} \leqslant A\|P\|_{L^1}.$$

It remains to choose now

$$q_\nu(z_1, z_2) := \sum_{\substack{0 \leqslant k_1 \leqslant m_1 \\ 0 \leqslant k_2 \leqslant m_2}} \widehat{f}_\nu(k_1, k_2) z_1^{k_1} z_2^{k_2}$$

and

$$r_\nu(z_1, z_2) := z_1^{m_1} z_2^{m_2} \overline{\left(\sum_{\substack{0 \leqslant k_1 \leqslant m_1 \\ 0 \leqslant k_2 \leqslant m_2}} \widehat{g}_\nu(k_1, k_2) z_1^{k_1} z_2^{k_2} \right)}, \qquad (z_1, z_2) \in \mathbb{T}^2.$$

$\square$

Similarly, one can find an L^1-norm-controlled weak factorization of parts of p corresponding to the corners R_{+-}, R_{-+}, and R_{++}. We obtain therefore a decomposition

$$p = p_{--} + p_{-+} + p_{+-} + p_{++} + p_1,$$

where all terms are trigonometric polynomials of degree at most $(2m_1, 2m_2)$, first four terms admit weak factorization controlled by the L^1-norm of p, and the last term p_1 satisfies

$$\widehat{p}_1(k_1, k_2) = 0 \qquad \text{if} \quad (k_1, k_2) \in R_{--} \cup R_{-+} \cup R_{+-} \cup R_{++}.$$

Let

$$\Psi_m(\theta) = 2\Phi_{2m-1}(\theta) - \Phi_{m-1}(\theta) = 1 + 2 \sum_{k=1}^{m} \cos k\theta + 4 \sum_{k=m+1}^{2m-1} \left(1 - \frac{k}{2m}\right) \cos k\theta$$

be a de la Vallée-Poussin kernel. We put

(3.3) $$p_2(e^{i\theta_1}, e^{i\theta_2}) := \int_{-\pi}^{\pi} p_1(e^{i\theta}, e^{i\theta_2}) \, \Psi_{m_1}(\theta_1 - \theta) \frac{d\theta}{2\pi}$$

or, briefly,

$$p_2 = p_1 * (\Psi_{m_1} \otimes \delta_0).$$

We have

$$\widehat{p}_2(k_1, k_2) = \widehat{p}_1(k_1, k_2) \qquad \text{if} \quad -m_1 \leqslant k_1 \leqslant m_1, \ -2m_2 \leqslant k_2 \leqslant 2m_2,$$

and hence a trigonometric polynomial

$$p_3 := p_1 - p_2$$

satisfies $\widehat{p}_3(k_1, k_2) = 0$ outside the rectangle

$$\{-2m_1 \leqslant k_1 \leqslant 2m_1, \ -m_2 \leqslant k_2 \leqslant m_2\}.$$

In particular,

$$p_3 = p_3 * (\delta_0 \otimes \Psi_{m_2})$$

or

$$(3.4) \qquad p_3(e^{i\theta_1}, e^{i\theta_2}) = \int_{-\pi}^{\pi} p_3(e^{i\theta_1}, e^{i\theta}) \Psi_{m_2}(\theta_2 - \theta) \frac{d\theta}{2\pi}.$$

Further, let $\varphi_n(z)$ be such an analytic polynomial of degree n that

$$\Phi_n(\theta) = |\varphi_n(e^{i\theta})|^2.$$

We have then a weak factorization

$$(3.5) \qquad \Psi_m(\theta) = 2|\varphi_{2m-1}(e^{i\theta})|^2 - |\varphi_{m-1}(e^{i\theta})|^2.$$

For each fixed θ, a one-dimensional trigonometric polynomial $p_1(e^{i\theta}, \cdot)$ admits a weak factorization

$$(3.6) \qquad p_1(e^{i\theta}, \cdot) = \sum_{\nu=1}^{4} q_\nu^\theta(\cdot) \overline{r_\nu^\theta(\cdot)}$$

with some analytic polynomials q_ν^θ, r_ν^θ of degree at most $2m_2$. Substituting (3.5) and (3.6) to (3.3), we obtain an integral weak factorization of p_2:

$$p_2(e^{i\theta_1}, e^{i\theta_2})$$

$$= 2 \int_{-\pi}^{\pi} \sum_{\nu=1}^{4} \left(q_\nu^\theta(e^{i\theta_2}) \varphi_{2m_1-1}(e^{i(\theta_1-\theta)}) \right) \overline{\left(r_\nu^\theta(e^{i\theta_2}) \varphi_{2m_1-1}(e^{i(\theta_1-\theta)}) \right)} \frac{d\theta}{2\pi}$$

$$- \int_{-\pi}^{\pi} \sum_{\nu=1}^{4} \left(q_\nu^\theta(e^{i\theta_2}) \varphi_{m_1-1}(e^{i(\theta_1-\theta)}) \right) \overline{\left(r_\nu^\theta(e^{i\theta_2}) \varphi_{m_1-1}(e^{i(\theta_1-\theta)}) \right)} \frac{d\theta}{2\pi}$$

with an obvious L^1-norm control.

A similar argument applies also to p_3 represented by (3.4) which accomplishes the proof of Theorem 3.1.

References

[1] S. Ferguson and M. Lacey, *A characterization of product BMO by commutators*, Acta Math. **189** (2002), no. 2, 143–160.

[2] A. R. Rajwade, *Squares*, London Mathematical Society Lecture Note Series, **171**. Cambridge University Press, Cambridge, 1993.

[3] M. Rosenblum and J. Rovnyak, *The factorization problem for nonnegative operator valued functions*, Bull. Amer. Math. Soc. **77** (1971), 287–318.

[4] W. Rudin, *The extension problem for positive-definite functions*, Illinois J. Math. **7** (1963), 532–539.

DEPARTMENT OF MATHEMATICS, ROYAL INSTITUTE OF TECHNOLOGY, 100 44 STOCKHOLM, SWEDEN

E-mail address: `shimorin@e.kth.se`

Contemporary Mathematics
Volume **404**, 2006

A Sharp Norm Estimate of the
Bergman Projection on L^p Spaces

Kehe Zhu

ABSTRACT. We show that the norm of the Bergman projection on L^p of the unit ball in $\mathbb{C}^n$ is comparable to $\csc(\pi/p)$ for $1 < p < \infty$.

1. Introduction

Throughout the paper, we fix a positive integer n and let $\mathbb{B}$ denote the open unit ball in $\mathbb{C}^n$. For $-1 < \alpha < \infty$, let

$$dv_\alpha(z) = c_\alpha(1 - |z|^2)^\alpha \, dv(z),$$

where dv is the normalized volume measure on $\mathbb{B}$ and c_α is a positive constant chosen so that dv_α is a probability measure.

For $0 < p < \infty$, let

$$A_\alpha^p = H(\mathbb{B}) \cap L^p(\mathbb{B}, dv_\alpha)$$

denote the weighted Bergman space on $\mathbb{B}$ with standard radial weights. Here $H(\mathbb{B})$ is the space of all holomorphic functions in $\mathbb{B}$. It is easy to see that A_α^p is closed in $L^p(\mathbb{B}, dv_\alpha)$. We will use $\| \ \|_{p,\alpha}$ for the norm in $L^p(\mathbb{B}, dv_\alpha)$.

We use P_α to denote the orthogonal projection from $L^2(\mathbb{B}, dv_\alpha)$ onto A_α^2. It is well-known that P_α is an integral operator on $L^2(\mathbb{B}, dv_\alpha)$,

$$P_\alpha f(z) = \int_\mathbb{B} K_\alpha(z, w) f(w) \, dv_\alpha(w),$$

where the integral kernel is given by

$$K_\alpha(z, w) = \frac{1}{(1 - \langle z, w \rangle)^{n+1+\alpha}}.$$

It is also well-known that, for $1 < p < \infty$, the Bergman projection P_α maps $L^p(\mathbb{B}, dv_\alpha)$ boundedly onto A_α^p. See, for example, Section 7.1 of [**3**].

The purpose of this paper is to give a sharp estimate of the norm of P_α on $L^p(\mathbb{B}, dv_\alpha)$. Our main result is the following theorem.

2000 *Mathematics Subject Classification.* Primary 32A36, 32A25.

This work was done while I was visiting the University of Marseille I in France. I wish to thank the Centre de Mathématiques et d'Informatique at Marseille I, and Professor Hassan Youssfi in particular, for a very nice visit.

THEOREM. *For any* $-1 < \alpha < \infty$, *there exists a constant* $C > 0$, *depending on* α *and* n *but not on* p, *such that the norm of the operator*

$$P_\alpha : L^p(\mathbb{B}, dv_\alpha) \to A_\alpha^p$$

satisfies the estimate

$$C^{-1} \csc \frac{\pi}{p} \le \|P_\alpha\|_p \le C \csc \frac{\pi}{p}$$

for all $1 < p < \infty$.

It is easy to see that the quantity $\csc(\pi/p)$ is comparable to p as $p \to \infty$ and comparable to $1/(p-1)$ as $p \to 1$.

The corresponding result for the Riesz projection onto the Hardy spaces of the unit disk can be found in [2]. I wish to thank Dechao Zheng for bringing this paper to my attention.

2. Preliminaries

The proof of our main result still depends on the two traditional tools, namely, Forelli-Rudin type estimates for certain integrals on the ball and Schur's test for the boundedness of integral operators on L^p spaces. But three new ingredients are necessary here. First, we need a more precise version of the Forelli-Rudin estimates, namely, how the estimates depend on various parameters. Second, we need to find the right test function for Schur's lemma in order to control the parameters in the Forelli-Rudin estimates. And finally, we need to show that our estimates are sharp in a certain sense.

LEMMA 1. *For any* $T > 0$, *there exists a constant* $C > 0$, *depending on* n *and* T *but not on* t, *such that*

$$\int_{\mathbb{S}} \frac{d\sigma(\zeta)}{|1 - \langle z, \zeta \rangle|^{n+t}} \le \frac{C\Gamma(t)}{(1-|z|^2)^t}$$

for all $z \in \mathbb{B}$ *and* $0 < t < T$, *where* $\mathbb{S}$ *is the unit sphere in* $\mathbb{C}^n$ *and* σ *is the normalized Lebesgue measure on* $\mathbb{S}$.

PROOF. By the proof of Proposition 1.4.10 in [3],

$$(1) \qquad \int_{\mathbb{S}} \frac{d\sigma(\zeta)}{|1 - \langle z, \zeta \rangle|^{n+t}} = \frac{\Gamma(n)}{\Gamma^2(\lambda)} \sum_{k=0}^{\infty} \frac{\Gamma^2(k+\lambda)}{\Gamma(k+1)\Gamma(k+n)} |z|^{2k},$$

where $\lambda = (n+t)/2$. Also,

$$(2) \qquad \frac{1}{(1-|z|^2)^t} = \sum_{k=0}^{\infty} \frac{\Gamma(k+t)}{\Gamma(k+1)\Gamma(t)} |z|^{2k}.$$

As t goes from 0 to T, the parameter λ goes from $n/2$ to $(n+T)/2$, so the quotient $\Gamma(n)/\Gamma^2(\lambda)$ is bounded below away from 0 and bounded above away from infinity. We now use Stirling's formula to compare (1) and (2).

For each non-negative k, let

$$a_k = a_k(t) = \frac{\Gamma^2(k+\lambda)}{\Gamma(k+n)\Gamma(k+t)}.$$

It is obvious that if $k > 0$ then a_k is positive and bounded in $t \in (0, T)$; this is also true when $k = 0$, because the gamma function is bounded away from 0 on the interval $(0, \infty)$. We need to show that there exists a positive constant C (depending only on n and T) such that $a_k \leq C$ for all $k \geq 0$ and $0 < t < T$.

By Stirling's formula, there exist positive constants C_1 and M such that

$$C_1^{-1} \leq \frac{\Gamma(x)}{x^{x-\frac{1}{2}} e^{-x}} \leq C_1$$

for all $x \geq M$. It then follows easily that there exists a constant $C_2 > 0$, depending only on n and T, such that

$$a_k \leq C_2 \left[1 + \frac{\lambda - n}{k + n}\right]^{k+n} \left[1 + \frac{\lambda - t}{k + t}\right]^{k+t} \frac{\sqrt{(k+n)(k+t)}}{k + \lambda}.$$

As $k \to +\infty$, the right hand side above approaches C_2 uniformly for $t \in (0, T)$ (when n is fixed), because

$$\lim_{y \to \infty} \left(1 + \frac{x}{y}\right)^y = e^x,$$

and the convergence is uniform for x in any finite interval. This proves the desired estimate. $\qquad\square$

LEMMA 2. *Given any* $T > 0$ *and* $A > -1$, *there exists a constant* $C > 0$, *depending on* n, T, *and* A, *but not on* t *and* α, *such that*

$$\int_{\mathbb{B}} \frac{(1 - |w|^2)^\alpha \, dv(w)}{|1 - \langle z, w \rangle|^{n+1+\alpha+t}} \leq \frac{C\Gamma(\alpha + 1)\Gamma(t)}{(1 - |z|^2)^t}$$

for all $-1 < \alpha < A$, $0 < t < T$, *and* $z \in \mathbb{B}$.

PROOF. Let I denote the integral concerned. By the proof of Proposition 1.4.10 in [3],

(3) $$I = \frac{\Gamma(n+1)\Gamma(\alpha+1)}{\Gamma^2(\lambda)} \sum_{k=0}^{\infty} \frac{\Gamma^2(k+\lambda)}{\Gamma(k+1)\Gamma(n+1+\alpha+k)} |z|^{2k},$$

where $\lambda = (n + 1 + \alpha + t)/2$. The desired result then follows from (2), (3), and Stirling's formula. The details are exactly the same as in the proof of Lemma 1. $\quad\square$

LEMMA 3. *Suppose* μ *is a positive measure on a space* X *and* $H(x, y)$ *is a positive kernel on* X. *If there exists a constant* $C > 0$ *and a positive function* $h(x)$ *on* X *such that*

$$\int_X H(x, y)h(y)^q d\mu(y) \leq Ch(x)^q$$

for all x *in* X, *and*

$$\int_X H(x, y)h(x)^p d\mu(x) \leq Ch(y)^p$$

for all y *in* X, *then the integral operator*

$$Sf(x) = \int_X H(x, y)f(y) \, d\mu(y)$$

is bounded on $L^p(X, \mu)$ *with norm not exceeding* C. *Here* $1 < p < \infty$ *and* $1/p + 1/q = 1$.

PROOF. See, for example, [4]. $\qquad\square$

3. Proof of the Main Result

We now prove the main result of the paper.

THEOREM 4. *For any* $-1 < \alpha < \infty$, *there exists a constant* $C > 0$, *depending on* α *and* n *but not on* p, *such that the norm of the operator*

$$P_\alpha : L^p(\mathbb{B}, dv_\alpha) \to A_\alpha^p$$

satisfies the estimate

$$C^{-1} \csc \frac{\pi}{p} \leq \|P_\alpha\|_p \leq C \csc \frac{\pi}{p}$$

for all $1 < p < \infty$.

PROOF. Fix $1 < p < \infty$ and let q be the conjugate exponent,

$$\frac{1}{p} + \frac{1}{q} = 1.$$

Consider the function

$$h(z) = (1 - |z|^2)^{-(\alpha+1)/(pq)}$$

on $\mathbb{B}$ and the operator

$$Sf(z) = \int_{\mathbb{B}} \frac{f(w)\, dv_\alpha(w)}{|1 - \langle z, w \rangle|^{n+1+\alpha}}$$

on $L^p(\mathbb{B}, dv_\alpha)$. By Lemma 2, with $T = \alpha + 1$ and $A = \alpha$, there exists a constant $C > 0$, independent of p, such that

$$
\begin{aligned}
\int_{\mathbb{B}} \frac{h(w)^q dv_\alpha(w)}{|1 - \langle z, w \rangle|^{n+1+\alpha}} &= (\alpha+1) \int_{\mathbb{B}} \frac{(1 - |w|^2)^{\frac{\alpha+1}{q} - 1} dv(w)}{|1 - \langle z, w \rangle|^{n+1+\frac{\alpha+1}{q} - 1 + \frac{\alpha+1}{p}}} \\
&\leq \frac{C(\alpha+1)\Gamma\left(\frac{\alpha+1}{q}\right)\Gamma\left(\frac{\alpha+1}{p}\right)}{(1 - |z|^2)^{(\alpha+1)/p}} \\
&= C(\alpha+1)\Gamma\left(\frac{\alpha+1}{q}\right)\Gamma\left(\frac{\alpha+1}{p}\right) h(z)^q.
\end{aligned}
$$

Similarly,

$$\int_{\mathbb{B}} \frac{h(z)^p dv_\alpha(z)}{|1 - \langle z, w \rangle|^{n+1+\alpha}} \leq C(\alpha+1)\Gamma\left(\frac{\alpha+1}{p}\right)\Gamma\left(\frac{\alpha+1}{q}\right) h(w)^p.$$

It follows from Lemma 3 that the norm of the operator S on $L^p(\mathbb{B}, dv_\alpha)$, and hence the norm of P_α on $L^p(\mathbb{B}, dv_\alpha)$ does not exceed

$$C(\alpha+1)\Gamma\left(\frac{\alpha+1}{q}\right)\Gamma\left(\frac{\alpha+1}{p}\right).$$

If $\alpha = 0$, a well-known property of the gamma function gives

$$\Gamma\left(\frac{1}{p}\right)\Gamma\left(\frac{1}{q}\right) = \frac{\pi}{\sin\frac{\pi}{p}};$$

see [1].

If $\alpha \neq 0$, we can find a constant $C_1 > 0$, independent of p but dependent on α, such that

$$\Gamma\left(\frac{\alpha+1}{q}\right)\Gamma\left(\frac{\alpha+1}{p}\right) \leq \frac{C_1}{\sin\frac{\pi}{p}}.$$

In fact, because of the symmetry of the sine function and the conjugacy between p and q, we only need to consider the case in which p is very large. In this case, the factor $\Gamma((\alpha + 1)/q)$ is bounded from above and from below, and

$$\Gamma\left(\frac{\alpha + 1}{p}\right) \sim p \sim \frac{1}{\sin\frac{\pi}{p}},$$

because

$$x\Gamma(x) = \Gamma(x + 1) \sim 1$$

when x is a small positive number.

To prove that the norm estimate

$$\|P_\alpha\|_p \leq \frac{C}{\sin\frac{\pi}{p}}$$

is sharp, we only need to consider the case when $p > 2$; the case when $1 < p < 2$ then follows from duality and the symmetry of the function $\sin(\pi/p)$. Note again that for $p > 2$ the constant $\sin(\pi/p)$ is comparable to $1/p$.

So we assume that $p > 2$ and consider the function

$$\begin{aligned} f(z) &= \log(1 - z_1) - \overline{\log(1 - z_1)} \\ &= 2i\arg(1 - z_1). \end{aligned}$$

It is clear that $|f(z)| \leq 2\pi$ for all $z \in \mathbb{B}$, so that the norm of f in $L^p(\mathbb{B}, dv_\alpha)$ does not exceed 2π. On the other hand, it is easy to see that

$$P_\alpha f(z) = \log(1 - z_1);$$

see Theorem 7.1.4(b) in [**3**].

It is well-known that every function g in A_α^p satisfies the pointwise estimate

$$(4) \qquad |g(z)| \leq \frac{\|g\|_{p,\alpha}}{(1 - |z|^2)^{(n+1+\alpha)/p}}, \qquad z \in \mathbb{B}.$$

In fact, for any fixed z in $\mathbb{B}$, a change of variables shows that the functions g and G have the same norm in A_α^p, where

$$G(w) = g \circ \varphi_z(w) \left[\frac{(1 - |z|^2)^{n+1+\alpha}}{(1 - \langle w, z\rangle)^{2(n+1+\alpha)}}\right]^{\frac{1}{p}}, \qquad w \in \mathbb{B},$$

and φ_z is the involutive automorphism of $\mathbb{B}$ that interchanges the origin and the point z; see Section 2.2 in [**3**]. The obvious estimate $|G(0)| \leq \|G\|_{p,\alpha}$ then leads to (4).

Let $g = P_\alpha f$ and $z = (r, 0, \cdots, 0)$ in (4), where $0 < r < 1$. We obtain

$$\begin{aligned} \|P_\alpha f\|_{p,\alpha} &\geq (1 - r^2)^{\frac{n+1+\alpha}{p}} \log\frac{1}{1 - r} \\ &\geq (1 - r)^{\frac{n+1+\alpha}{p}} \log\frac{1}{1 - r}. \end{aligned}$$

In particular, if $r = 1 - e^{-p}$, then

$$\|P_\alpha f\|_{p,\alpha} \geq pe^{-(n+1+\alpha)}.$$

This shows that

$$\frac{\|P_\alpha f\|_{p,\alpha}}{\|f\|_{p,\alpha}} \geq \frac{p}{2\pi e^{n+1+\alpha}},$$

so the norm of P_α on $L^p(\mathbb{B}, dv_\alpha)$ is at least $p/(2\pi e^{n+1+\alpha})$. This completes the proof of the theorem.   □

Note that the main result can be restated as follows: there exists a constant $C > 0$ such that

$$C^{-1} \frac{p^2}{p-1} \leq \|P_\alpha\|_p \leq C \frac{p^2}{p-1}$$

for every $p \in (1, \infty)$. The quotient $p^2/(p-1)$ can also be replaced by pq or $p+q$, where $1/p + 1/q = 1$.

4. Further Remarks and Questions

The referee of the paper noticed that Lemmas 1 and 2 could be proved alternatively as follows. First, one observes that

$$|1 - \langle z, \zeta \rangle| \geq C \max\{1 - |z|, |1 - \langle z/|z|, \zeta \rangle|\}$$

and

$$\sigma(\zeta \in \mathbb{S} : |1 - \langle \zeta, \zeta_0 \rangle| \leq h) \leq Ch^n,$$

where C is independent of h and $\zeta_0 \in \mathbb{S}$. Then immediately,

$$\int_{\mathbb{S}} \frac{d\sigma(\zeta)}{|1 - \langle z, \zeta \rangle|^{n+t}} \leq C \sum_{k=0}^{\infty} \frac{[2^k(1-|z|^2)]^n}{[2^k(1-|z|^2)]^{n+t}} \leq \frac{C}{\min(1,t)(1-|z|^2)^t},$$

which proves Lemma 1. The constant $1/\min(1, t)$ is actually more precise than $\Gamma(t)$. Lemma 2 then follows from integration on concentric spheres:

$$
\begin{aligned}
I &\leq \int_0^1 \frac{C(1-s^2)^\alpha}{(1-|z|^2 s^2)^{t+\alpha+1}} \cdot \frac{ds}{\min(1, t+\alpha+1)} \\
&= \int_{|z|}^1 + \int_0^{|z|} \cdots \\
&\leq C\left(\frac{1}{t} + \frac{1}{\alpha+1}\right) \cdot \frac{1}{\min(1, t+\alpha+1)} \cdot \frac{1}{(1-|z|^2)^t} \\
&\leq \frac{C}{\min(1, t)\min(1, \alpha+1)} \cdot \frac{1}{(1-|z|^2)^t}.
\end{aligned}
$$

Once again, this is more precise than the estimate obtained in Lemma 2. Since

$$\frac{1}{\min(1, (\alpha+1)/p)} \cdot \frac{1}{\min(1, (\alpha+1)/q)} \leq C(p+q),$$

where C is independent of p, the rest of the proof of the main theorem still works.

Our main result shows how fast the norm of the Bergman projection P_α on $L^p(\mathbb{B}, dv_\alpha)$ grows as p increases to infinity or as p decreases to 1, when α is fixed. A related question is to determine how the norm of P_α on $L^p(\mathbb{B}, dv_\alpha)$ depends on α when p is fixed. In particular, we are interested in estimates of this norm when p is fixed and when α approaches -1. We conjecture that the norm of P_α on $L^p(\mathbb{B}, dv_\alpha)$ remains bounded if p is fixed in $(1, \infty)$ and when α approaches -1. A direct proof of this, such as the one in the previous section, will give a proof for the boundedness of the Cauchy-Szëgo projection on L^p spaces of the unit sphere when $1 < p < \infty$.

It is of course well-known that the Cauchy-Szëgo projection Q is bounded on L^p spaces of the unit sphere when $1 < p < \infty$. However, we are not aware of any estimates for the norm of Q on L^p.

Another natural problem is to find sharp norm estimates for the Bergman projection on L^p spaces of other domains, such as strongly pseudo-convex domains in $\mathbb{C}^n$.

References

[1] J.B. Conway, *Functions of One Complex Variable*, Springer-Verlag, New York, 1978.

[2] S.K. Pichorides, *On the best values of the constants in the theorems of M. Riesz, Zygmund, and Kolmogorov*, Studia Math. **44** (1972), 165-179.

[3] W. Rudin, *Function Theory in the Unit Ball of $\mathbb{C}^n$*, Springer-Verlag, New York, 1980.

[4] K. Zhu, *Operator Theory in Function Spaces*, Marcel Dekker, New York, 1990.

DEPARTMENT OF MATHEMATICS, SUNY, ALBANY, NY 12222, USA
E-mail address: kzhu@math.albany.edu